Introduction to
Matrix Analysis

Introduction to Matrix Analysis

RICHARD BELLMAN

Professor of Mathematics, Electrical Engineering, and Medicine, University of Southern California

SECOND EDITION

McGRAW-HILL BOOK COMPANY

New York St. Louis San Francisco Düsseldorf London Mexico Panama Sydney Toronto

INTRODUCTION TO MATRIX ANALYSIS

 Library of Congress Catalog Card Number 76-96425

04416

1234567890 MAMM 7543210

To Nina

Preface

Our aim in this volume is to introduce the reader to the study of matrix theory, a field which with a great deal of justice may be called the arithmetic of higher mathematics.

Although this is a rather sweeping claim, let us see if we can justify it. Surveying any of the classical domains of mathematics, we observe that the more interesting and significant parts are characterized by an interplay of factors. This interaction between individual elements manifests itself in the appearance of functions of several variables and correspondingly in the shape of variables which depend upon several functions. The analysis of these functions leads to transformations of multidimensional type.

It soon becomes clear that the very problem of describing the problems that arise is itself of formidable nature. One has only to refer to various texts of one hundred years ago to be convinced that at the outset of any investigation there is a very real danger of being swamped by a sea of arithmetical and algebraical detail. And this is without regard of many conceptual and analytic difficulties that multidimensional analysis inevitably conjures up.

It follows that at the very beginning a determined effort must be made to devise a useful, sensitive, and perceptive notation. Although it would certainly be rash to attempt to assign a numerical value to the dependence of successful research upon well-conceived notation, it is not difficult to cite numerous examples where the solutions become apparent when the questions are appropriately formulated. Conversely, a major effort and great ingenuity would be required were a clumsy and unrevealing notation employed. Think, for instance, of how it would be to do arithmetic or algebra in terms of Roman numerals.

A well-designed notation attempts to express the essence of the underlying mathematics without obscuring or distracting.

With this as our introduction, we can now furnish a very simple syllogism. Matrices represent the most important of transformations, the linear transformations; transformations lie at the heart of mathematics; consequently, our first statement.

This volume, the first of a series of volumes devoted to an exposition of the results and methods of modern matrix theory, is intended to acquaint the reader with the fundamental concepts of matrix theory. Subsequent volumes will expand the domain in various directions. Here we shall pay particular attention to the field of analysis, both from the standpoint of motivation and application.

In consequence, the contents are specifically slanted toward the needs and aspirations of analysts, mathematical physicists, engineers of all shadings, and mathematical economists.

It turns out that the analytical theory of matrices, at the level at which we shall treat it, falls rather neatly into three main categories: the theory of symmetric matrices, which invades all fields, matrices and differential equations, of particular concern to the engineer and physicist, and positive matrices, crucial in the areas of probability theory and mathematical economics.

Although we have made no attempt to tie our exposition in with any actual applications, we have consistently tried to show the origin of the principal problems we consider.

We begin with the question of determining the maximum or minimum of a function of several variables. Using the methods of calculus, we see that the determination of a local maximum or minimum leads to the corresponding question for functions of much simpler form, namely functions which are quadratic in all the variables, under reasonable assumptions concerning the existence of a sufficient number of partial derivatives.

In this fashion, we are led to consider quadratic forms, and thus symmetric matrices.

We first treat the case of functions of two variables where the usual notation suffices to derive all results of interest. Turning to the higher dimensional case, it becomes clear a better notation will prove of value. Nevertheless, a thorough understanding of the two-dimensional case is quite worthwhile, since all the methods used in the multidimensional case are contained therein.

We turn aside, then, from the multidimensional maximization problem to develop matrix notation. However, we systematically try to introduce at each stage only those new symbols and ideas which are necessary for the problem at hand. It may surprise the reader, for example, to see how far into the theory of symmetric matrices one can penetrate without the concept of an inverse matrix.

Consistent with these ideas, we have not followed the usual approach of deluging the novice with a flood of results concerning the solutions of linear systems of equations. Without for one moment attempting to minimize the importance of this study, it is still true that a significant number of interesting and important results can be presented without slogging down this long and somewhat wearisome road. The concept of linear independence is introduced in connection with the orthogonalization process, where it plays a vital role. In the appendix we present a proof of the fundamental result concerning the solutions of linear systems, and a discussion of some of the principal results concerning the rank of a matrix.

This concept of much significance in many areas of matrix theory is not as important as might be thought in the regions we explore here. Too often, in many parts of mathematics, the reader is required to swallow on faith a large quantity of predigested material before being allowed to chew over any meaty questions. We have tried to avoid this. Once it has been seen that a real problem exists, then there is motivation for introducing more sophisticated concepts. This is the situation that the mathematician faces in actual research and in many applications.

Although we have tried throughout to make the presentation logical, we have not belabored the point. Logic, after all, is a trick devised by the human mind to solve certain types of problems. But mathematics is more than logic, it is logic plus the creative process. How the logical devices that constitute the tools of mathematics are to be combined to yield the desired results is not necessarily logical, no more than the writing of a symphony is a logical exercise, or the painting of a picture an exercise in syllogisms.

Having introduced square matrices, the class of greatest importance for our work here, we turn to the problem of the canonical representation of real quadratic forms, or alternatively of a real symmetric matrix. The most important result of this analysis, and one that is basic for all subsequent development of the theory of symmetric matrices, is the equivalence, in a sense that will be made precise below, of every real symmetric matrix with a diagonal matrix.

In other words, multidimensional transformations of this type to a very great extent can be regarded as a number of one-dimensional transformations performed simultaneously.

The results of these preliminary chapters are instructive for several reasons. In the first place, they show what a great simplification in proof can be obtained by making an initial assumption concerning the simplicity of the characteristic roots. Secondly, they show that two methods can frequently be employed to circumvent the difficulties attendant upon multiplicity of roots. Both are potent tools of the analyst. The first is induction, the second is continuity.

Of the two, continuity is the more delicate method, and requires for its rigorous use, quite a bit more of sophistication than is required elsewhere in the book. Consequently, although we indicate the applicability of this technique wherever possible, we leave it to the ambitious reader to fill in the details.

Once having obtained the diagonal representation, we are ready to derive the min-max properties of the characteristic roots discovered by Courant and Fischer. The extension of these results to the more general operators arising from partial differential equations by Courant is a fundamental result in the domain of analysis.

Having reached this point, it is now appropriate to introduce some other properties of matrices. We turn then to a brief study of some of the important matrix functions. The question of defining a general function of a matrix is quite a bit more complicated than might be imagined, and we discuss this only briefly. A number of references to the extensive literature on the subject are given.

We now return to our original stimulus, the question of the range of values of a quadratic form. However, we complicate the problem to the extent of adding certain linear constraints. Not only is the problem of interest in itself, but it also supplies a good reason for introducing rectangular matrices. Having gone this far, it also turns out to be expedient to discuss matrices whose elements are themselves matrices. This further refinement of the matrix notation is often exceedingly useful.

Following this, we consider a number of interesting inequalities relating to characteristic roots and various functions of the characteristic roots. This chapter is rather more specialized than any of the others in the volume and reflects perhaps the personal taste of the author more than the needs of the reader.

The last chapter in the part devoted to symmetric matrices deals with the functional equation technique of dynamic programming. A number of problems related to maximization and minimization of quadratic forms and the solution of linear systems are treated in this fashion. The analytic results are interesting in their dependence upon parameters which are usually taken to be constant, while the recurrence relations obtained in this way yield algorithms which are sometimes of computational value.

In the second third of the volume, we turn our attention to the application of matrix theory to the study of linear systems of differential equations. No previous knowledge of differential equations is assumed or required. The requisite existence and uniqueness theorems for linear systems will be demonstrated in the course of the discussion.

The first important concept that enters in the study of linear systems with constant coefficients is that of the matrix exponential. In terms of this matrix function, we have an explicit solution of the differential equation. The case of variable coefficients does not permit resolution in this easy fashion. To obtain an analogous expression, it is necessary to introduce the product integral, a concept we shall not enter into here. The product integral plays an important role in modern quantum mechanics.

Although the construction of the exponential matrix solves in a very elegant fashion the problem of constructing an explicit solution of the linear equation with constant coefficients, it does not yield a useful representation for the individual components of the vector solution. For this purpose, we employ a method due to Euler for finding particular solutions of exponential type. In this way we are once again led to the problem of determining characteristic roots and vectors of a matrix.

Since the matrix is in general no longer symmetric, the problem is very much more complicated than before. Although there are again a number of canonical forms, none of these are as convenient as that obtained for the case of the symmetric or hermitian matrix.

The representation of the solution as a sum of exponentials, and limiting cases, permits us to state a necessary and sufficient condition that all solutions of a homogeneous system tend to the zero vector as the time becomes arbitrarily large. This leads to a discussion of stability and the problem of determining in a simple fashion when a given system is stable. The general problem is quite complicated.

Having obtained a variety of results for general, not necessarily symmetric, matrices, we turn to what appears to be a problem of rather specialized interest. Given a matrix A, how do we determine a matrix whose characteristic roots are specified functions of the characteristic roots of A? If we ask that these functions be certain symmetric functions of the characteristic roots of A, then in a very natural fashion we are led to one of the important concepts of the algebraic side of matrix theory, the Kronecker product of two matrices. As we shall see, however, in the concluding part of the book, this function of two matrices arises also in the study of stochastic matrices.

The final part of the volume is devoted to the study of matrices all of whose elements are non-negative. Matrices of this apparently specialized type arise in two important ways. First in the study of Markoff processes, and secondly in the study of various economic processes.

A consideration of the physical origin of these matrices makes intuitively clear a number of quite important and interesting limit theorems associated with the names of Markoff, Perron, and Frobenius. In particular, a variational representation due to Wielandt plays a fundamental role in much of the theory of positive matrices.

A brief chapter is devoted to the study of stochastic matrices. This area dealing with the multiplicative, and hence noncommutative, aspect of stochastic processes, rather than the additive, and hence commutative, aspect, is almost completely unexplored. The results we present are quite elementary and introductory.

Finally, in a series of appendices, we have added some results which were either tangential to the main exposition, or else of quite specialized interest. The applications of Selberg, Hermite, and Fischer of the theory of symmetric matrices are, however, of such singular elegance that we felt that it was absolutely imperative that they be included.

Now a few words to the reader first entering this fascinating field. As stated above, this volume is designed to be an introduction to the theory of matrix analysis. Although all the chapters are thus introductory in this sense, to paraphrase Orwell, some are more introductory than others.

Consequently, it is not intended that the chapters be read consecutively. The beginner is urged to read Chaps. 1 through 5 for a general introduction to operations with matrices, and for the rudiments of the theory of symmetric matrices. At this point, it would be appropriate to jump to the study of general square matrices, using Chaps. 10 and 11 for this purpose. Finally, Chaps. 14 and 16 should be studied in order to understand the origins of Markoff matrices and non-negative matrices in general. Together with the working out of a number of the exercises, this should cover a semester course at an undergraduate senior, or first-year graduate level.

The reader who is studying matrix theory on his own can follow the same program.

Having attained this level, the next important topic is that of the minimum-maximum characterization of the characteristic roots. Following this, we would suggest Kronecker products. From here, Chap. 6 seems appropriate. At this point, the remaining chapters can be taken in any order.

Now a few comments concerning the exercises. Since the purpose of mathematics is to solve problems, it is impossible to judge one's progress without breaking a lance on a few problems from stage to stage. What we have attempted to do is to provide problems of all levels of difficulty, starting from those which merely illustrate the text and ending with those which are of considerable difficulty. These latter have usually been lifted from current research papers.

Generally, the problems immediately following the sections are routine and included for drill purpose. A few are on a higher level, containing results which we shall use at a subsequent time. Since they can be established without too much effort we felt that it would be better to use them as exercises than to expand the text unduly by furnishing the proofs. In any case, the repetition of the same technique would make for tedious reading.

On the other hand, the problems at the end of the chapter in the Miscellaneous Exercises are usually of a higher level, with some of greater complexity than others. Although the temptation is to star those problems which we consider more difficult, a little thought shows that this is bad pedagogy. After all, the purpose of a text such as this is to prepare the student to solve problems on his own, as they appear in research in the fields of analysis, mathematical physics, engineering, mathematical economics, and so forth. It is very seldom the case that a problem arising in this fashion is neatly stated with a star attached. Furthermore, it is important for the student to observe that the complication of a problem can practically never be gauged by its formulation. One very quickly learns that some very simple statements can conceal major difficulties. Taking this all into account, we have mixed problems of all levels together in a fairly random fashion and with no warning as to their complexity. It follows that in attacking these problems, one should do those that can be done, struggle for a while with the more obdurate ones, and then return to these from time to time, as maturity and analytic ability increase.

Next, a word about the general plan of the volume. Not only do we wish to present a number of the fundamental results, but far more importantly, we wish to present a variety of fundamental methods. In order to do this, we have at several places presented alternative proofs of theorems, or indicated alternative proofs in the exercises. It is essential to realize that there are many different approaches to matrix theory, as indeed to all parts of mathematics. The importance of having all approaches available lies in the fact that different extensions may require different approaches. Indeed, some extensions may be routine with one approach and formidable, or impossible, following other routes.

In this direction, although it is quite elegant and useful to derive many of the results of matrix theory from results valid for general operators, it is also important to point out various special techniques and devices which are particularly valuable in dealing with finite-dimensional transformations. It is for this reason that we have tried to rescue from oblivion a number of simple and powerful approaches which were used fifty to seventy-five years ago in the days of halcyon innocence.

The human mind being what it is, repetition and cross-sections from different angles are powerful pedagogical devices. In this connection, it is appropriate to quote Lewis Carroll in the *Hunting of the Snark*, Fit the First—"I have said it thrice: What I tell you three times is true."—The Bellman.

Let us now briefly indicate some of the many fundamental aspects of matrix theory which have reluctantly not been discussed in this volume.

In the first place, we have omitted any discussion of the theory of computational treatment of matrices. This theory has vastly expanded in recent years, stimulated by the extraordinary abilities of modern computers and those that are planned for the immediate future, and by the extraordinary demands of the current physical and economic scenes.

The necessity for the development of new techniques in the numerical treatment of matrices lies in the fact that the problem of solving systems of linear equations involving a large number of variables, or that of finding the characteristic roots and vectors of a matrix of high dimension, cannot be treated in any routine fashion. Any successful treatment of these problems requires the use of new and subtle techniques. So much work has been done in this field that we thought it wise to devote a separate volume of this series to its recital. This volume will be written by George Forsythe.

In a different direction, the combinatorial theory of matrices has mushroomed in recent years. The mathematical theory of games of Borel and von Neumann, the theory of linear programming, and the mathematical study of scheduling theory have all combined to create virtually a new field of matrix theory. Not only do novel and significant analytic and algebraic problems arise directly in this way, but the pressure of obtaining computational results also has resulted in the development of still further techniques and concepts. A volume devoted to this field will be written by Alan Hoffman.

Classically, topological aspects of matrix theory entered by way of the study of electrical networks. Here we encounter a beautiful blend of analytic, algebraic, and geometric theory. A volume on this important field is to be written by Louis Weinberg.

We have, of course, barely penetrated the domain of matrix theory in the foregoing enumeration of topics.

On the distant horizon, we foresee a volume on the advanced theory of matrix analysis. This would contain, among other results, various aspects of the theory of functions of matrices, the Loewner theory, the Siegel theory of modular functions of matrices, and the R matrices of Wigner. In the more general theory of functionals of matrices, the Baker-Campbell-Hausdorff theory leads to the study of product integrals. These theories have assumed dominant roles in many parts of modern mathematical physics.

Another vast field of matrix analysis has developed in connection with the study of multivariate analysis in mathematical statistics. Again we have felt that results in this domain are best presented with the background and motivation of statistics.

In a common ground between analysis and algebra, we meet the theory of group representation with its many beautiful applications to algebra, analysis, and mathematical physics. This subject also requires a separate volume.

In the purely algebraic domain, there is the treatment of ideal theory by way of matrices due to Poincaré. We have not mentioned this because of the necessity for the introduction of rather sophisticated concepts. In the exercises, however, there are constant reminders of the interconnection between complex numbers, quaternions, and matrices, and hints of more general connections.

Closely related is the study of matrices with integer elements, a number theory of matrices. Despite the attractiveness of these topics, we felt that it would be best to have them discussed in detail in a separate volume.

It is important to note that not even the foregoing enumeration exhausts the many roles played by matrices in modern mathematics and its applications.

We have included in the discussions at the ends of the chapters, and occasionally in the exercises, a large number of references to original papers, current research papers, and various books on matrix theory. The reader who becomes particularly interested in a topic may thus pursue it thoroughly. Nonetheless, we make no pretension to an exhaustive bibliography, and many significant papers are not mentioned.

Finally, I wish to express my heartfelt appreciation of the efforts of a number of friends who devotedly read through several drafts of the manuscript. Through their comments and criticisms, a number of significant improvements resulted in content, in style, and in the form of many interesting exercises. To Paul Brock, Ky Fan, and Olga Taussky, thanks.

My sincere thanks are also due to Albert Madansky and Ingram Olkin who read several chapters and furnished a number of interesting exercises.

I would like especially to express my gratitude to The RAND Corporation for its research policies which have permitted generous support of my work in the basic field of matrix theory. This has been merely one aspect of the freedom offered by RAND to pursue those endeavors which simultaneously advance science and serve the national interest.

At the last, a vote of thanks to my secretary, Jeanette Hiebert, who unflinchingly typed hundreds of pages of equations, uncomplainingly made revision after revision, and devotedly helped with the proofreading.

Richard Bellman

Preface to Second Edition

Since the publication of the First Edition of this book in 1960, the field of matrix theory has expanded at a furious rate. In preparing a Second Edition, the question arises of how to take into account this vast proliferation of topics, methods, and results. Clearly, it would be impossible to retain any reasonable size and also to cover the developments in new and old areas in some meaningful fashion.

A compromise is, therefore, essential. We decided after much soul-searching to add some new results in areas already covered by including exercises at the ends of chapters and by updating references. New areas, on the other hand, are represented by three additional chapters devoted to control theory, invariant imbedding, and numerical inversion of the Laplace transform.

As in the First Edition, we wish to emphasize the manner in which mathematical investigations of new scientific problems generate novel and interesting questions in matrix theory. Secondly, we hope to illustrate how the scientific background provides valuable clues as to the results to expect and even as to the methods of proof to employ. Many new areas have not been touched upon at all, or have been just barely mentioned: graph theory, scheduling theory, network theory, linear inequalities, and numerical analysis. We have preferred, for obvious reasons, to concentrate on topics to which a considerable amount of personal effort has been devoted.

Matrix theory is replete with fascinating results and elegant techniques. It is a domain constantly stimulated by interactions with the outside world and it contributes to all areas of the physical and social sciences. It represents a healthy trend in modern mathematics.

Richard Bellman
University of Southern California

Contents

1

Maximization, Minimization, and Motivation

1. Introduction. The purpose of this opening chapter is to show how the question of ascertaining the range of values of a homogeneous quadratic function of two variables enters in a very natural way in connection with the problem of determining the maximum or minimum of a general function of two variables.

We shall treat the problem of determining the extreme values of a quadratic function of two variables, or as we shall say, a quadratic form in two variables, in great detail. There are several important reasons for doing so. In the first place, the three different techniques we employ, algebraic, analytic, and geometric, can all, suitably interpreted, be generalized to apply to the multidimensional cases we consider subsequently. Even more important from the pedagogical point of view is the fact that the algebraic and analytic detail, to some extent threatening in the two-dimensional case but truly formidable in the N-dimensional case, rather pointedly impels us to devise a new notation.

A detailed examination of this case thus furnishes excellent motivation for the introduction of new concepts.

2. Maximization of Functions of One Variable. Let $f(x)$ be a real function of the real variable x for x in the closed interval $[a,b]$, and let us suppose that it possesses a convergent Taylor series of the form

$$f(x) = f(c) + (x - c)f'(c) + \frac{(x - c)^2}{2!} f''(c) + \cdots \tag{1}$$

around each point in the open interval (a,b).

Let c be a *stationary point* of $f(x)$, which is to say a point where $f'(x) = 0$, and let it be required to determine whether c is a point at which $f(x)$ is a relative maximum, a relative minimum, or a stationary point of more complicated nature.

If c is a stationary point, the expansion appearing in (1) takes the simpler form

$$f(x) = f(c) + \frac{(x - c)^2}{2!} f''(c) + \cdots \tag{2}$$

If $f''(c) = 0$, we must consider further terms in the expansion. If, however, $f''(c) \neq 0$, its sign tells the story. When $f''(c) > 0$, $f(x)$ has a relative minimum at $x = c$; when $f''(c) < 0$, $f(x)$ has a relative maximum at $x = c$.

EXERCISE

1. If $f''(c) = 0$, what are sufficient conditions that c furnish a relative maximum?

3. Maximization of Functions of Two Variables. Let us now pursue the same question for a function of two variables, $f(x,y)$, defined over the rectangle $a_1 \leq x \leq b_1$, $a_2 \leq y \leq b_2$, and possessing a convergent Taylor series around each point (c_1,c_2) inside this region. Thus, for $|x - c_1|$ and $|y - c_2|$ sufficiently small, we have

$$\begin{aligned} f(x,y) = f(c_1,c_2) &+ (x - c_1) \frac{\partial f}{\partial c_1} + (y - c_2) \frac{\partial f}{\partial c_2} + \frac{(x - c_1)^2}{2} \frac{\partial^2 f}{\partial c_1{}^2} \\ &+ (x - c_1)(y - c_2) \frac{\partial^2 f}{\partial c_1 \partial c_2} + \frac{(y - c_2)^2}{2} \frac{\partial^2 f}{\partial c_2{}^2} + \cdots \end{aligned} \tag{1}$$

Here

$$\begin{aligned} \frac{\partial f}{\partial c_1} &= \frac{\partial f}{\partial x} \text{ at } x = c_1 \qquad y = c_2 \\ \frac{\partial f}{\partial c_2} &= \frac{\partial f}{\partial y} \text{ at } x = c_1 \qquad y = c_2 \end{aligned} \tag{2}$$

and so on.

Let (c_1,c_2) be a stationary point of $f(x,y)$ which means that we have the equations

$$\frac{\partial f}{\partial c_1} = 0 \qquad \frac{\partial f}{\partial c_2} = 0 \tag{3}$$

Then, as in the foregoing section, the nature of $f(x,y)$ in the immediate neighborhood of (c_1,c_2) depends upon the behavior of the quadratic terms appearing in the expansion in (1), namely,

$$Q_2(x,y) = a(x - c_1)^2 + 2b(x - c_1)(y - c_2) + c(y - c_2)^2 \tag{4}$$

where to simplify the notation we have set

$$a = \frac{1}{2} \frac{\partial^2 f}{\partial c_1{}^2} \qquad 2b = \frac{\partial^2 f}{\partial c_1 \partial c_2} \qquad c = \frac{1}{2} \frac{\partial^2 f}{\partial c_2{}^2} \tag{5}$$

To simplify the algebra still further, let us set

$$x - c_1 = u \qquad y - c_2 = v \tag{6}$$

and consider the homogeneous quadratic expression

$$Q(u,v) = au^2 + 2buv + cv^2 \tag{7}$$

An expression of this type will be called a *quadratic form*, specifically a quadratic form in the two variables u and v.

Although we are interested only in the behavior of $Q(u,v)$ in the vicinity of $u = v = 0$, the fact that $Q(u,v)$ is homogeneous permits us to examine, if we wish, the range of values of $Q(u,v)$ as u and v take all real values, or, if it is more convenient, the set of values assumed on $u^2 + v^2 = 1$.

The fact that $Q(ku,kv) = k^2Q(u,v)$ for any value of k shows that the set of values assumed on the circle $u^2 + v^2 = k^2$ is related in a very simple way to the values assumed on $u^2 + v^2 = 1$.

If $Q(u,v) > 0$ for all u and v distinct from $u = v = 0$, $f(x,y)$ will have a relative minimum at $x = c_1$, $y = c_2$; if $Q(u,v) < 0$, there will be a relative maximum. If $Q(u,v)$ can assume both negative and positive values, we face a stationary point of more complicated type—*a saddle point.*

Although a number of quite interesting algebraic and geometric questions arise in connection with saddle points, we shall not be concerned with these matters in this volume.

If $Q(u,v)$ is identically zero, the problem is, of course, even more complicated, but one, fortunately, of no particular importance.

EXERCISE

1. Can the study of the positivity or negativity of a homogeneous form of the fourth degree, $Q(u,v) = a_0u^4 + a_1u^3v + a_2u^2v^2 + a_3uv^3 + a_4v^4$, be reduced to the study of the corresponding problem for quadratic forms?

4. Algebraic Approach. Let us now see if we can obtain some simple relations connecting the coefficients a, b, and c which will tell us which of the three situations described above actually occurs for any given quadratic form, $au^2 + 2buv + cv^2$, with real coefficients.

In order to determine the sign of $Q(u,v)$, we complete the square in the expression $au^2 + 2buv$ and write $Q(u,v)$ in the following form:

$$Q(u,v) = a\left(u + \frac{bv}{a}\right)^2 + \left(c - \frac{b^2}{a}\right)v^2 \tag{1}$$

provided that $a \neq 0$.

If $a = 0$, but $c \neq 0$, we carry out the same type of transformation, reversing the roles of u and v. If $a = c = 0$, then $Q(u,v)$ reduces to

$2buv$. If $b \neq 0$, it is clear that $Q(u,v)$ can assume both negative and positive values. If $b = 0$, the quadratic form disappears.

Let us then henceforth assume that $a \neq 0$, since otherwise the problem is readily resolved.

From (1) it follows that $Q(u,v) > 0$ for all *nontrivial* u and v (i.e., for all u and v distinct from the pair (0,0). We shall employ this expression frequently below), provided that

$$a > 0 \qquad c - \frac{b^2}{a} > 0 \tag{2}$$

Similarly, $Q(u,v) < 0$ for all nontrivial u and v, provided that we have the inequalities

$$a < 0 \qquad c - \frac{b^2}{a} < 0 \tag{3}$$

Conversely, if Q is positive for all nontrivial u and v, then the two inequalities in (2) must hold, with a similar result holding for the case where Q is negative for all nontrivial u and v.

We have thus proved the following theorem.

Theorem 1. *A set of necessary and sufficient conditions that $Q(u,v)$ be positive for all nontrivial u and v is that*

$$a > 0 \qquad \begin{vmatrix} a & b \\ b & c \end{vmatrix} > 0 \tag{4}$$

Observe that we say a set of necessary and sufficient conditions, since there may be, and actually are, a number of alternative, but, of course, equivalent, sets of necessary and sufficient conditions. We usually try to obtain as many alternative sets as possible, since some are more convenient to apply than others in various situations.

EXERCISE

1. Show that a set of necessary and sufficient conditions that $Q(u,v)$ be positive is that $c > 0$, $ac - b^2 > 0$.

5. Analytic Approach. As noted above, to determine the range of values of $Q(u,v)$ it is sufficient to examine the set of values assumed by $Q(u,v)$ on the circle $u^2 + v^2 = 1$. If Q is to be positive for all nontrivial values of u and v, we must have

$$\min_{u^2+v^2=1} Q(u,v) > 0 \tag{1}$$

while the condition

$$\max_{u^2+v^2=1} Q(u,v) < 0 \tag{2}$$

is the required condition that $Q(u,v)$ be negative for all u and v on the unit circumference.

To treat these variational problems in a symmetric fashion, we employ a Lagrange multiplier. Consider the problem of determining the stationary points of the new quadratic expression

$$R(u,v) = au^2 + 2buv + cv^2 - \lambda(u^2 + v^2) \tag{3}$$

The conditions $\partial R/\partial u = \partial R/\partial v = 0$ yield the two linear expressions

$$\begin{aligned} au + bv - \lambda u &= 0 \\ bu + cv - \lambda v &= 0 \end{aligned} \tag{4}$$

Eliminating u and v from these two equations, we see that λ satisfies the determinantal equation

$$\begin{vmatrix} a - \lambda & b \\ b & c - \lambda \end{vmatrix} = 0 \tag{5}$$

or

$$\lambda^2 - (a + c)\lambda + ac - b^2 = 0 \tag{6}$$

Since the discriminant is

$$(a + c)^2 - 4(ac - b^2) = (a - c)^2 + 4b^2 \tag{7}$$

clearly non-negative, we see that the roots of (6) which we shall call λ_1 and λ_2 are always real. Unless $a = c$ and $b = 0$, these roots are distinct. Let us consider the case of *distinct roots* in detail.

If $b = 0$, the roots of the quadratic in (6) are $\lambda_1 = a$, $\lambda_2 = c$. In the first case, $\lambda_1 = a$, the equations in (4) are

$$(a - \lambda_1)u = 0 \qquad (c - \lambda_1)v = 0 \tag{8}$$

which leaves u arbitrary and $v = 0$, if $a \neq c$. Since we are considering only the case of distinct roots, this must be so.

If $b \neq 0$, we obtain the nontrivial solutions of (4) by using one of the equations and discarding the other. Thus u and v are connected by the relation

$$(a - \lambda_1)u = -bv \tag{9}$$

In order to talk about a particular solution, let us add the requirement that $u^2 + v^2 = 1$. This is called a *normalization*. The values of u and v determined in this way are

$$\begin{aligned} u_1 &= -b/(b^2 + (a - \lambda_1)^2)^{1/2} \\ v_1 &= (a - \lambda_1)/(b^2 + (a - \lambda_1)^2)^{1/2} \end{aligned} \tag{10}$$

with another set (u_2,v_2) determined in a similar fashion when λ_2 is used in place of λ_1.

6. Analytic Approach—II. Once λ_1 and λ_2 have been determined by way of (5.6),[1] u_i and v_i are determined by the formulas of (5.10). These values, when substituted, yield the required minimum and maximum values of $au^2 + 2buv + cv^2$ on $u^2 + v^2 = 1$.

It turns out, however, that we can proceed in a very much more adroit fashion. Returning to the linear equations in (5.4), and multiplying the first by u and the second by v, we obtain

$$au^2 + 2buv + cv^2 - \lambda(u^2 + v^2) = 0 \tag{1}$$

This result is not unexpected; it is a special case of Euler's theorem concerning homogeneous functions, i.e.,

$$u\frac{\partial Q}{\partial u} + v\frac{\partial Q}{\partial v} = 2Q \tag{2}$$

if $Q(u,v)$ is homogeneous of degree 2. Since $u_i^2 + v_i^2 = 1$, for $i = 1, 2$, we see that

$$\begin{aligned} \lambda_1 &= au_1^2 + 2bu_1v_1 + cv_1^2 \\ \lambda_2 &= au_2^2 + 2bu_2v_2 + cv_2^2 \end{aligned} \tag{3}$$

Hence one solution of the quadratic equation in (5.6) is the required maximum, with the other the required minimum.

We observe then the remarkable fact that the maximum and minimum values of $Q(u,v)$ can be obtained without any explicit calculation of the points at which they are obtained. Nonetheless, as we shall see, these points have important features of their own.

Let us now derive an important property of the points (u_i,v_i), still without using their explicit values.

As we have seen in the foregoing sections, these points are determined by the sets of equations

$$\begin{aligned} au_1 + bv_1 - \lambda_1u_1 &= 0 & \qquad au_2 + bv_2 - \lambda_2u_2 &= 0 \\ bu_1 + cv_1 - \lambda_1v_1 &= 0 & \qquad bu_2 + cv_2 - \lambda_2u_2 &= 0 \\ u_1^2 + v_1^2 &= 1 & \qquad u_2^2 + v_2^2 &= 1 \end{aligned} \tag{4}$$

Considering the first set, we have, upon multiplying the first equation by u_2, the second by v_2, and adding,

$$\begin{aligned} &u_2(au_1 + bv_1 - \lambda_1u_1) + v_2(bu_1 + cv_1 - \lambda_1v_1) \\ &\qquad = au_1u_2 + b(u_2v_1 + u_1v_2) + cv_1v_2 - \lambda_1(u_1u_2 + v_1v_2) = 0 \end{aligned} \tag{5}$$

Similarly, the second set yields

$$au_1u_2 + b(u_2v_1 + u_1v_2) + cv_1v_2 - \lambda_2(u_1u_2 + v_1v_2) = 0 \tag{6}$$

[1] Double numbers in parentheses refer to equations in another section of the chapter.

Subtracting, we have

$$(\lambda_1 - \lambda_2)(u_1u_2 + v_1v_2) = 0 \tag{7}$$

Since $\lambda_1 \neq \lambda_2$ (by assumption), we obtain the result that

$$u_1u_2 + v_1v_2 = 0 \tag{8}$$

The geometric significance of this relation will be discussed below.

Let us also note that the quantity $u_1v_2 - u_2v_1$ is nonzero. For assume that it were zero. Then, together with (8), we would have the two linear equations in u_1 and v_1,

$$\begin{aligned} u_1u_2 + v_1v_2 &= 0 \\ u_1v_2 - v_1u_2 &= 0 \end{aligned} \tag{9}$$

Since u_1 and v_1 are not both zero, a consequence of the normalization conditions in (4), we must have the determinantal relation

$$\begin{vmatrix} u_2 & v_2 \\ v_2 & -u_2 \end{vmatrix} = 0 \tag{10}$$

or $u_2^2 + v_2^2 = 0$, contradicting the last relation in (4).

EXERCISE

1. Show that for any two sets of values (u_1,u_2) and (v_1,v_2) we have the relation

$$(u_1^2 + v_1^2)(u_2^2 + v_2^2) = (u_1u_2 + v_1v_2)^2 + (u_1v_2 - u_2v_1)^2,$$

and thus again that $u_1v_2 - u_2v_1 \neq 0$ if the u_i and v_i are as above.

7. A Simplifying Transformation. Armed with a knowledge of the properties of the (u_i,v_i) contained in (6.4) and (6.8), let us see what happens if we make the change of variable

$$\begin{aligned} u &= u_1u' + u_2v' \\ v &= v_1u' + v_2v' \end{aligned} \tag{1}$$

This is a one-to-one transformation since the determinant $u_1v_2 - u_2v_1$ is nonzero, as noted at the end of Sec. 6.

In the first place, we see that

$$u^2 + v^2 = (u_1^2 + v_1^2)u'^2 + (u_2^2 + v_2^2)v'^2 + 2(u_1u_2 + v_1v_2)u'v' = u'^2 + v'^2 \tag{2}$$

It follows that the set of values assumed by $Q(u,v)$ on the circle $u^2 + v^2 = 1$ is the same as the set of values assumed by $Q(u_1u' + u_2v', v_1u' + v_2v')$ on the circle $u'^2 + v'^2 = 1$.

Let us now see what the expression for Q looks like in terms of the new variables. We have, upon collecting terms,

$$Q(u_1u' + u_2v', v_1u' + v_2v') \\ = (au_1^2 + 2bu_1v_1 + cv_1^2)u'^2 + (au_2^2 + 2bu_2v_2 + cv_2^2)v'^2 \\ + 2(au_1u_2 + b(u_1v_2 + u_2v_1) + cv_1v_2)u'v' \quad (3)$$

Referring to (6.3), (6.6), and (6.8), we see that this reduces to

$$\lambda_1 u'^2 + \lambda_2 v'^2 \quad (4)$$

The effect of the change of variable has been to eliminate the cross-product term $2buv$.

This is quite convenient for various algebraic, analytic and geometric purposes, as we shall have occasion to observe in subsequent chapters where a similar transformation will be applied in the multidimensional case.

As a matter of fact, the principal part of the theory of quadratic forms rests upon the fact that an analogous result holds for quadratic forms in any number of variables.

EXERCISE

1. Referring to (4), determine the conic section described by the equation $Q(u,v) = 1$ in the following cases:

(*a*) $\lambda_1 > 0, \lambda_2 > 0$
(*b*) $\lambda_1 > 0, \lambda_2 < 0$
(*c*) $\lambda_1 = \lambda_2 > 0$
(*d*) $\lambda_1 = 0, \lambda_2 > 0$

8. Another Necessary and Sufficient Condition. Using the preceding representation, we see that we can make the following statement.

Theorem 2. *A necessary and sufficient condition that $Q(u,v)$ be positive for all nontrivial u and v is that the roots of the determinantal equation*

$$\begin{vmatrix} a - \lambda & b \\ b & c - \lambda \end{vmatrix} = 0 \quad (1)$$

be positive.

EXERCISE

1. Show directly that the condition stated above is equivalent to that given in Theorem 1.

9. Definite and Indefinite Forms. Let us now introduce some terminology. If $Q(u,v) = au^2 + 2buv + cv^2 > 0$ for all nontrivial u and v, we shall say that $Q(u,v)$ is *positive definite*. If $Q(u,v) < 0$ for these values of u and v, we shall call $Q(u,v)$ *negative definite*. If $Q(u,v)$ can be of either sign, we shall say that it is *indefinite*. If we merely have the inequality $Q(u,v) \geq 0$ for all nontrivial u and v, we say that Q is *non-negative definite*, with non-positive definite defined analogously. Occasionally, the term *positive indefinite* is used in place of non-negative definite.

EXERCISES

1. Show that if $a_1u^2 + 2b_1uv + c_1v^2$ and $a_2u^2 + 2b_2uv + c_2v^2$ are both positive definite, then $a_1a_2u^2 + 2b_1b_2uv + c_1c_2v^2$ is positive definite.

2. Under what conditions is $(a_1u_1 + a_2u_2)^2 + (b_1u_1 + b_2u_2)^2$ positive definite?

3. In terms of the foregoing notation, how can one tell whether $au^2 + 2buv + cv^2 = 1$ represents an ellipse, a hyperbola, or a parabola?

10. Geometric Approach. Let us now consider a variant of the foregoing method which will yield a valuable insight into the geometrical significance of the roots, λ_1 and λ_2, and the values (u_i,v_i).

Assume that the equation

$$au^2 + 2buv + cv^2 = 1 \tag{1}$$

represents an ellipse, as pictured:

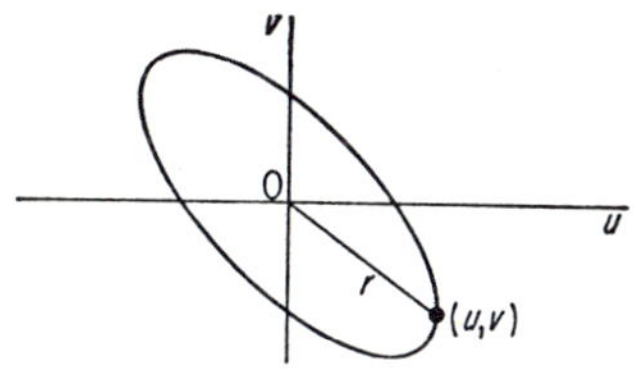

The quantity $r = (u^2 + v^2)^{1/2}$ denotes the length of the radius vector from the origin to a point (u,v) on the ellipse.

Let us use the fact that the problem of determining the maximum of $Q(u,v)$ on $u^2 + v^2 = 1$ is equivalent to that of determining the minimum of $u^2 + v^2$ on the curve $Q(u,v) = 1$.

The Lagrange multiplier formalism as before leads to the equations

$$\begin{aligned} u - \lambda(au + bv) &= 0 \\ v - \lambda(bu + cv) &= 0 \end{aligned} \tag{2}$$

These yield the equation

$$\begin{vmatrix} 1 - a\lambda & -b\lambda \\ -b\lambda & 1 - c\lambda \end{vmatrix} = 0 \tag{3}$$

or

$$\begin{vmatrix} a - \dfrac{1}{\lambda} & b \\ b & c - \dfrac{1}{\lambda} \end{vmatrix} = 0 \tag{4}$$

If λ_i is a root of this equation and (u_i,v_i) the corresponding extremum point, we see as above that

$$u_i^2 + v_i^2 = \lambda_i \tag{5}$$

From this we conclude that one root of (4) is the square of the minimum distance from the origin to the ellipse and that the other is the square of the maximum distance. We observe then that the variational problem we have posed yields in the course of its solution the lengths of the major and minor axes of the ellipse. The condition of (6.8) we now recognize as the well-known perpendicularity or *orthogonality* of the principal axes of an ellipse.

The linear transformation of (7.1) is clearly a rotation, since it preserves both the origin and distance from the origin. We see that it is precisely the rotation which aligns the coordinate axes and the axes of the ellipse.

EXERCISES

1. From the foregoing facts, conclude that the area of the ellipse is given by $\pi/(ac - b^2)^{1/2}$.

2. Following the algebraic approach, show that necessary and sufficient conditions that the form $Q = au^2 + 2buv + cv^2 + 2duw + ew^2 + 2fvw$ be positive definite are that

$$a > 0 \qquad \begin{vmatrix} a & b \\ b & c \end{vmatrix} > 0 \qquad \begin{vmatrix} a & b & d \\ b & c & f \\ d & f & e \end{vmatrix} > 0$$

3. Similarly, following the analytic approach, show that a necessary and sufficient condition that Q be positive definite is that all the roots of the determinantal equation

$$\begin{vmatrix} a - \lambda & b & d \\ b & c - \lambda & f \\ d & f & e - \lambda \end{vmatrix} = 0$$

be positive.

4. If Q is positive definite, show that the equation $Q(u,v,w) = 1$ represents an ellipsoid and determine its volume.

11. Discussion. We have pursued the details of the two-dimensional case, details which are elementary but whose origins are perhaps obscure, in order that the reader will more readily understand the need for a better notation and appreciate its advantages. The results which seem so providential here, and which have essentially been verified by direct calculation, will be derived quite naturally in the general case.

The basic ideas, however, and the basic devices are all contained in the preceding discussions.

MISCELLANEOUS EXERCISES

1. For what values of x_1 and x_2 is the quadratic form $(a_{11}x_1 + a_{12}x_2 - b_1)^2 + (a_{21}x_1 + a_{22}x_2 - b_2)^2$ a minimum, and what is the minimum value?

2. Show that $(x_1^2 + y_1^2)(x_2^2 + y_2^2)$ can be written in the form

$$(a_1x_1x_2 + a_2x_1y_2 + a_3x_2y_1 + a_4y_1y_2)^2 + (b_1x_1x_2 + b_2x_1y_2 + b_3x_2y_1 + b_4y_1y_2)^2$$

for values of a_i and b_i which are independent of x_i and y_i, and determine all such values.

3. Show that there exists no corresponding result for

$$(x_1^2 + y_1^2 + z_1^2)(x_2^2 + y_2^2 + z_2^2)$$

4. If $x_1u^2 + 2x_2uv + x_3v^2$ and $y_1u^2 + 2y_2uv + y_3v^2$ are positive definite, then

$$\begin{vmatrix} x_1y_1 & x_2y_2 \\ x_2y_2 & x_3y_3 \end{vmatrix} \geq \begin{vmatrix} x_1 & x_2 \\ x_2 & x_3 \end{vmatrix} \begin{vmatrix} y_1 & y_2 \\ y_2 & y_3 \end{vmatrix}$$

5. Utilize this result to treat Exercise 1 of Sec. 9.

6. Establish the validity of the Lagrange multiplier formalism by considering the maximum and minimum values of $(au^2 + 2buv + cv^2)/(u^2 + v^2)$.

7. What linear transformations leave the quadratic form $Q(x_1,x_2) = \lambda(x_1^2 + x_2^2) + (1 - \lambda)(x_1 + x_2)^2$ invariant? Here $0 \leq \lambda \leq 1$.

Bibliography

§1. A general discussion of the maxima and minima of functions of several variables can be found in any of a number of books on advanced calculus. A particularly thorough discussion may be found in

H. Hancock, *The Theory of Maxima and Minima*, Ginn & Company, Boston, 1907.

In a later chapter we shall study the more difficult problem of constrained maxima and minima.

The problem of determining the relations between the number of relative maxima, relative minima, and stationary points of other types is one which belongs to the domain of topology and will not be discussed here; see M. Morse, *The Calculus of Variations in the Large*, Amer. Math. Soc., 1934, vol. 18.

§9. Other locutions also appear frequently. Schwerdtfeger and Mirsky use *positive semidefinite;* Halmos uses *non-negative semidefinite.*

2

Vectors and Matrices

1. Introduction. In Chap. 1 we studied the question of determining the local maxima and minima of a function of two variables. If we consider the corresponding problem for functions of N variables, and proceed as before, we immediately encounter the problem of determining simple necessary and sufficient conditions which ensure the positivity of a quadratic form in N variables,

$$Q(x_1,x_2, \ldots ,x_N) = \sum_{i,j=1}^{N} a_{ij}x_ix_j \tag{1}$$

As we shall see later, in Chap. 5, the algebraic method presented in the previous chapter yields a simple and elegant solution of this problem. However, since we really want a much deeper understanding of quadratic forms than merely that required for this particular problem, we shall pretend here that this solution does not exist.

Our objective in this chapter then is to develop a notation which will enable us to pursue the analytic approach with a very minimum of arithmetic or analytic calculation. Pursuant to this aim, we want a notation as independent of dimension as possible.

Oddly enough, the study of quadratic functions of the form appearing above is enormously simplified by a notation introduced initially to study linear transformations of the form

$$y_i = \sum_{j=1}^{N} a_{ij}x_j \qquad i = 1, 2, \ldots , N \tag{2}$$

2. Vectors. Let us begin by defining a *vector*, a set of N complex numbers which we shall write in the form

$$x = \begin{bmatrix} x_1 \\ x_2 \\ \cdot \\ \cdot \\ \cdot \\ x_N \end{bmatrix} \tag{1}$$

A vector of this type is called a *column vector*. If N numbers are arranged in horizontal array,

$$x = (x_1, x_2, \ . \ . \ . \ , x_N) \tag{2}$$

x is called a *row vector*.

Since we can do all that we wish to do here working with column vectors, we shall henceforth reserve the term "vector" to denote a quantity having the form in (1).

Lower-case letters such as x, y, z or a, b, c will be employed throughout to designate vectors. When discussing a particular set of vectors, we shall use superscripts, thus x^1, x^2, etc.

The quantities x_i are called the *components* of x, while N is called the *dimension* of the vector x. One-dimensional vectors are called *scalars*. These are the usual quantities of analysis.

By $\bar{x}$ we shall denote the vector whose components are the complex conjugates of the elements of x. If the components of x are all real, we shall say that x is real.

3. Vector Addition. Let us now proceed to develop an algebra of vectors, which is to say a set of rules for manipulating these quantities. Since arbitrarily many sets of rules can be devised, the justification for those we present will and must lie in the demonstration that they permit us to treat some important problems in a straightforward and elegant fashion.

Two vectors x and y are said to be *equal* if and only if their components are equal, $x_i = y_i$ for $i = 1, 2, \ . \ . \ . \ , N$. The simplest operation acting on two vectors is *addition*. The sum of two vectors, x and y, is written $x + y$ and defined to be the vector

$$x + y = \begin{bmatrix} x_1 + y_1 \\ x_2 + y_2 \\ \cdot \\ \cdot \\ \cdot \\ x_N + y_N \end{bmatrix} \tag{1}$$

It should be noted that the plus sign connecting x and y is not the same as the sign connecting x_i and y_i. However, since, as we readily see, it enjoys the same analytic properties, there is no harm in using the same symbol in both cases.

EXERCISES

1. Show that we have *commutativity*, $x + y = y + x$, and *associativity*, $x + (y + z) = (x + y) + z$.

2. Hence, show that $x^1 + x^2 + \cdots + x^M$ is an unambiguous vector.

3. Define subtraction of two vectors directly, and in terms of addition; that is, $x - y$ is a vector z such that $y + z = x$.

4. Scalar Multiplication. Multiplication of a vector x by a scalar c_1 is defined by means of the relation

$$c_1 x = x c_1 = \begin{bmatrix} c_1 x_1 \\ c_1 x_2 \\ \cdot \\ \cdot \\ \cdot \\ c_1 x_N \end{bmatrix} \tag{1}$$

EXERCISES

1. Show that $(c_1 + c_2)(x + y) = c_1 x + c_1 y + c_2 x + c_2 y$.

2. Define the *null* vector, written 0, to be the vector all of whose components are zero. Show that it is uniquely determined as the vector 0 for which $x + 0 = x$ for all x.

3. Show that it is uniquely determined by the condition that $c_1 0 = 0$ for all scalars c_1.

5. The Inner Product of Two Vectors. We now introduce a most important scalar function of two vectors x and y, the *inner product*. This function will be written (x,y) and defined by the relation

$$(x,y) = \sum_{i=1}^{N} x_i y_i \tag{1}$$

The following properties of the inner product are derived directly from the definition:

$$(x,y) = (y,x) \tag{2a}$$
$$(x + y, z + w) = (x,z) + (x,w) + (y,z) + (y,w) \tag{2b}$$
$$(c_1 x, y) = c_1 (x,y) \tag{2c}$$

This is one way of "multiplying" two vectors. However, there are other ways, which we shall not use here.

The importance of the inner product lies in the fact that (x,x) can be considered to represent the square of the "length" of the real vector x. We thus possess a method for evaluating these non-numerical quantities.

EXERCISES

1. If x is real, show that $(x,x) > 0$ unless $x = 0$.

2. Show that $(ux + vy, ux + vy) = u^2(x,x) + 2uv(x,y) + v^2(y,y)$ is a non-negative definite quadratic form in the scalar variables u and v if x and y are real and hence that

$$(x,y)^2 \leq (x,x)(y,y)$$

for any two real vectors x and y (Cauchy's inequality).

3. What is the geometric interpretation of this result?

4. Using this result, show that

$$(x + y, x + y)^{1/2} \leq (x,x)^{1/2} + (y,y)^{1/2}$$

for any two real vectors.

5. Why is this inequality called the "triangle inequality"?

6. Show that for any two complex vectors x and y we have

$$|(x,y)|^2 \leq (x,\bar{x})(y,\bar{y})$$

6. Orthogonality. If two real vectors are connected by the relation

$$(x,y) = 0 \tag{1}$$

they are said to be *orthogonal*. The importance of this concept derives to a great extent from the following easily established result.

Theorem 1. *Let $\{x^i\}$ be a set of real vectors which are mutually orthogonal; that is, $(x^i,x^j) = 0$, $i \neq j$, and normalized by the condition that*

$$(x^i,x^i) = 1$$

Then if a particular vector x has the representation

$$x = \sum_{i=1}^{M} c_i x^i \tag{2}$$

we have the following values for the coefficients

$$c_i = (x,x^i) \tag{3}$$

and, in addition,

$$(x,x) = \sum_{i=1}^{M} c_i^2 \tag{4}$$

EXERCISES

1. Why is the proper concept of orthogonality for complex vectors the relation

$$(x,\bar{y}) = 0$$

and the proper normalization the condition $(x,\bar{x}) = 1$?

2. Show that $(x,\bar{y}) = \overline{(\bar{x},y)}$.

3. In N-dimensional Euclidean space, consider the coordinate vectors

$$y^1 = \begin{bmatrix} 1 \\ 0 \\ 0 \\ \cdot \\ \cdot \\ \cdot \\ 0 \end{bmatrix} \qquad y^2 = \begin{bmatrix} 0 \\ 1 \\ 0 \\ \cdot \\ \cdot \\ \cdot \\ 0 \end{bmatrix} \qquad \ldots, \qquad y^N = \begin{bmatrix} 0 \\ 0 \\ 0 \\ \cdot \\ \cdot \\ \cdot \\ 1 \end{bmatrix}$$

Show that the y^i are mutually orthogonal, and are, in addition, normalized. A set of this type is called *orthonormal*. If $x = \sum_{i=1}^{N} c_i y^i$, determine the values of the c_i, and discuss the geometric meaning of the result.

7. Matrices. Let us now introduce the concept of a *matrix*. An array of complex numbers written in the form

$$A = \begin{bmatrix} a_{11} & a_{12} & \cdots & a_{1N} \\ a_{21} & a_{22} & \cdots & a_{2N} \\ \cdot & & & \\ \cdot & & & \\ \cdot & & & \\ a_{N1} & a_{N2} & \cdots & a_{NN} \end{bmatrix} \tag{1}$$

will be called a *square matrix*. Since these are the only types of matrices we shall consider to any extent, we shall reserve the term "matrix" for these entities. When other types of matrices are introduced subsequently, we shall use an appropriate modifying adjective.

The quantities a_{ij} are called the *elements* of A and the integer N is the dimension. The quantities $a_{i1}, a_{i2}, \ldots, a_{iN}$ are said to constitute the ith *row* of A, and the quantities $a_{1j}, a_{2j}, \ldots, a_{Nj}$ are said to constitute the jth *column*. Throughout, matrices will be denoted by upper-case letters, X, Y, Z and A, B, C. From time to time we shall use the shorthand notation

$$A = (a_{ij}) \tag{2}$$

The determinant associated with the array in (1) will be written $|A|$ or $|a_{ij}|$.

The simplest relation between matrices is that of equality; two matrices are called equal if and only if their elements are equal. Following the same path as for vectors, we next define the operation of addition. The sum of two matrices A and B is written $A + B$ and defined by the notation

$$A + B = (a_{ij} + b_{ij}) \tag{3}$$

Multiplication of a matrix A by a scalar c_1 is defined by the relation

$$c_1 A = A c_1 = (c_1 a_{ij}) \tag{4}$$

Finally, by $\bar{A}$ we shall mean the matrix whose elements are the complex conjugates of A. When the elements of A are real, we shall call A a real matrix.

EXERCISES

1. Show that 0, the null matrix defined by the condition that all its elements are zero, is uniquely determined by the condition that $A + 0 = A$ for all A.

2. Show that matrix addition is commutative and associative.
3. Show that $A_1 + A_2 + \cdots + A_N$ is unambiguously defined.

8. Matrix Multiplication—Vector by Matrix. In order to make our algebra more interesting, we shall define some multiplicative operations. In order to render these concepts reasonable, we return to linear transformations of the form

$$y_i = \sum_{j=1}^{N} a_{ij}x_j, \qquad i = 1, 2, \ldots, N \tag{1}$$

where the coefficients a_{ij} are complex quantities. Since, after all, the whole purpose of our introduction of vector-matrix notation is to facilitate the study of these transformations, it is only fair that in deriving the fundamental properties of vectors and matrices we turn occasionally to the defining equations for guidance.

The point we wish to make is that the definitions of addition and multiplication of vectors and matrices are not arbitrary, but, on the contrary, are essentially forced upon us by the analytic and geometric properties of the entities we are studying.

Given two vectors x and y related as in (1), we write

$$y = Ax \tag{2}$$

This relation defines multiplication of a vector x by a matrix A. Observe carefully the order in which the product is written.

EXERCISES

1. Show that $(A + B)(x + y) = Ax + Ay + Bx + By$.
2. Consider the *identity matrix* I defined by the tableau

$$I = \begin{bmatrix} 1 & & & & 0 \\ & 1 & & & \\ & & \cdot & & \\ & & & \cdot & \\ & & & & \cdot \\ 0 & & & & 1 \end{bmatrix}$$

Explicitly, $I = (\delta_{ij})$, where δ_{ij} is the Kronecker delta symbol defined by the relation

$$\begin{aligned} \delta_{ij} &= 0, \qquad i \neq j \\ &= 1, \qquad i = j \end{aligned}$$

Show that

$$\delta_{ij} = \sum_{k=1}^{N} \delta_{ik}\delta_{kj}$$

3. Show that $Ix = x$ for all x, and that this relation uniquely determines I.

4. Show that

$$(Ax,Ax) = \sum_{i=1}^{N} \left(\sum_{j=1}^{N} a_{ij}x_j \right)^2$$

9. Matrix Multiplication—Matrix by Matrix. Now let us see how to define the product of a matrix by a matrix. Consider a second linear transformation

$$z = By \tag{1}$$

which converts the components of y into the components of z. In order to express the components of z in terms of the components of x, where, as above, $y = Ax$, we write

$$\begin{aligned} z_i &= \sum_{k=1}^{N} b_{ik}y_k = \sum_{k=1}^{N} b_{ik} \left(\sum_{j=1}^{N} a_{kj}x_j \right) \\ &= \sum_{j=1}^{N} \left(\sum_{k=1}^{N} b_{ik}a_{kj} \right) x_j \end{aligned} \tag{2}$$

If we now introduce a new matrix $C = (c_{ij})$ defined by the relations

$$c_{ij} = \sum_{k=1}^{N} b_{ik}a_{kj} \qquad i, j = 1, 2, \ldots, N \tag{3}$$

we may write

$$z = Cx \tag{4}$$

Since, formally,

$$z = By = B(Ax) = (BA)x \tag{5}$$

we are led to define the product of A by B,

$$C = BA \tag{6}$$

where C is determined by (3). Once again, note carefully the order in which the product is written.

EXERCISES

1. Show that $(A + B)(C + D) = AC + AD + BC + BD$.

2. If

$$T(\theta) = \begin{bmatrix} \cos\theta & -\sin\theta \\ \sin\theta & \cos\theta \end{bmatrix}$$

show that

$$T(\theta_1)T(\theta_2) = T(\theta_2)T(\theta_1) = T(\theta_1 + \theta_2)$$

3. Let A be a matrix with the property that $a_{ij} = 0$, if $j \neq i$, a *diagonal* matrix, and let B be a matrix of similar type. Show that AB is again a diagonal matrix, and that $AB = BA$.

4. Let A be a matrix with the property that $a_{ij} = 0,\ j > i$, a *triangular* or *semi-diagonal* matrix, and let B be a matrix of similar type. Show that AB is again a triangular matrix, but that $AB \neq BA$, in general.

5. Let

$$A = \begin{bmatrix} a_1 & a_2 \\ -a_2 & a_1 \end{bmatrix} \qquad B = \begin{bmatrix} b_1 & b_2 \\ -b_2 & b_1 \end{bmatrix}$$

Show that AB is again a matrix of the same type, and that $AB = BA$. (As we shall subsequently see, these matrices are equivalent to complex numbers of the form $a_1 + ia_2$ if a_1 and a_2 are real.)

6. Use the relation $|AB| = |A|\,|B|$ to show that

$$(a_1^2 + a_2^2)(b_1^2 + b_2^2) = (a_1b_1 - a_2b_2)^2 + (a_2b_1 + a_1b_2)^2$$

7. Let

$$A = \begin{bmatrix} a_1 + ia_3 & a_2 + ia_4 \\ -a_2 + ia_4 & a_1 - ia_3 \end{bmatrix}$$

and B be a matrix of similar type. Show that AB is a matrix of similar type, but that $AB \neq BA$ in general.

8. Use the relation $|AB| = |A|\,|B|$ to express $(a_1^2 + a_2^2 + a_3^2 + a_4^2)(b_1^2 + b_2^2 + b_3^2 + b_4^2)$ as a sum of four squares.

9. Let

$$A = \begin{bmatrix} a_1 & a_2 & a_3 & a_4 \\ -a_2 & a_1 & -a_4 & a_3 \\ -a_3 & a_4 & a_1 & a_2 \\ -a_4 & a_3 & -a_2 & a_1 \end{bmatrix}$$

and B be a matrix of similar type. Show that AB is a matrix of similar type, but that $AB \neq BA$ in general. Evaluate $|A|$. (These matrices are equivalent to quaternions.)

10. Consider the linear fractional transformation

$$w = \frac{a_1z + b_1}{c_1z + d_1} = T_1(z)$$

If $T_2(z)$ is a similar transformation, with coefficients a_1, b_1, c_1, d_1 replaced by a_2, b_2, c_2, d_2, show that $T_1(T_2(z))$ and $T_2(T_1(z))$ are again transformations of the same type.

11. If the expression $a_1d_1 - b_1c_1 \neq 0$, show that $T_1^{-1}(z)$ is a transformation of the same type. If $a_1d_1 - b_1c_1 = 0$, what type of a transformation is $T_1(z)$?

12. Consider a correspondence between $T_1(z)$ and the matrix

$$A_1 = \begin{bmatrix} a_1 & b_1 \\ c_1 & d_1 \end{bmatrix}$$

written $A_1 \sim T_1(z)$. Show that if $A_1 \sim T_1(z)$ and $A_2 \sim T_2(z)$, then $A_1A_2 \sim T_1(T_2(z))$. What then is the condition that $T_1(T_2(z)) = T_2(T_1(z))$ for all z?

13. How can the foregoing results be used to obtain a representation for the iterates of $T_1(z)$?

14. Show that

$$\begin{bmatrix} -1 & -1 \\ 1 & 0 \end{bmatrix}^3 = I$$

15. If $X = (x_{ij})$, where $x_{ij} = (-1)^{N-j}\binom{N-j}{i-1}$ (the binomial coefficient), then $X^3 = I$.

16. From the fact that we can establish a correspondence

$$1 \sim \begin{bmatrix} 1 & 0 \\ 0 & 1 \end{bmatrix} \qquad -1 \sim \begin{bmatrix} 0 & 1 \\ 1 & 0 \end{bmatrix}$$

deduce that arithmetic can be carried out without the aid of negative numbers.

10. Noncommutativity. What makes the study of matrices so fascinating, albeit occasionally thorny, is the fact that multiplication is *not* commutative. In other words, in general,

$$AB \neq BA \tag{1}$$

A simple example is furnished by the 2×2 matrices

$$A = \begin{bmatrix} 1 & 2 \\ 3 & 4 \end{bmatrix} \qquad B = \begin{bmatrix} 2 & 1 \\ 4 & 3 \end{bmatrix} \tag{2}$$

and more interesting examples by the matrices appearing in Exercises 7 and 9 of Sec. 9. Then,

$$AB = \begin{bmatrix} 10 & 7 \\ 22 & 15 \end{bmatrix} \qquad BA = \begin{bmatrix} 5 & 8 \\ 13 & 20 \end{bmatrix} \tag{3}$$

If $AB = BA$, we shall say that A and B *commute*.

The theory of matrices yields a very natural transition from the tame domain of scalars and their amenable algebra to the more interesting and realistic world where many different species of algebras abound, each with its own singular and yet compensating properties.

EXERCISES

1. Show that $AI = IA = A$ for all A, and that this relation uniquely determines the identity matrix.

2. Show that $A0 = 0A = 0$ for all A, and that this relation uniquely determines the null matrix.

3. Let the rows of A be considered to consist, respectively, of the components of the vectors $a^1, a^2, \ldots, a^N$, and the columns of B to consist of the components of $b^1, b^2, \ldots, b^N$. Then we may write

$$AB = [(a^i,b^j)]$$

4. If $AX = XA$ for *all* X, then A is a scalar multiple of I.

11. Associativity. Fortunately, although commutativity does not hold, associativity of multiplication is preserved in this new algebra. In other words, for all A, B, and C we have

$$(AB)C = A(BC) \tag{1}$$

This means that the product ABC is unambiguously defined without the aid of parentheses. To establish this result most simply, let us employ the "dummy index" convention, which asserts that any index which is repeated is to be summed over all of its admissible values. The ijth element in AB may then be written

$$a_{ik}b_{kj} \tag{2}$$

Employing this convention and the definition of multiplication given above, we see that

$$\begin{aligned} (AB)C &= [(a_{ik}b_{kl})c_{lj}] \\ A(BC) &= [a_{ik}(b_{kl}c_{lj})] \end{aligned} \tag{3}$$

which establishes the equality of $(AB)C$ and $A(BC)$.

EXERCISES

1. Show that $ABCD = A(BCD) = (AB)(CD) = (ABC)D$, and generally that $A_1A_2 \cdots A_N$ has a unique significance.

2. Show that $A^n = A \cdot A \cdot (n \text{ times}) A$ is unambiguously defined, and that $A^{m+n} = A^mA^n$, $m,n = 1, 2, \ldots$.

3. Show that A^m and B^n commute if A and B commute.

4. Write

$$X^n = \begin{bmatrix} x_1(n) & x_2(n) \\ x_3(n) & x_4(n) \end{bmatrix} \qquad \text{where } X = \begin{bmatrix} x_1 & x_2 \\ x_3 & x_4 \end{bmatrix}$$

a given matrix. Using the relation $X^{n+1} = XX^n$, derive recurrence relations for the $x_i(n)$ and thus derive the analytic form of the $x_i(n)$.

5. Use these relations for the case where

$$X = \begin{bmatrix} x_1 & x_2 \\ -x_2 & x_1 \end{bmatrix} \qquad x_1, x_2 \text{ real}$$

or

$$X = \begin{bmatrix} x_1 & x_2 \\ -\bar{x}_2 & \bar{x}_1 \end{bmatrix} \qquad x_1, x_2 \text{ complex}$$

6. Use these relations to find explicit representations for the elements of X^n where

$$X = \begin{bmatrix} \lambda & 1 \\ 0 & \lambda \end{bmatrix} \qquad X = \begin{bmatrix} \lambda & 1 & 0 \\ 0 & \lambda & 1 \\ 0 & 0 & \lambda \end{bmatrix}$$

7. Find all 2×2 matrix solutions of $X^2 = X$.

8. Find all 2×2 matrix solutions of $X^n = X$, where n is a positive integer.

12. Invariant Vectors. Proceeding as in Sec. 5 of Chap. 1, we see that the problem of determining the maximum and minimum values of $Q = \sum_{i,j=1}^{N} a_{ij}x_ix_j$ for x_i satisfying the relation $\sum_{i=1}^{N} x_i^2 = 1$ can be reduced to the problem of determining the values of λ for which the linear homogeneous equations

$$\sum_{j=1}^{N} a_{ij}x_j = \lambda x_i \qquad i = 1, 2, \ldots, N \tag{1}$$

possess nontrivial solutions.

In vector-matrix terms, we can write these equations in the form of a single equation

$$Ax = \lambda x \tag{2}$$

Written in this way, the equation has a very simple significance. We are looking for those vectors x which are transformed into scalar multiples of themselves by the matrix A. Thinking of x as representing a direction indicated by the N direction numbers $x_1, x_2, \ldots, x_N$, we are searching for invariant directions.

We shall pursue this investigation vigorously in the following chapters. In the meantime, we wish to introduce a small amount of further notation which will be useful to us in what follows.

13. Quadratic Forms as Inner Products. Let us now present another justification of the notation we have introduced. The quadratic form $Q(u,v) = au^2 + 2buv + cv^2$ can be written in the form

$$u(au + bv) + v(bu + cv) \tag{1}$$

Hence, if the vector x and the matrix A are defined by

$$x = \begin{bmatrix} u \\ v \end{bmatrix} \qquad A = \begin{bmatrix} a & b \\ b & c \end{bmatrix} \tag{2}$$

we see that

$$Q(u,v) = (x,Ax) \tag{3}$$

Similarly, given the N-dimensional quadratic form

$$Q(x) = \sum_{i,j=1}^{N} a_{ij}x_ix_j \tag{4}$$

where without loss of generality we may take $a_{ij} = a_{ji}$, we can write

$$\begin{aligned} Q(x) &= x_1 \left[\sum_{j=1}^{N} a_{1j}x_j \right] + x_2 \left[\sum_{j=1}^{N} a_{2j}x_j \right] + \cdots + x_N \left[\sum_{j=1}^{N} a_{Nj}x_j \right] \\ &= (x,Ax) \end{aligned} \tag{5}$$

where x has the components x_i and $A = (a_{ij})$.

EXERCISES

1. Does $(x,Ax) = (x,Bx)$ for all x imply that $A = B$?
2. Under what conditions does $(x,Ax) = 0$ for all x?

14. The Transpose Matrix. Let us now define a most important matrix function of A, the *transpose* matrix, by means of the relation

$$A' = (a_{ji}) \tag{1}$$

The rows of A' are the columns of A and the rows of A are the columns of A'.

We are led to consider this new matrix in the following fashion. Consider a set of vectors $\{x\}$ and another set $\{y\}$, and form all inner products (x,y) composed of one vector from one set and one from the other.

Suppose now that we transform the set $\{x\}$ by means of matrix multiplication of A, obtaining the new set $\{Ax\}$. Forming inner products as before, we obtain the set of values (Ax,y).

Observe that

$$(Ax,y) = y_1 \left[\sum_{j=1}^{N} a_{1j}x_j \right] + y_2 \left[\sum_{j=1}^{N} a_{2j}x_j \right] + \cdots + y_N \left[\sum_{j=1}^{N} a_{Nj}x_j \right] \tag{2}$$

or, rearranging,

$$\begin{aligned}(Ax,y) &= x_1 \left[\sum_{i=1}^{N} a_{i1}y_i \right] + x_2 \left[\sum_{i=1}^{N} a_{i2}y_i \right] + \cdots + x_N \left[\sum_{i=1}^{N} a_{iN}y_i \right] \\ &= (x,A'y)\end{aligned} \tag{3}$$

In other words, as far as the inner product is concerned, the effect of the transformation A on the set of x's is equivalent to the transformation A' on the set of y's. We can then regard A' as an *induced transformation* or *adjoint transformation.* This simple, but powerful, idea pervades much of classical and modern analysis.

15. Symmetric Matrices. From the foregoing discussion, it is plausible that matrices satisfying the condition

$$A = A' \tag{1}$$

should enjoy special properties and play an important role in the study of quadratic forms. This is indeed the case. These matrices are called *symmetric,* and are characterized by the condition that

$$a_{ij} = a_{ji} \tag{2}$$

The first part of this volume will be devoted to a study of the basic properties of this class of matrices for the case where the a_{ij} are *real.*

Henceforth, we shall use the term "symmetric" to denote *real symmetric*. When there is any danger of confusion, we shall say real symmetric or complex symmetric, depending upon the type of matrix we are considering.

EXERCISES

1. Show that $(A')' = A$.

2. Show that $(A + B)' = A' + B'$, $(AB)' = B'A'$, $(A_1A_2 \dots A_n)' = A'_n \dots A'_2A'_1$, $(A^n)' = (A')^n$.

3. Show that AB is not necessarily symmetric if A and B are.

4. Show that $A'BA$ is symmetric if B is symmetric.

5. Show that $(Ax,By) = (x,A'By)$.

6. Show that $|A| = |A'|$.

7. Show that when we write $Q(x) = \sum_{i,j=1}^{N} a_{ij}x_ix_j$, there is no loss of generality in assuming that $a_{ij} = a_{ji}$.

16. Hermitian Matrices. As we have mentioned above, the important scalar function for complex vectors turns out not to be the usual inner product, but the expression $(x,\bar{y})$. If we note that

$$(Ax,\bar{y}) = (x,\bar{z}) \tag{1}$$

where $z = \overline{A'}y$, we see that the induced transformation is now $\overline{A'}$, the complex conjugate of the transform of A. Matrices for which

$$A = \overline{A'} \tag{2}$$

are called *Hermitian*, after the great French mathematician Charles Hermite.

We shall write A^* in place of $\overline{A'}$ for simplicity of notation.

As we shall see, all the vital properties of symmetric matrices have immediate analogues for Hermitian matrices. Furthermore, if we had wished, we could have introduced a notation

$$[x,y] = (x,\bar{y}) \tag{3}$$

in terms of which the properties of both types of matrices can be derived simultaneously. There are advantages to both procedures, and the reader can take his choice, once he has absorbed the basic techniques.

EXERCISES

1. A real Hermitian matrix is symmetric.

2. $(A^*)^* = A$, $(AB)^* = B^*A^*$, $(A_1A_2 \cdots A_n)^* = A_n^* \cdots A_2^*A_1^*$.

3. If $A + iB$ is Hermitian, A,B real, then $A' = A$, $B' = -B$.

17. Invariance of Distance—Orthogonal Matrices. Taking as our guide the Euclidean measure of distance, we introduced the quantity (x,x) as a measure of the magnitude of the real vector x.

It is a matter of some curiosity, and importance too as we shall see, to determine the linear transformations $y = Tx$ which leave (x,x) unchanged. In other words, we wish to determine T so that the equation

$$(x,x) = (Tx,Tx) \tag{1}$$

is satisfied for *all* x. Since

$$(Tx,Tx) = (x,T'Tx) \tag{2}$$

and $T'T$ is symmetric, we see that (1) yields the relation

$$T'T = I \tag{3}$$

A real matrix T possessing this property is called *orthogonal.*

EXERCISES

1. Show that T' is orthogonal whenever T is.
2. Show that every 2×2 orthogonal matrix with determinant $+1$ can be written in the form

$$\begin{bmatrix} \cos\theta & -\sin\theta \\ \sin\theta & \cos\theta \end{bmatrix}$$

What is the geometrical significance of this result?

3. Show that the columns of T are orthogonal vectors.
4. Show that the product of two orthogonal matrices is again an orthogonal matrix
5. Show that the determinant of an orthogonal matrix is ± 1.
6. Let T_N be an orthogonal matrix of dimension N, and form the $(N + 1)$-dimensional matrix

$$T_{N+1} = \begin{bmatrix} 1 & 0 & \cdots & 0 \\ 0 & & & \\ \cdot & & & \\ \cdot & & T_N & \\ \cdot & & & \\ 0 & & & \end{bmatrix}$$

Show that T_{N+1} is orthogonal.

7. Show that if T is orthogonal, $x = Ty$ implies that $y = T'x$.
8. If $AB = BA$, then TAT' and TBT' commute if T is orthogonal.

18. Unitary Matrices. Since the appropriate measure for a complex vector is $(x,\bar{x})$, we see that the analogue of the invariance condition of (17.3) is

$$T^*T = I \tag{1}$$

Matrices possessing this property are called *unitary,* and play the same role in the treatment of Hermitian matrices that orthogonal matrices enjoy in the theory of symmetric matrices.

EXERCISES

1. Show that T^* is unitary if T is.

2. Show that the product of two unitary matrices is again unitary.

3. Show that the determinant of a unitary matrix has absolute value 1.

4. Show that if T is unitary, $x = Ty$ implies that $y = T^*x$.

5. Obtain a result corresponding to that given in Exercise 2 of Sec. 17 for the elements of a 2×2 unitary matrix. (Analogous, but more complicated, results hold for the representation of the elements of 3×3 orthogonal matrices in terms of elliptic functions. See F. Caspary, Zur Theorie der Thetafunktionen mit zwei Argumenten, *Kronecker J.*, XCIV, pp. 74–86; and F. Caspary, Sur les systèmes orthogonaux, formés par les fonctions théta, *Comptes Rendus de Paris*, CIV, pp. 490–493.)

6. Is a real unitary matrix orthogonal?

7. Is a complex orthogonal matrix unitary?

8. Every 3×3 orthogonal matrix can be represented as the product of

$$\begin{bmatrix} \cos a & \sin a & 0 \\ -\sin a & \cos a & 0 \\ 0 & 0 & 1 \end{bmatrix} \begin{bmatrix} \cos b & 0 & -\sin b \\ 0 & 1 & 0 \\ \sin b & 0 & \cos b \end{bmatrix} \begin{bmatrix} 1 & 0 & 0 \\ 0 & \cos c & \sin c \\ 0 & -\sin c & \cos c \end{bmatrix}$$

What is the geometric origin of this result?

MISCELLANEOUS EXERCISES

1. Prove that

$$\begin{vmatrix} a_{11} & a_{12} & \cdots & a_{1N} & x_1 \\ a_{21} & a_{22} & \cdots & a_{2N} & x_2 \\ \cdot & & & & \cdot \\ \cdot & & & & \cdot \\ \cdot & & & & \cdot \\ a_{N1} & a_{N2} & \cdots & a_{NN} & x_N \\ x_1 & x_2 & \cdots & x_N & 0 \end{vmatrix} = -(a^{11}x_1^2 + \cdots (a^{ij} + a^{ji})x_ix_j + \cdots)$$

where a^{ij} is the cofactor of a_{ij} in $|a_{ij}|$.

2. Let I_{ij} denote the matrix obtained by interchanging the ith and jth rows of the unit matrix I. Prove that

$$I_{ij}^2 = I \qquad I_{ik}I_{kj}I_{ji} = I_{kj}$$

3. Show that $I_{ij}A$ is a matrix identical to A except that the ith and jth rows have been interchanged, while AI_{ij} yields the interchange of the ith and jth columns.

4. Let H_{ij} denote the matrix whose ijth element is h, and whose other elements are zero. Show that $(I + H_{ij})A$ yields a matrix identical with A except that the ith row has been replaced by the ith row plus h times the jth row, while $A(I + H_{ij})$ has a similar effect upon columns.

5. Let H_r denote the matrix equal to I except for the fact that the element one in the rr position is equal to k. What are H_rA and AH_r equal to?

6. If A is real and $AA' = 0$, then $A = 0$.

7. If $AA^* = 0$, then $A = 0$.

8. Show that if T is orthogonal, its elements are uniformly bounded. Similarly, if U is unitary, its elements are uniformly bounded in absolute value.

9. Let d_1 denote the determinant of the system of linear homogeneous equations derived from the relations

$$\left[\sum_{j=1}^{N} a_{ij}x_j\right]\left[\sum_{j=1}^{N} a_{kj}x_j\right] = 0 \qquad i, k = 1, 2, \ldots, N$$

regarding the $N(N+1)/2$ quantities x_ix_j, $i, j = 1, 2, \ldots, N$, as unknowns. Then $d_1 = |a_{ij}|^{N(N-1)/2}$ (*Schäfli*).

10. Show $\lim_{n\to\infty} A^n$, $\lim_{n\to\infty} B^n$ may exist as $n \to \infty$, without $\lim_{n\to\infty} (AB)^n$ existing. It is sufficient to take A and B two-dimensional.

11. Suppose that we are given the system of linear equations $\sum_{j=1}^{N} a_{ij}x_j = b_i$, $i = 1, 2, \ldots, N$. If it is possible to obtain the equations $x_1 = c_1$, $x_2 = c_2$, $\ldots$, $x_N = c_N$, by forming linear combinations of the given equations, then these equations yield a solution of the given equations, and the only solution (*R. M. Robinson*).

12. Introduce the Jacobi bracket symbol $[A,B] = AB - BA$, the *commutator* of A and B. Show, by direct calculation, that

$$[A,[B,C]] + [B,[C,A]] + [C,[A,B]] = 0$$

13. Let $r_1 = e^{2\pi i/n}$ be an irreducible root of unity, and let $r_k = e^{2\pi ik/n}$, $k = 1, 2, \ldots, n-1$. Consider the matrix

$$T = \begin{bmatrix} 1 & 1 & \cdots & 1 \\ 1 & r_1 & \cdots & r_1^{n-1} \\ \cdot & & & \\ \cdot & & & \\ \cdot & & & \\ 1 & r_{n-1} & \cdots & r_{n-1}^{n-1} \end{bmatrix}$$

Show that $T/n^{1/2}$ is unitary. Hence, if $x_k = \sum_{j=0}^{n-1} e^{2\pi ikj/n}y_j$, then $y_k = \frac{1}{n}\sum_{j=0}^{n-1} e^{-2\pi ikj/n}x_j$, and $\sum_{k=0}^{n-1} |x_k|^2 = n\sum_{k=0}^{n-1} |y_k|^2$. This transformation is called a *finite Fourier transform*.

14. Suppose that

$$\sum_{i,j=1}^{N} a_{ij}x_iy_j = \sum_{k=1}^{N} c_k \left\{\sum_{s=k}^{N} b_{ks}x_s\right\}\left\{\sum_{t=k}^{N} d_{kt}y_t\right\}$$

with $b_{kk} = d_{kk} = 1$ for all k. Then $|a_{ij}| = \prod_{k=1}^{N} c_k$. See J. L. Burchnall.[1]

15. Consider the Gauss transform $B = (b_{ij})$ of the matrix $A = (a_{ij})$,

$$b_{i1} = \delta_{i1}a_{i1}, \qquad b_{ik} = a_{11}^{-1}(a_{11}a_{ik} - a_{i1}a_{1k}), \qquad k > 1$$

Let $A_{11} = (a_{ij})$, $i, j = 2, \ldots, N$. Show that

$$|\lambda I - B| = a_{11}^{-1}[\lambda|\lambda I - A_{11}| - |\lambda I - A|]$$

(*D. M. Kotelyanskii*)

[1] J. L. Burchnall, *Proc. Edinburgh Math. Soc.*, (2), vol. 9, pp. 100–104, 1954.

16. Show that the matrices

$$A = \begin{bmatrix} 1 & a_1 & b_1 \\ 0 & 1 & 0 \\ 0 & 0 & 1 \end{bmatrix} \qquad B = \begin{bmatrix} 1 & -a_1 & -b_1 \\ 0 & 1 & 0 \\ 0 & 0 & 1 \end{bmatrix}$$

satisfy the relation $AB = I$.

17. Show that

$$\begin{vmatrix} x_1 & y_1 & z_1 \\ z_1 & x_1 & y_1 \\ y_1 & z_1 & x_1 \end{vmatrix} = (x_1 + y_1 + z_1)(x_1 + \omega y_1 + \omega^2 z_1)(x_1 + \omega^2 y_1 + \omega z_1)$$
$$= x_1^3 + y_1^3 + z_1^3 - 3x_1y_1z_1$$

where ω is a cube root of unity.

18. Hence, show that if $Q(x) = x_1^3 + x_2^3 + x_3^3 - 3x_1x_2x_3$, then $Q(x)Q(y) = Q(z)$, where z is a bilinear form in the x_i and y_i, that is, $z_i = \Sigma a_{ijk}x_jy_k$, where the coefficients a_{ijk} are independent of the x_i and y_j.

19. Show that

$$\begin{vmatrix} x_1 & x_2 & x_3 & x_4 \\ -x_2 & x_1 & -x_4 & x_3 \\ -x_3 & x_4 & x_1 & x_2 \\ -x_4 & x_3 & -x_2 & x_1 \end{vmatrix} = (x_1^2 + x_2^2 + x_3^2 + x_4^2)^2$$

and thus that

$$(x_1^2 + x_2^2 + x_3^2 + x_4^2)(y_1^2 + y_2^2 + y_3^2 + y_4^2) = z_1^2 + z_2^2 + z_3^2 + z_4^2,$$

where the z_i are bilinear forms in the x_i and y_i. It was shown by Hurwitz that a product of N squares multiplied by a product of N squares is a product of N squares in the above sense, only when $N = 1, 2, 4, 8$; see A. Hurwitz.[1] For an exposition of the theory of reproducing forms, see C. C. MacDuffee.[2,3]

20. Prove that a matrix A whose elements are given by the relations

$$\begin{aligned} a_{ij} &= (-1)^{i-1}\binom{j-1}{i-1} && i < j \\ &= (-1)^{i-1} && i = j \\ &= 0 && i > j \end{aligned}$$

satisfies the relation $A^2 = I$. Here $\binom{n}{k}$ is the binomial coefficient, $n!/k!(n-k)!$.

21. Let $y_i = y_i(x_1x_2, \ldots, x_N)$ be a set of N functions of the x_i, $i = 1, 2, \ldots, N$. The matrix $J = J(y,x) = (\partial y_i/\partial x_j)$ is called the Jacobian matrix and its determinant the Jacobian of the transformation. Show that

$$J(z,y)J(y,x) = J(z,x)$$

22. Consider the relation between the N^2 variables y_{ij} and the N^2 variables x_{ij} given by $Y = AXB$, where A and B are constant matrices. Show that $|J(Y,X)| = |A|^N|B|^N$.

[1] A. Hurwitz, Über die Komposition der quadratischen Formen von beliebig vielen Variablen, *Math. Werke*, bd II, Basel, 1933, pp. 565–571.

[2] C. C. MacDuffee, On the Composition of Algebraic Forms of Higher Degree, *Bull. Amer. Math. Soc.*, vol. 51 (1945), pp. 198–211.

[3] J. Radon, Lineare Scharen orthogonaler Matrizen, *Abh. Math. Sem. Hamb.*, vol. 1 (1921), pp. 1–14.

23. If $Y = XX'$, where X is triangular, then $|J(Y,X)| = 2^N \prod_{i=1}^{N} x_{ii}^{N-i+1}$.

24. Show that the problem of determining the maximum of the function $(x,Ax) - 2(x,b)$ leads to the vector equation $Ax = b$.

25. Similarly, show that the problem of minimizing $(x,Bx) + 2(x,Ay) + (y,By) - 2(a,x) - 2(b,y)$ over all x and y leads to the simultaneous equations $Bx + Ay = a$, $A'x + By = b$.

26. Let $f(x)$ be a function of the variable x which assumes only two values, $x = 0$, $x = 1$. Show that $f(x)$ may be written in the form $a + bx$, where $a = f(0)$, $b = f(1) - f(0)$.

27. Let $g(x)$ be a function of the same type, which itself assumes only two values 0 or 1. Then $f(g(x)) = a_1 + b_1 x$. Show that a_1 and b_1 are linear combinations of a and b, and thus that the effect of replacing x by $g(x)$ is equivalent to a matrix transformation of the vector whose components are a and b.

28. Let $f(x_1,x_2)$ be a function of the two variables x_1 and x_2 each of which assumes only the values 0 and 1. Show that we may write $f(x_1,x_2) = a_1 + a_2 x_1 + a_3 x_2 + a_4 x_1 x_2$.

29. Let $g_1(x_1,x_2)$, $g_2(x_1,x_2)$ be functions of the same type, each of which itself assumes only two values, 0 or 1. Then if we write

$$f(g_1,g_2) = a_1' + a_2' x_1 + a_3' x_2 + a_4' x_1 x_2$$

we have a matrix transformation

$$\begin{bmatrix} a_1' \\ a_2' \\ a_3' \\ a_4' \end{bmatrix} = M \begin{bmatrix} a_1 \\ a_2 \\ a_3 \\ a_4 \end{bmatrix}$$

For the case where $g_1(x_1,x_2) = x_1 x_2$, $g_2(x_1,x_2) = x_1(1 - x_2)$, evaluate M.

30. Generalize the foregoing results to the case where we have a function of N variables, $f(x_1,x_2, \ldots ,x_N)$, and to the case where the x_i can assume any of a given finite set of values.

(The foregoing results are useful in the study of various types of logical nets; see, for example, R. Bellman, J. Holland, and R. Kalaba, Dynamic Programming and the Synthesis of Logical Nets, *J. Assoc. Comp. Mach.*, 1959.)

31. If A is a given 2×2 matrix and X an unknown 2×2 matrix, show that the equation $AX - XA = I$ has no solution.

32. Show that the same result holds for the case where A and X are $N \times N$ matrices.

33. Consider the relation between the $N(N + 1)/2$ variables y_{ij} and the $N(N + 1)/2$ variables x_{ij} given by $Y = AXA'$, where X and Y are symmetric. Show that $|J(Y,X)| = |A|^{N+1}$. Two semiexpository papers discussing matters of this nature have been written by W. L. Deemer and I. Olkin.[1]

34. Construct a 4×4 symmetric orthogonal matrix whose elements are ± 1. This question, and extensions, is of interest in connection with Hadamard's inequality, see Sec. 7 of Chap. 8, and (amazingly!), in connection with the design of experiments. See R. E. A. C. Paley[2] and R. L. Plackett and J. P. Burman.[3]

[1] W. L. Deemer and I. Olkin, Jacobians of Matrix Transformations Useful in Multivariate Analysis, *Biometrika*, vol. 38, pp. 345–367, 1951.

I. Olkin, Note on "Jacobians of Matrix Transformations Useful in Multivariate Analysis," *Biometrika*, vol. 40, pp. 43–46, 1953.

[2] R. E. A. C. Paley, On Orthogonal Matrices, *J. Math. and Physics*, vol. 12, pp. 311–320, 1933.

[3] R. L. Plackett and J. P. Burman, The Design of Optimum Multifactorial Experiments, *Biometrika*, vol. 33, pp. 305–325, 1946.

Bibliography

The reader who is interested in the origins of matrix theory is urged to read the monograph by MacDuffee

C. C. MacDuffee, *The Theory of Matrices,* Chelsea Publishing Co., New York, 1946.

Although we shall occasionally refer to various theorems by the names of their discoverers, we have made no serious attempt to restore every result to its rightful owner. As in many other parts of mathematics, there is a certain amount of erroneous attribution. When aware of it, we have attempted to rectify matters.

§8. It should be pointed out that other physical situations may very well introduce other types of algebras with quite different "addition" and "multiplication." As I. Olkin points out, in the study of sociometrics, a multiplication of the form $A \cdot B = (a_{ij}b_{ij})$ is quite useful. Oddly enough, as will be seen subsequently, this Schur product crops up in the theory of partial differential equations.

§10. For an extensive discussion of commutativity, see

O. Taussky, Commutativity in Finite Matrices, *Amer. Math. Monthly,* vol. 64, pp. 229–235, 1957.

For an interesting extension of commutativity, see

B. Friedman, n-commutative Matrices, *Math. Annalen,* vol. 136, 1958, pp. 343–347.

For a discussion of the solution of equations in X of the form $XA = AX$, $XA = A'X$, see

H. O. Foukes, *J. London Math. Soc.,* vol. 17, pp. 70–80, 1942.

§11. Other algebras can be defined in which commutativity and associativity of multiplication both fail. A particularly interesting example is the Cayley algebra. For a discussion of these matters, see

A. A. Albert, *Modern Higher Algebra,* University of Chicago Press, Chicago, 1937.

G. A. Birkhoff and S. Maclane, *Survey of Modern Algebra,* The Macmillan Company, New York, 1958.

§14. In the case of more general transformations or operators, A' is often called the *adjoint* transformation or operator. The reason for its importance resides in the fact that occasionally the induced transformation may be simpler to study than the original transformation. Furthermore, in many cases, certain properties of A are only simply expressed when stated in terms of A'.

In our discussion of Markoff matrices in Chap. **14**, we shall see an example of this.

§16. The notation H^* for $\overline{H'}$ appears to be due to Ostrowski:

A. Ostrowski, Über die Existenz einer endlichen Basis bei gewissen Funktionensystemen, *Math. Ann.*, vol. 78, pp. 94–119, 1917.

For an interesting geometric interpretation of the Jacobi bracket identity of Exercise 12 in the Miscellaneous Exercises, see

W. A. Hoffman, *Q. Appl. Math.*, 1968.

Some remarkable results have been obtained recently concerning the number of multiplications required for matrix multiplication. See

S. Winograd, The Number of Multiplications Involved in Computing Certain Functions, *Proc. IFIP Congress*, 1968.

For some extensions of scalar number theory, see

F. A. Ficken, Rosser's Generalization of the Euclidean Algorithm, *Duke Math. J.*, vol. 10, pp. 355–379, 1943.

A paper of importance is

M. R. Hestenes, A Ternary Algebra with Applications to Matrices and Linear Transformations, *Archive Rat. Mech. Anal.*, vol. 11, pp. 138–194, 1962.

3

Diagonalization and Canonical Forms for Symmetric Matrices

1. Recapitulation. Our discussion of the problem of determining the stationary values of the quadratic form $Q(x) = \sum_{i,j=1}^{N} a_{ij}x_ix_j$ on the sphere $x_1^2 + x_2^2 + \cdots + x_N^2 = 1$ led us to the problem of finding nontrivial solutions of the linear homogeneous equations

$$\sum_{j=1}^{N} a_{ij}x_j = \lambda x_i \qquad i = 1, 2, \ldots, N \tag{1}$$

We interrupted our story at this point to introduce vector-matrix notation, stating in extenuation of our excursion that this tool would permit us to treat this and related questions in a simple and elegant fashion.

Observe that the condition of symmetry $a_{ij} = a_{ji}$ is automatically satisfied in (1) in view of the origins of these equations. This simple but significant property will permit us to deduce a great deal of information concerning the nature of the solutions. On the basis of this knowledge, we shall transform $Q(x)$ into a simpler form which plays a paramount role in the higher theory of matrices and quadratic forms.

Now to resume our story!

2. The Solution of Linear Homogeneous Equations. We require the following fundamental result.

Lemma. A necessary and sufficient condition that the linear system

$$\sum_{j=1}^{N} b_{ij}x_j = 0 \qquad i = 1, 2, \ldots, N \tag{1}$$

possess a nontrivial solution is that we have the determinantal relation

$$|b_{ij}| = 0 \tag{2}$$

As usual, by "nontrivial" we mean that at least one x_i is nonzero.

This result is actually a special case of a more general result concerning linear systems in which the number of equations is not necessarily equal to the number of variables, a matter which is discussed in Appendix A. Here, however, we shall give a simple inductive proof which establishes all that we need.

Proof of Lemma. The necessity is clear. If $|b_{ij}| \neq 0$, we can solve by Cramer's rule, obtaining thereby the unique solution $x_1 = x_2 = \cdots = 0$.

Let us concentrate then on the sufficiency. It is clear that the result is true for $N = 1$. Let us then see if we can establish its truth for N, assuming its validity for $N - 1$. Since at least one of the b_{ij} is nonzero, or else the result is trivially true, assume that one of the elements in the first row is nonzero, and then, without loss of generality, that this element is b_{11}.

Turning to the linear system in (1), let us eliminate x_1 between the first and second equations, the first and third, and so on. The resulting system has the form

$$\begin{aligned} \left(b_{22} - \frac{b_{21}b_{12}}{b_{11}}\right) x_2 + \cdots + \left(b_{2N} - \frac{b_{21}b_{1N}}{b_{11}}\right) x_N &= 0 \\ &\vdots \\ \left(b_{N2} - \frac{b_{N1}b_{12}}{b_{11}}\right) x_2 + \cdots + \left(b_{NN} - \frac{b_{N1}b_{1N}}{b_{11}}\right) x_N &= 0 \end{aligned} \tag{3}$$

Let us obtain a relation between the determinant of this system and the original $N \times N$ determinant, $|b_{ij}|$, in the following way.

Subtracting b_{21}/b_{11} times the first row of $|b_{ij}|$ from the second, b_{31}/b_{11} times the first row from the third row and so on, we obtain the relation

$$\begin{vmatrix} b_{11} & b_{12} & \cdots & b_{1N} \\ b_{21} & b_{22} & \cdots & b_{2N} \\ \cdot & \cdot & & \cdot \\ \cdot & \cdot & & \cdot \\ \cdot & \cdot & & \cdot \\ b_{N1} & b_{N2} & \cdots & b_{NN} \end{vmatrix} = \begin{vmatrix} b_{11} & & \cdots & \\ 0 & \left(b_{22} - \frac{b_{21}b_{12}}{b_{11}}\right) & \cdots & \left(b_{2N} - \frac{b_{21}b_{1N}}{b_{11}}\right) \\ \cdot & & & \\ \cdot & & & \\ \cdot & & & \\ 0 & \left(b_{N2} - \frac{b_{N1}b_{12}}{b_{11}}\right) & \cdots & \left(b_{NN} - \frac{b_{N1}b_{1N}}{b_{11}}\right) \end{vmatrix} \tag{4}$$

Hence, the determinant of $(N - 1)$-dimensional system in (3) is zero, since by assumption $b_{11} \neq 0$ and $|b_{ij}| = 0$. From our inductive hypothesis it follows that there exists a nontrivial solution of (3), $x_2, x_3, \ldots, x_N$. Setting

$$x_1 = -\sum_{j=2}^{N} b_{1j}x_j/b_{11} \tag{5}$$

we obtain, thereby, the desired nontrivial solution of (1), $x_1, x_2, \ldots, x_N$.

EXERCISE

1. Show that if A is real, the equation $Ax = 0$ always possesses a *real* nontrivial solution if it possesses *any* nontrivial solution.

3. Characteristic Roots and Vectors. Setting x equal to the vector whose components are x_i and $A = (a_{ij})$, we may write (1.1) in the form

$$Ax = \lambda x \tag{1}$$

Referring to the lemma of Sec. 2, we know that a necessary and sufficient condition that there exist a nontrivial vector x satisfying this equation is that λ be a root of the determinantal equation

$$|a_{ij} - \lambda\delta_{ij}| = 0 \tag{2}$$

or, as we shall usually write,

$$|A - \lambda I| = 0 \tag{3}$$

This equation is called the *characteristic equation* of A. As a polynomial equation in λ, it possesses N roots, distinct or not, which are called the *characteristic roots* or *characteristic values.* If the roots are distinct, we shall occasionally use the term *simple,* as opposed to *multiple.*

The hybrid word *eigenvalue* appears with great frequency in the literature, a bilingual compromise between the German word "Eigenwerte" and the English expression given above. Despite its ugliness, it seems to be too firmly entrenched to dislodge.

Associated with each distinct characteristic value λ, there is a *characteristic vector,* determined up to a scalar multiple. This characteristic vector may be found via the inductive route sketched in Sec. 2, or following the path traced in Appendix A. Neither of these is particularly attractive for large values of N, since they involve a large number of arithmetic operations. In actuality, there are no easy methods for obtaining the characteristic roots and characteristic vectors of matrices of large dimension.

As stated in the Preface, we have deliberately avoided in this volume any references to computational techniques which can be employed to determine numerical values for characteristic roots and vectors.

If λ is a multiple root, there may or may not be an equal number of associated characteristic vectors if A is an arbitrary square matrix. These matters will be discussed in the second part of the book, devoted to the study of general, not necessarily symmetric, matrices.

For the case of symmetric matrices, multiple roots cause a certain amount of inconvenience, but nothing of any moment. We will show that a real symmetric matrix of order N has N distinct characteristic vectors.

EXERCISES

1. A and A' have the same characteristic values.
2. $T'AT - \lambda I = T'(A - \lambda I)T$ if T is orthogonal. Hence, A and $T'AT$ have the same characteristic values if T is orthogonal.
3. A and T^*AT have the same characteristic values if T is unitary.
4. SAT and A have the same characteristic values if $ST = I$.
5. Show by direct calculation for A and B, 2×2 matrices, that AB and BA have the same characteristic equation.
6. Does the result hold generally?
7. Show that any scalar multiple apart from zero of a characteristic vector is also a characteristic vector. Hence, show that we can always choose a characteristic vector x so that $(x,\bar{x}) = 1$.
8. Show, by considering 2×2 matrices, that the characteristic roots of $A + B$ cannot be obtained in general as sums of characteristic roots of A and of B.
9. Show that a similar comment is true for the characteristic roots of AB.
10. For the 2×2 case, obtain the relation between the characteristic roots of A and those of A^2.
11. Does a corresponding relation hold for the characteristic roots of A and A^n for $n = 3, 4, \ldots$?

4. Two Fundamental Properties of Symmetric Matrices. Let us now give the simple proofs of the two fundamental results upon which the entire analysis of real symmetric matrices hinges.

Although we are interested only in symmetric matrices whose elements are real, we shall insert the word "real" here and there in order to emphasize this fact and prevent any possible confusion.

Theorem 1. *The characteristic roots of a real symmetric matrix are real.*

Proof. Assume the contrary. Since A is a real matrix, it follows from the characteristic equation $|A - \lambda I| = 0$ that the conjugate of any complex characteristic root λ is also a characteristic root. We obtain this result and further information from the fact that if the equation

$$Ax = \lambda x \tag{1}$$

holds, then the relation

$$A\bar{x} = \bar{\lambda}\bar{x} \tag{2}$$

is also valid. From these equations, we obtain the further relations

$$(\bar{x},Ax) = \lambda(\bar{x},x)$$
$$(x,A\bar{x}) = \bar{\lambda}(x,\bar{x}) \qquad (3)$$

Since A is symmetric, which implies that $(\bar{x},Ax) = (A\bar{x},x) = (x,A\bar{x})$, the foregoing relations yield

$$0 = (\lambda - \bar{\lambda})(x,\bar{x}), \qquad (4)$$

whence $\lambda = \bar{\lambda}$, a contradiction.

This means that the characteristic vectors of a real symmetric matrix A can always be taken to be real, and we shall consistently do this.

The second result is:

Theorem 2. *Characteristic vectors associated with distinct characteristic roots of a real symmetric matrix A are orthogonal.*

Proof. From

$$Ax = \lambda x$$
$$Ay = \mu y \qquad (5)$$

$\lambda \neq \mu$, we obtain

$$(y,Ax) = \lambda(y,x)$$
$$(x,Ay) = \mu(x,y) \qquad (6)$$

Since $(x,Ay) = (Ax,y) = (y,Ax)$, subtraction yields

$$0 = (\lambda - \mu)(x,y) \qquad (7)$$

whence $(x,y) = 0$.

This result is of basic importance. Its generalization to more general operators is one of the cornerstones of classical analysis.

EXERCISES

1. A characteristic vector cannot be associated with two distinct characteristic values.

2. Show by means of a 2×2 matrix, however, that two distinct vectors can be associated with the same characteristic root.

3. Show by means of an example that there exist 2×2 symmetric matrices A and B with the property that $|A - \lambda B|$ is identically zero. Hence, under what conditions on B can we assert that all roots of $|A - \lambda B| = 0$ are real?

4. Show that if A and B are real symmetric matrices, and if B is positive definite, the roots of $|A - \lambda B| = 0$ are all real.

5. Show that the characteristic roots of a Hermitian matrix are real and that the characteristic vectors corresponding to distinct characteristic roots are orthogonal using the generalized inner product $(x,\bar{y})$.

6. Let the elements a_{ij} of A depend upon a parameter t. Show that the derivatives of $|A|$ with respect to t can be written as the sum of N determinants, where the kth determinant is obtained by differentiating the elements of the kth row and leaving the others unaltered.

7. Show that the derivative of $|A - \lambda I|$ with respect to λ is equal to $-\sum_{k=1}^{N} |A_k - \lambda I|$, where A_k is the $(N-1) \times (N-1)$ matrix obtained from A by striking out the kth row and column.

8. From this, conclude that if λ is a simple root of A, then at least one of the determinants $|A_k - \lambda I|$ is nonzero.

9. Use this result to show that if λ is a simple root of A, a characteristic vector x associated with λ can always be taken to be a vector whose components are polynomials in λ and the elements of A.

5. Reduction to Diagonal Form—Distinct Characteristic Roots. We can obtain an important result quite painlessly at this juncture, if we agree to make the simplifying assumption that A has *distinct* characteristic roots $\lambda_1, \lambda_2, \ldots, \lambda_N$. Let $x^1, x^2, \ldots, x^N$ be an associated set of characteristic vectors, normalized by the condition that

$$(x^i,x^i) = 1 \qquad i = 1, 2, \ldots, N \tag{1}$$

Consider the matrix T formed upon using the vectors x^i as columns. Schematically,

$$T = (x^1,x^2, \ldots ,x^N) \tag{2}$$

Then T' is the matrix obtained using the x^i as rows,

$$T' = \begin{bmatrix} x^1 \\ x^2 \\ \cdot \\ \cdot \\ \cdot \\ x^N \end{bmatrix} \tag{3}$$

Since

$$T'T = ((x^i,x^j)) = (\delta_{ij}) \tag{4}$$

(in view of the orthogonality of the x^i as characteristic vectors associated with distinct characteristic roots of the symmetric matrix A), we see that T is an orthogonal matrix.

We now assert that the product AT has the simple form

$$AT = (\lambda_1 x^1,\lambda_2 x^2, \ldots ,\lambda_N x^N) \tag{5}$$

by which we mean that the matrix AT has as its ith column the vector $\lambda_i x^i$.

It follows that

$$T'AT = (\lambda_i(x^i,x^j)) = (\lambda_i\delta_{ij}) \tag{6}$$

(recalling once again the orthogonality of the x_i),

$$= \begin{bmatrix} \lambda_1 & & & & 0 \\ & \lambda_2 & & & \\ & & \cdot & & \\ & & & \cdot & \\ & & & & \cdot \\ 0 & & & & \lambda_N \end{bmatrix}$$

The matrix on the right-hand side has as its main diagonal the characteristic values $\lambda_1, \lambda_2, \ldots, \lambda_N$, and zeros every place else. A matrix of this type is, as earlier noted, called a *diagonal* matrix.

Multiplying on the right by T' and on the left by T, and using the fact that $TT' = I$, we obtain the important result

$$A = T \begin{bmatrix} \lambda_1 & & & & 0 \\ & \lambda_2 & & & \\ & & \cdot & & \\ & & & \cdot & \\ & & & & \cdot \\ 0 & & & & \lambda_N \end{bmatrix} T' \tag{7}$$

This process is called *reduction to diagonal form.* As we shall see, this representation plays a fundamental role in the theory of symmetric matrices. Let us use the notation

$$\Lambda = \begin{bmatrix} \lambda_1 & & & & 0 \\ & \lambda_2 & & & \\ & & \cdot & & \\ & & & \cdot & \\ & & & & \cdot \\ 0 & & & & \lambda_N \end{bmatrix} \tag{8}$$

EXERCISES

1. Show that $\Lambda^k = (\lambda_i^k \delta_{ij})$, and that $A^k = T\Lambda^k T'$, for $k = 1, 2, \ldots$.

2. Show that if A has distinct characteristic roots, then A satisfies its own characteristic equation. This is a particular case of a more general result we shall establish later on.

3. If A has distinct characteristic roots, obtain the set of characteristic vectors associated with the characteristic roots of A^k, $k = 2, 3, \ldots$.

6. Reduction of Quadratic Forms to Canonical Form. Let us now show that this matrix transformation leads to an important transformation of $Q(x)$. Setting $x = Ty$, where T is as defined in (2) of Sec. 5, we have

$$(x,Ax) = (Ty,ATy) = (y,T'ATy) = (y,\Lambda y) \tag{1}$$

or the fundamental relation

$$\sum_{i,j=1}^{N} a_{ij}x_ix_j = \sum_{i=1}^{N} \lambda_i y_i^2 \tag{2}$$

Since T is orthogonal, we see that $x = Ty$ implies that

$$T'x = T'Ty = y \tag{3}$$

Hence to each value of x corresponds precisely one value of y and conversely.

We thus obtain the exceedingly useful result that the set of values assumed by $Q(x)$ on the sphere $(x,x) = 1$ is identical with the set of values assumed by $(y,\Lambda y)$ on the sphere $(y,y) = 1$.

So far we have only established this for the case where the λ_i are all distinct. As we shall see in Chap. 4, it is true in general, constituting the foundation stone of the theory of quadratic forms.

EXERCISES

1. Let A have distinct characteristic roots which are all positive. Use the preceding result to compute the volume of the N-dimensional ellipsoid $(x,Ax) = 1$.

2. Prove along the preceding lines that the characteristic roots of Hermitian matrices are real and that characteristic vectors associated with distinct characteristic roots are orthogonal in terms of the notation $[x,y]$ of Sec. 16 of Chap. 2.

3. Show that if the characteristic roots of a Hermitian matrix A are distinct, we can find a unitary matrix T such that $A = T\Lambda T^*$. This again is a particular case of a more general result we shall prove in the following chapter.

4. Let A be a real matrix with the property that $A' = -A$, a *skew-symmetric* matrix. Show that the characteristic roots are either zero or pure imaginary.

5. Let T be an orthogonal matrix. Show that all characteristic roots have absolute value one.

6. Let T be a unitary matrix. Show that all characteristic roots have absolute value one.

7. Suppose that we attempt to obtain the representation of (5.2) under no restriction on the characteristic roots of the symmetric matrix A in the following way. To begin with, we assert that we can always find a symmetric matrix B, with elements arbitrarily small, possessing the property that $A + B$ has simple characteristic roots. We do not dwell upon this point since the proof is a bit more complicated than might be suspected; see Sec. 16 of Chap. 11. Let $\{\mu_i\}$ be the characteristic roots of $A + B$. Then, as we know, there exists an orthogonal matrix S such that

$$A + B = S \begin{bmatrix} \mu_1 & & & 0 \\ & \mu_2 & & \\ & & \ddots & \\ 0 & & & \mu_N \end{bmatrix} S'$$

Since S is an orthogonal matrix, its elements are uniformly bounded. Let $\{B_n\}$ be a sequence of matrices approaching 0 such that the corresponding sequence of orthogonal matrices $\{S_n\}$ converges. The limit matrix must then be an orthogonal matrix, say T. Since $\lim_{n\to\infty} \mu_i = \lambda_i$, we have

$$A = \lim_{n\to\infty} (A + B_n) = \lim_{n\to\infty} S_n \begin{bmatrix} \lim_{n\to\infty} \mu_1 & & & \\ & \cdot & & \\ & & \cdot & \\ & & & \lim_{n\to\infty} \mu_N \end{bmatrix} \lim_{n\to\infty} S_n'$$

Since this proof relies upon a number of analytic concepts which we do not wish to introduce until much later, we have not put it into the text. It is an illustration of a quite useful metamathematical principle that results valid for general real symmetric matrices can always be established by first considering matrices with distinct characteristic roots and then passing to the limit.

7. Positive Definite Quadratic Forms and Matrices. In Sec. 9 of Chap. 1, we introduced the concept of a positive definite quadratic form in two variables, and the allied concepts of positive indefinite or non-negative definite. Let us now extend this to N-dimensional quadratic forms. If $A = (a_{ij})$ is a real symmetric matrix, and

$$Q_N(x) = \sum_{i,j=1}^{N} a_{ij}x_ix_j > 0$$

for all real nontrivial x_i, we shall say that $Q_N(x)$ is *positive definite* and that A is *positive definite*. If $Q_N(x) \geq 0$, we shall say that $Q_N(x)$ and A are *positive indefinite*.

Similarly, if H is a Hermitian form and $P_N(x) = \sum_{i,j=1}^{N} h_{ij}x_i\bar{x}_j > 0$ for all complex nontrivial x_i, we shall say that $P_N(x)$ and H are positive definite.

EXERCISES

1. If A is a symmetric matrix with distinct characteristic roots, obtain a set of necessary and sufficient conditions that A be positive definite.

2. There exists a scalar c_1 such that $A + c_1I$ is positive definite, given any symmetric matrix A.

3. Show that we can write A in the form

$$A = \sum_{i=1}^{N} \lambda_i E_i$$

where the E_i are non-negative definite matrices. Then

$$A^k = \sum_{i=1}^{N} \lambda_i^k E_i$$

for $k = 1, 2, \ldots$.

4. If $p(\lambda)$ is a polynomial in λ with scalar coefficients, $p(\lambda) = \sum_{k=0}^{m} c_k \lambda^k$, let $p(A)$ denote the matrix $p(A) = \sum_{k=0}^{m} c_k A^k$. Show that $p(A) = \sum_{i=1}^{N} p(\lambda_i) E_i$.

MISCELLANEOUS EXERCISES

1. Let A and B be two symmetric matrices. Then the roots of $|A - \lambda B| = 0$ are all real if B is positive definite. What can be said of the vectors satisfying the relations $Ax = \lambda_i Bx$?

2. Every matrix is uniquely expressible in the form $A = H + S$, where H is Hermitian and S is skew-Hermitian, that is, $S^* = -S$.

3. As an extension of Theorem 1, show that if λ is a characteristic root of a real matrix A, then $|\text{Im}\ (\lambda)| \leq d(N(N-1)/2)^{1/2}$, where

$$d = \max_{1 \leq i,j \leq N} |a_{ij} - a_{ji}|/2 \qquad (I.\ Bendixson)$$

4. Generally, let A be complex; then if

$$d_1 = \max_{i,j} |a_{ij}| \qquad d_2 = \max_{i,j} |a_{ij} + \overline{a_{ji}}|/2$$
$$d_3 = \max |a_{ij} - \overline{a_{ji}}|/2$$

we have

$$|\lambda| \leq Nd_1 \qquad |\text{Re}\ (\lambda)| \leq Nd_2 \qquad |\text{Im}\ (\lambda)| \leq Nd_3\dagger$$

5. Show that for any complex matrix A, we have the inequalities

$$\sum_{i=1}^{N} |\lambda_i|^2 \leq \sum_{i,j=1}^{N} |a_{ij}|^2$$

$$\sum_{i=1}^{N} |\text{Re}\ (\lambda_i)|^2 \leq \sum_{i,j=1}^{N} |(a_{ij} + \overline{a_{ji}})/2|^2$$

$$\sum_{i=1}^{N} |\text{Im}\ (\lambda_i)|^2 \leq \sum_{i,j=1}^{N} |(a_{ij} - \overline{a_{ji}})/2|^2 \qquad (I.\ Schur)$$

† Far more sophisticated results are available. See the papers by A. Brauer in *Duke Math. J.*, 1946, 1947, 1948, where further references are given. See also W. V. Parker, Characteristic Roots and Fields of Value of a Matrix, *Bull. Am. Math. Soc.*, vol. 57, pp. 103–108, 1951.

6. So far, we have not dismissed the possibility that a characteristic root may have several associated characteristic vectors, not all multiples of a particular characteristic vector. As we shall see, this can happen if A has multiple characteristic roots. For the case of distinct characteristic roots, this cannot occur. Although the simplest proof uses concepts of the succeeding chapter, consider a proof along the following lines:

(*a*) Let x^1 and y be two characteristic vectors associated with λ_1 and suppose that $y \neq c_1 x^1$ for any scalar c_1. Then x^1 and $z = y - x^1(x^1,y)/(x^1,x^1)$ are characteristic vectors and x^1 and z are orthogonal.

(*b*) Let z^1 be the normalized multiple of z. Then

$$S = \begin{bmatrix} x^1 & z & x^3 & \cdots & x^N \\ & & & & \\ & & & & \end{bmatrix}$$

is an orthogonal transformation.

(*c*) $A = SDS'$, where

$$D = \begin{bmatrix} \lambda_1 & & & & 0 \\ & \lambda_1 & & & \\ & & \lambda_3 & & \\ & & & \ddots & \\ 0 & & & & \lambda_N \end{bmatrix}$$

(*d*) It follows that the transformation $x = Sy$ changes (x,Ax) into (y,Dy).

(*e*) Assume that A is positive definite (if not, consider $A + c_1 I$), then, on one hand, the volume of the ellipsoid $(x,Ax) = 1$ is equal to the volume of the ellipsoid

$$\lambda_1 y_1^2 + \lambda_2 y_2^2 + \cdots + \lambda_N y_N^2 = 1,$$

and, on the other hand, from what has just preceded, is equal to the volume of

$$\lambda_1 y_1^2 + \lambda_1 y_1^2 + \cdots + \lambda_N y_N^2 = 1$$

This is a contradiction if $\lambda_1 \neq \lambda_2$.

Bibliography

§2. This proof is taken from the book by L. Mirsky,

L. Mirsky, *Introduction to Linear Algebra*, Oxford University Press, New York, 1955.

§3. The term "latent root" for characteristic value is due to Sylvester. For the reason, and full quotation, see

N. Dunford and J. T. Schwartz, *Linear Operators*, part I, Interscience Publishers, New York, 1958, pp. 606–607.

The term "spectrum" for set of characteristic values is due to Hilbert.

§5. Throughout the volume we shall use this device of examining the case of distinct characteristic roots before treating the general case. In many cases, we can employ continuity techniques to deduce the general case from the special case, as in Exercise 7, Sec. 6. Apart from the fact that the method must be used with care, since occasionally there is a vast difference between the behaviors of the two types of matrices, we have not emphasized the method because of its occasional dependence upon quite sophisticated analysis.

It is, however, a most powerful technique, and one that is well worth acquiring.

4

Reduction of General Symmetric Matrices to Diagonal Form

1. Introduction. In this chapter, we wish to demonstrate that the results obtained in Chap. 3 under the assumption of simple characteristic roots are actually valid for general symmetric matrices. The proof we will present will afford us excellent motivation for discussing the useful concept of linear dependence and for demonstrating the Gram-Schmidt orthogonalization technique.

Along the way we will have opportunities to discuss some other interesting techniques, and finally, to illustrate the inductive method for dealing with matrices of arbitrary order.

2. Linear Dependence. Let $x^1, x^2, \ldots, x^k$ be a set of k N-dimensional vectors. If a set of scalars, $c_1, c_2, \ldots, c_k$, at least one of which is nonzero, exists with the property that

$$c_1x^1 + c_2x^2 + \cdots + c_kx^k = 0 \tag{1}$$

where 0 represents the null vector, we say that the vectors are *linearly dependent.* If no such set of scalars exist, we say that the vectors are *linearly independent.*

Referring to the results concerning linear systems established in Appendix A, we see that this concept is only of interest if $k \leq N$, since any k vectors, where $k > N$, are related by a relation of the type given in (1).

EXERCISES

1. Show that any set of mutually orthogonal nontrivial vectors is linearly independent.

2. Given any nontrivial vector in N-dimensional space, we can always find $N - 1$ vectors which together with the given N-dimensional vector constitute a linearly independent set.

3. Gram-Schmidt Orthogonalization. Let $x^1, x^2, \ldots, x^N$ be a set of N real linearly independent N-dimensional vectors. We wish to show that we can form suitable linear combinations of these base vectors which will constitute a set of mutually orthogonal vectors.

The procedure we follow is inductive. We begin by defining two new vectors as follows.

$$
\begin{aligned}
y^1 &= x^1 \\
y^2 &= x^2 + a_{11}x^1
\end{aligned}
\tag{1}
$$

where a_{11} is a scalar to be determined by the condition that y^1 and y^2 are orthogonal. The relation

$$(y^1,y^2) = (x^1, x^2 + a_{11}x^1) = 0 \tag{2}$$

yields the value

$$a_{11} = -(x^1,x^2)/(x^1,x^1) \tag{3}$$

Since the set of x^i is by assumption linearly independent, we cannot have x^1 equal to the null vector, and thus (x^1,x^1) is not equal to zero.

Let us next set

$$y^3 = x^3 + a_{21}x^1 + a_{22}x^2 \tag{4}$$

where now the two scalar coefficients a_{21} and a_{22} are to be determined by the conditions of orthogonality

$$(y^3,y^1) = 0 \qquad (y^3,y^2) = 0 \tag{5}$$

These conditions are easier to employ if we note that (1) shows that the equation in (5) is equivalent to the relations

$$(y^3,x^1) = 0 \qquad (y^3,x^2) = 0 \tag{6}$$

These equations reduce to the simultaneous equations

$$
\begin{aligned}
(x^3,x^1) + a_{21}(x^1,x^1) + a_{22}(x^2,x^1) &= 0 \\
(x^3,x^2) + a_{21}(x^1,x^2) + a_{22}(x^2,x^2) &= 0
\end{aligned}
\tag{7}
$$

which we hope determine the unknown coefficients a_{21} and a_{22}. We can solve for these quantities, using Cramer's rule, provided that the determinant

$$D_2 = \begin{vmatrix} (x^1,x^1) & (x^1,x^2) \\ (x^1,x^2) & (x^2,x^2) \end{vmatrix} \tag{8}$$

is not equal to zero.

To show that D_2 is not equal to zero, we can proceed as follows. If $D_2 = 0$, we know from the lemma established in Sec. 2 of Chap. 3 there are two scalars r_1 and s_1, not both equal to zero, such that the linear equations

$$
\begin{aligned}
r_1(x^1,x^1) + s_1(x^1,x^2) &= 0 \\
r_1(x^1,x^2) + s_1(x^2,x^2) &= 0
\end{aligned}
\tag{9}
$$

are satisfied. These equations may be written in the form

$$\begin{aligned}(x^1, r_1x^1 + s_1x^2) &= 0\\(x^2, r_1x^1 + s_1x^2) &= 0\end{aligned} \tag{10}$$

Multiplying the first equation by r_1 and the second by s_1 and adding, we obtain the resultant equation

$$(r_1x^1 + s_1x^2, r_1x^1 + s_1x^2) = 0 \tag{11}$$

This equation, however, in view of our assumption of reality, can hold only if $r_1x^1 + s_1x^2 = 0$, contradicting the assumed linear independence of the x^i. Hence $D_2 \neq 0$, and there is a unique solution for the quantities a_{21} and a_{22}.

At the next step, we introduce the vector

$$y^4 = x^4 + a_{31}x^1 + a_{32}x^2 + a_{33}x^3 \tag{12}$$

As above, the conditions of mutual orthogonality yield the equations

$$(y^4,x^1) = (y^4,x^2) = (y^4,x^3) = 0 \tag{13}$$

which lead to the simultaneous equations

$$(x^4,x^i) + a_{31}(x^1,x^i) + a_{32}(x^2,x^i) + a_{33}(x^3,x^i) = 0 \qquad i = 1, 2, 3 \tag{14}$$

We can solve for the coefficients a_{31}, a_{32}, and a_{33}, provided that the determinant

$$D_3 = |(x^i,x^j)| \qquad i, j = 1, 2, 3 \tag{15}$$

is nonzero. The proof that D_3 is nonzero is precisely analogous to that given above in the two-dimensional case. Hence, we may continue this procedure, step by step, until we have obtained a complete set of vectors $\{y^i\}$ which are mutually orthogonal.

These vectors can then be normalized by the condition $(y^i,y^i) = 1$. We then say they form an *orthonormal* set. The determinants D_k are called *Gramians.*

EXERCISES

1. Consider the interval $[-1,1]$ and define the inner product of two real functions $f(t)$ and $g(t)$ in the following way:

$$(f,g) = \int_{-1}^{1} f(t)g(t)\,dt$$

Let $P_0(t) = \frac{1}{2}$, and define the other elements of the sequence of real polynomials $\{P_n(t)\}$ by the condition that $P_n(t)$ is of degree n and $(P_n,P_m) = 0$, $m \neq n$, $(P_n,P_n) = 1$. Prove that we have $(P_n,t^m) = 0$, for $0 \leq m \leq n - 1$, and construct the first few members of the sequence in this way.

2. With the same definition of an inner product as above, prove that

$$\left(\left(\frac{d}{dt}\right)^n (1 - t^2)^n, t^m\right) = 0,$$

$0 \leq m \leq n - 1$, and thus express P_n in terms of the expression $\left(\frac{d}{dt}\right)^n (1 - t^2)^n$. These polynomials are, apart from constant factors, the classical Legendre polynomials.

3. Consider the interval $(0, \infty)$ and define the inner product $(f,g) = \int_0^\infty e^{-t}f(t)g(t)\,dt$ for any two real polynomials $f(t)$ and $g(t)$. Let the sequence of polynomials $\{L_n(t)\}$ be determined by the conditions $L_0(t) = 1$, $L_n(t)$ is a polynomial of degree n, and $(L_n,L_m) = 0$, $n \neq m$, $(L_n,L_n) = 1$. Prove that these conditions imply that $(L_n,t^m) = 0$, $0 \leq m \leq n - 1$, and construct the first few terms of the sequence using these relations.

4. Prove that $(e^t(d/dt)^n(e^{-t}t^n), t^m) = 0$, $0 \leq m \leq n - 1$, and thus express L_n in terms of $e^t(d/dt)^n(e^{-t}t^n)$. These polynomials are apart from constant factors the classical Laguerre polynomials.

5. Consider the interval $(-\infty, \infty)$ and define the inner product

$$(f,g) = \int_{-\infty}^{\infty} e^{-t^2}f(t)g(t)\,dt$$

for any two real polynomials $f(t)$ and $g(t)$. Let the sequence of polynomials $\{H_n(t)\}$ be determined as follows. $H_0(t) = 1$, $H_n(t)$ is a polynomial of degree n, and $(H_m,H_n) = 0$, $m \neq n$, $(H_n,H_n) = 1$. Prove that $(H_n,t^m) = 0$, $0 \leq m \leq n - 1$, and construct the first few terms of the sequence in this fashion.

6. Prove that $(e^{t^2}(d/dt)^n(e^{-t^2}t^n), t^m) = 0$, $0 \leq m \leq n - 1$ and hence express H_n in terms of $e^{t^2}(d/dt)^n(e^{-t^2}t^n)$. These polynomials are apart from constant factors the classical Hermite polynomials.

7. Show that given a real N-dimensional vector x^1, normalized by the condition that $(x^1,x^1) = 1$, we can find $(N - 1)$ additional vectors $x^2, x^3, \ldots, x^N$ with the property that the matrix $T = (x^1,x^2, \ldots, x^N)$ is orthogonal.

8. Obtain the analogue of the Gram-Schmidt method for complex vectors.

4. On the Positivity of the D_k. All that was required in the foregoing section on orthogonalization was the nonvanishing of the determinants D_k. Let us now show that a slight extension of the previous argument will enable us to conclude that actually the D_k are positive. The result is not particularly important at this stage, but the method is an important and occasionally useful one.

Consider the quadratic form

$$\begin{aligned} Q(u_1,u_2, \ldots, u_k) &= (u_1x^1 + u_2x^2 + \cdots + u_kx^k, \\ &\qquad u_1x^1 + u_2x^2 + \cdots + u_kx^k) \\ &= \sum_{i,j=1}^{k} (x^i,x^j)u_iu_j \end{aligned} \tag{1}$$

where the u_i are real quantities, and the x^i as above constitute a system of real linearly independent vectors.

In view of the linear independence of the vectors x^i, we see that $Q > 0$ for all nontrivial sets of values of the u_i. Consequently, the result we wish to establish can be made a corollary of the more general result that the determinant

$$D = |a_{ij}| \tag{2}$$

of any positive definite quadratic form

$$Q = \sum_{i,j=1}^{N} a_{ij}u_iu_j \tag{3}$$

is positive. We shall show that this positivity is a simple consequence of the fact that it is nonzero.

All this, as we shall see, can easily be derived from results we shall obtain further on. It is, however, interesting and profitable to see the depth of various results, and to note how far one can go by means of fairly simple reasoning.

Let us begin by observing that D is never zero. For, if $D = 0$, there is a nontrivial solution of the system of linear equations

$$\sum_{j=1}^{N} a_{ij}u_j = 0 \qquad i = 1, 2, \ldots, N \tag{4}$$

From these equations, we conclude that

$$Q = \sum_{i=1}^{N} u_i \left(\sum_{j=1}^{N} a_{ij}u_j\right) = 0 \tag{5}$$

a contradiction. This is essentially the same proof we used in Sec. 3.

The novelty arises in the proof that D is positive. Consider the family of quadratic forms defined by the relation

$$P(\lambda) = \lambda Q + (1 - \lambda) \sum_{i=1}^{N} u_i^2 \tag{6}$$

where λ is a scalar parameter ranging over the interval [0,1]. For all λ in this interval, it is clear that $P(\lambda)$ is positive for all nontrivial u_i. Consequently, the determinant of the quadratic form $P(\lambda)$ is never zero.

For $\lambda = 0$, the determinant has the simple form

$$\begin{vmatrix} 1 & & & & & & \\ & \cdot & & & & 0 & \\ & & \cdot & & & & \\ & & & 1 & & & \\ & & & & \cdot & & \\ & 0 & & & & \cdot & \\ & & & & & & 1 \end{vmatrix} = 1 \tag{7}$$

clearly positive. Since the determinant of $P(\lambda)$ is continuous in λ for $0 \le \lambda \le 1$, and never zero in this range, it follows that positivity at $\lambda = 0$ implies positivity at $\lambda = 1$, the desired result.

5. An Identity. An alternative method of establishing the positivity of the D_k when the vectors x^i are linearly independent depends upon the following identity, which we shall call upon again in a later chapter devoted to inequalities.

Theorem 1. *Let x^i, $i = 1, 2, \ldots, k$, be a set of N-dimensional vectors, $N \ge k$. Then*

$$|(x^i,x^j)|_{i,j=1,2,\ldots,k} = \frac{1}{k!}\sum_{\{i\}} \begin{vmatrix} x_{i_1}{}^1 & x_{i_2}{}^1 & \cdots & x_{i_k}{}^1 \\ x_{i_1}{}^2 & x_{i_2}{}^2 & \cdots & x_{i_k}{}^2 \\ \cdot & & & \\ \cdot & & & \\ \cdot & & & \\ x_{i_1}{}^k & x_{i_2}{}^k & \cdots & x_{i_k}{}^k \end{vmatrix}^2 \tag{1}$$

where the sum is over all sets of integers $\{i_k\}$ with $1 \le i_1 \le i_2 \le \cdots < i_k \le N$.

Proof. Before presenting the proof, let us observe that the case $k = 2$ is the identity of Lagrange

$$\left(\sum_{i=1}^{N} x_i^2\right)\left(\sum_{i=1}^{N} y_i^2\right) - \left(\sum_{i=1}^{N} x_iy_i\right)^2 = \tfrac{1}{2}\sum_{i,j=1}^{N} (x_iy_j - x_jy_i)^2 \tag{2}$$

It is sufficient to consider the case $k = 3$ in order to indicate the method of proof, and it is helpful to employ a less barbarous notation. We wish to demonstrate that

$$\begin{vmatrix} \sum_{i=1}^{N} x_i^2 & \sum_{i=1}^{N} x_iy_i & \sum_{i=1}^{N} x_iz_i \\ \sum_{i=1}^{N} x_iy_i & \sum_{i=1}^{N} y_i^2 & \sum_{i=1}^{N} y_iz_i \\ \sum_{i=1}^{N} x_iz_i & \sum_{i=1}^{N} y_iz_i & \sum_{i=1}^{N} z_i^2 \end{vmatrix} = \frac{1}{3!}\sum_{1\le k,l,m\le N} \begin{vmatrix} x_k & x_l & x_m \\ y_k & y_l & y_m \\ z_k & z_l & z_m \end{vmatrix}^2 \tag{3}$$

Let us begin with the result

$$\begin{vmatrix} x_k & x_l & x_m \\ y_k & y_l & y_m \\ z_k & z_l & z_m \end{vmatrix}^2 = \begin{vmatrix} x_k^2 + x_l^2 + x_m^2 & x_ky_k + x_ly_l + x_my_m & x_kz_k + x_lz_l + x_mz_m \\ x_ky_k + x_ly_l + x_my_m & y_k^2 + y_l^2 + y_m^2 & y_kz_k + y_lz_l + y_mz_m \\ x_kz_k + x_lz_l + x_mz_m & y_kz_k + y_lz_l + y_mz_m & z_k^2 + z_l^2 + z_m^2 \end{vmatrix} \tag{4}$$

which we obtain from the rule for multiplying determinants. The last determinant may be written as the sum of three determinants, of which the first is typical,

$$\begin{vmatrix} x_k^2 & x_ky_k + x_ly_l + x_my_m & x_kz_k + x_lz_l + x_mz_m \\ x_ky_k & y_k^2 + y_l^2 + y_m^2 & y_kz_k + y_lz_l + y_mz_m \\ x_kz_k & y_kz_k + y_lz_l + y_mz_m & z_k^2 + z_l^2 + z_m^2 \end{vmatrix} \tag{5}$$

Subtracting y_k/x_k times the first column from the second column, and z_k/x_k times the first column from the third column, we are left with the determinant

$$\begin{vmatrix} x_k^2 & x_ly_l + x_my_m & x_lz_l + x_mz_m \\ x_ky_k & y_l^2 + y_m^2 & y_lz_l + y_mz_m \\ x_kz_k & y_lz_l + y_mz_m & z_l^2 + z_m^2 \end{vmatrix} \tag{6}$$

Writing this determinant as a sum of two determinants, of which the following is typical, we are left with

$$\begin{vmatrix} x_k^2 & x_ly_l & x_lz_l + x_mz_m \\ x_ky_k & y_l^2 & y_lz_l + y_mz_m \\ x_kz_k & y_lz_l & z_l^2 + z_m^2 \end{vmatrix} \tag{7}$$

and subtracting z_l/y_l times the second column from the third column, we obtain the determinant

$$\begin{vmatrix} x_k^2 & x_ly_l & x_mz_m \\ x_ky_k & y_l^2 & y_mz_m \\ x_kz_k & y_lz_l & z_m^2 \end{vmatrix} \tag{8}$$

Summing over all k, l, and m, we obtain the left side of (3). Observing that the above procedures reduce us to a determinant of the form given in (8) in 3×2 ways, we see that a factor of $1/3!$ is required.

6. The Diagonalization of General Symmetric Matrices—Two Dimensional. In order to show that a general real symmetric matrix can be reduced to diagonal form by means of an orthogonal transformation, we shall proceed inductively, considering the 2×2 case first.

Let

$$A_2 = \begin{bmatrix} a_{11} & a_{12} \\ a_{21} & a_{22} \end{bmatrix} = \begin{bmatrix} a^1 \\ a^2 \end{bmatrix} \tag{1}$$

be a symmetric matrix, and let λ_1 and x^1 be an associated characteristic root and characteristic vector. This statement is equivalent to the relations

$$Ax^1 = \lambda_1 x^1 \qquad \text{or} \qquad (a^1,x^1) = \lambda_1 x_{11} \qquad (a^2,x^1) = \lambda_1 x_{12} \tag{2}$$

where x_{11} and x_{12} are the components of x^1, which we take to be normalized by the condition $(x^1,x^1) = 1$.

As we know, we can form a 2×2 orthogonal matrix T_2 one of whose columns is x^1. Let the other column be designated by x^2.

We wish to show that

$$T_2'AT_2 = \begin{bmatrix} \lambda_1 & 0 \\ 0 & \lambda_2 \end{bmatrix} \tag{3}$$

where λ_1 and λ_2 are the two characteristic values of A, which need not be distinct. This, of course, constitutes the difficulty of the general case. We already have a very simple proof for the case where the roots are distinct.

Let us show first that

$$T_2'AT_2 = \begin{bmatrix} \lambda_1 & b_{12} \\ 0 & b_{22} \end{bmatrix} \tag{4}$$

where b_{12} and b_{22} are as yet unknown parameters. The significant fact is that the element below the main diagonal is zero. We have

$$T_2'AT_2 = T_2' \begin{bmatrix} \lambda_1 x_{11} & (a^1,x^2) \\ \lambda_1 x_{12} & (a^2,x^2) \end{bmatrix} \tag{5}$$

upon referring to the equations in (2). Since $T_2'T_2 = I$, we obtain, upon carrying through the multiplication,

$$T_2' \begin{bmatrix} \lambda_1 x_{11} & (a^1,x^2) \\ \lambda_1 x_{12} & (a^2,x^2) \end{bmatrix} = \begin{bmatrix} \lambda_1 & b_{12} \\ 0 & b_{22} \end{bmatrix} \tag{6}$$

where b_{12} and b_{22} are parameters whose value we shall determine below.

Let us begin with b_{12}, and show that it has the value zero. This follows from the fact that $T_2'AT_2$ is a symmetric matrix, for

$$(T_2'AT_2)' = T_2'A'(T_2')' = T_2'A'T_2 = T_2'AT_2 \tag{7}$$

Finally, let us show that b_{22} must be the other characteristic root of A. This follows from the fact already noted that the characteristic roots of A are identical with the characteristic roots of $T_2'AT_2$. Hence $b_{22} = \lambda_2$.

This completes the proof of the two-dimensional case. Not only is this case essential for our induction, but it is valuable because it contains in its treatment precisely the same ingredients we shall use for the general case.

7. N-dimensional Case. Let us proceed inductively. Assume that for each k, $k = 1, 2, \ldots, N$, we can determine an orthogonal matrix T_k which reduces a real symmetric matrix $A = (a_{ij})$, $i, j = 1, 2, \ldots, k$, to diagonal form,

$$T_k'AT_k = \begin{bmatrix} \lambda_1 & & & & 0 \\ & \lambda_2 & & & \\ & & \cdot & & \\ & & & \cdot & \\ 0 & & & & \lambda_k \end{bmatrix} \tag{1}$$

The elements of the main diagonal λ_i must then be characteristic roots of A. Under this assumption, which we know to be valid for $N = 2$, let us show that the same reduction can be carried out for $N + 1$.

Let

$$A_{N+1} = (a_{ij}) = \begin{bmatrix} a^1 \\ a^2 \\ \cdot \\ \cdot \\ \cdot \\ a^{N+1} \end{bmatrix} \tag{2}$$

and let λ_1 and x^1 be, respectively, an associated characteristic value and normalized characteristic vector of A_{N+1}.

Proceeding as in the two-dimensional case, we form an orthogonal matrix T_1 whose first column is x^1. Let the other columns be designated as $x^2, x^3, \ldots, x^{N+1}$, so that T_1 has the form

$$T_1 = (x^1,x^2, \ldots ,x^{N+1}) \tag{3}$$

Then as before, the first step consists of showing that the matrix $T_1'A_{N+1}T_1$ has the representation

$$T_1'A_{N+1}T_1 = \begin{bmatrix} \lambda_1 & b_{12} & b_{13} & \cdots & b_{1N} \\ 0 & & & & \\ \cdot & & & & \\ \cdot & & A_N & & \\ \cdot & & & & \\ 0 & & & & \end{bmatrix} \tag{4}$$

where

a. The elements of the first column are all zero except for the first which is λ_1
b. The quantities $b_{12}, b_{13}, \ldots, b_{1N}$ will be determined below (5)
c. The matrix A_N is an $N \times N$ matrix

Carrying out the multiplication, we have

$$\begin{aligned} A_{N+1}T_1 &= \begin{bmatrix} (a^1,x^1) & (a^1,x^2) & \cdots & (a^1,x^{N+1}) \\ (a^2,x^1) & (a^2,x^2) & \cdots & (a^2,x^{N+1}) \\ \cdot & \cdot & & \\ \cdot & \cdot & & \\ \cdot & \cdot & & \\ (a^{N+1},x^1) & (a^{N+1},x^2) & \cdots & (a^{N+1},x^{N+1}) \end{bmatrix} \\ &= \begin{bmatrix} \lambda_1 x_{11} & (a^1,x^2) & \cdots & (a^1,x^{N+1}) \\ \lambda_1 x_{12} & (a^2,x^2) & \cdots & (a^2,x^{N+1}) \\ \cdot & \cdot & & \\ \cdot & \cdot & & \\ \cdot & \cdot & & \\ \lambda_1 x_{1,N+1} & (a^{N+1},x^2) & \cdots & (a^{N+1},x^{N+1}) \end{bmatrix} \end{aligned} \tag{6}$$

Since T_1 is an orthogonal matrix, we see that

$$T_1'A_{N+1}T_1 = \begin{bmatrix} \lambda_1 & b_{12} & \cdots & b_{1,N+1} \\ 0 & & & \\ \cdot & & & \\ \cdot & & A_N & \\ \cdot & & & \\ 0 & & & \end{bmatrix} \tag{7}$$

To determine the values of the quantities b_{1i}, we use the fact that $T_1'A_{N+1}T_1$ is symmetric. This shows that these quantities are zero and simultaneously that the matrix A_N must be symmetric.

The result of all this is to establish the fact that there is an orthogonal matrix T_1 with the property that

$$T_1'A_{N+1}T_1 = \begin{bmatrix} \lambda_1 & 0 & \cdots & 0 \\ 0 & & & \\ \cdot & & & \\ \cdot & & A_N & \\ \cdot & & & \\ 0 & & & \end{bmatrix} \tag{8}$$

with A_N a symmetric matrix.

Finally, let us note that the characteristic roots of the matrix A_N must be $\lambda_2, \lambda_3, \ldots, \lambda_{N+1}$, the remaining N characteristic roots of the matrix A_{N+1}. This follows from what we have already observed concerning the identity of the characteristic roots of A_{N+1} and $T_1'A_{N+1}T_1$, and the fact that the characteristic equation of the matrix appearing on the right-hand side of (7) has the form $|\lambda_1 - \lambda|\,|A_N - \lambda I| = 0$.

Let us now employ our inductive hypothesis. Let T_N be an orthogonal matrix which reduces A_N to diagonal form. Form the $(N + 1)$-dimensional matrix

$$S_{N+1} = \begin{bmatrix} 1 & 0 & \cdots & 0 \\ 0 & & & \\ \cdot & & & \\ \cdot & & T_N & \\ \cdot & & & \\ 0 & & & \end{bmatrix} \tag{9}$$

which, as we know, is also orthogonal.

It is readily verified that

$$S_{N+1}'(T_1'A_{N+1}T_1)S_{N+1} = \begin{bmatrix} \lambda_1 & & & & 0 \\ & \lambda_2 & & & \\ & & \cdot & & \\ & & & \cdot & \\ & & & & \cdot \\ 0 & & & & \lambda_{N+1} \end{bmatrix} \tag{10}$$

Since we may write

$$S'_{N+1}(T'_1 A_{N+1} T_1) S_{N+1} = (T_1 S_{N+1})' A_{N+1} (T_1 S_{N+1}) \tag{11}$$

we see that $T_1 S_{N+1}$ is the required diagonalizing orthogonal matrix for A_{N+1}.

We have thus proved the following basic result.

Theorem 2. *Let A be a real symmetric matrix. Then it may be transformed into diagonal form by means of an orthogonal transformation, which is to say, there is an orthogonal matrix T such that*

$$T'AT = \begin{bmatrix} \lambda_1 & & & & 0 \\ & \lambda_2 & & & \\ & & \cdot & & \\ & & & \cdot & \\ 0 & & & & \lambda_N \end{bmatrix} \tag{12}$$

where λ_i are the characteristic roots of A.

Equivalently,

$$(x,Ax) = \sum_{i=1}^{N} \lambda_i y_i^2 \tag{13}$$

where $y = T'x$.

EXERCISE

1. What can be said if A is a complex symmetric matrix?

8. A Necessary and Sufficient Condition for Positive Definiteness. The previous result immediately yields Theorem 3.

Theorem 3. *A necessary and sufficient condition that A be positive definite is that all the characteristic roots of A be positive.*

Similarly, we see that a necessary and sufficient condition that A be positive indefinite, or non-negative definite, is that all characteristic roots of A be non-negative.

EXERCISES

1. $A = BB'$ is positive definite if B is real and $|B| \neq 0$.
2. $H = CC^*$ is positive definite if $|C| \neq 0$.
3. If A is symmetric, then $I + \epsilon A$ is positive definite if ϵ is sufficiently small.

9. Characteristic Vectors Associated with Multiple Characteristic Roots. If the characteristic roots of A are distinct, it follows from the orthogonality of the associated characteristic vectors that these vectors are linearly independent.

Let us now examine the general case. Suppose that λ_1 is a root of multiplicity k. Is it true that there exist k linearly independent characteristic vectors with λ_1 as the associated characteristic root? If so, is it true that every characteristic vector associated with λ_1 is a linear combination of these k vectors?

To answer the question, let us refer to the representation in (7.12), and suppose that $\lambda_1 = \lambda_2 = \cdots = \lambda_k$, but that $\lambda_i \neq \lambda_1$ for

$$i = k + 1, \ldots, N$$

Retracing our steps and writing

$$AT = T \begin{bmatrix} \lambda_1 & & & & 0 \\ & \lambda_2 & & & \\ & & \cdot & & \\ & & & \cdot & \\ & & & & \cdot \\ 0 & & & & \lambda_N \end{bmatrix} \tag{1}$$

it follows that the jth column of T is a characteristic vector of A with the characteristic value λ_j. Since T is orthogonal, its columns are linearly independent. Hence, if λ_1 is a root of multiplicity k, it possesses k linearly independent characteristic vectors.

It remains to show that any other characteristic vector y associated with λ_1 is a linear combination of these k vectors. Let y be written as a linear combination of the N columns of T,

$$y = \sum_{i=1}^{N} c_i x^i \tag{2}$$

The coefficients c_i can be determined by Cramer's rule since the determinant is nonzero as a consequence of the linear independence of the x^i.

Since characteristic vectors associated with distinct characteristic roots are orthogonal, we see that $c_i = 0$ unless x^i is a characteristic vector associated with λ_1. This shows that y is a linear combination of the characteristic vectors associated with λ_1, obtained from the columns of T.

10. The Cayley-Hamilton Theorem for Symmetric Matrices. From (7.12), we see that for any polynomial $p(\lambda)$ we have the relation

$$p(A) = T \begin{bmatrix} p(\lambda_1) & & & & 0 \\ & p(\lambda_2) & & & \\ & & \cdot & & \\ 0 & & & \cdot & \\ & & & & \cdot \\ & & & & p(\lambda_N) \end{bmatrix} T' \tag{1}$$

If, in particular, we choose $p(\lambda) = |A - \lambda I|$, the characteristic polynomial of A, we see that $p(A) = 0$. This furnishes a proof of the following special case of a famous result of Cayley and Hamilton.

Theorem 4. *Every symmetric matrix satisfies its characteristic equation.*

As we shall see subsequently, this result can be extended to arbitrary square matrices.

EXERCISE

1. Use the method of continuity to derive the Cayley-Hamilton theorem for general symmetric matrices from the result for symmetric matrices for simple roots.

11. Simultaneous Reduction to Diagonal Form. Having seen that we can reduce a real symmetric matrix to diagonal form by means of an orthogonal transformation, it is natural to ask whether or not we can simultaneously reduce two real symmetric matrices to diagonal form. The answer is given by the following result.

Theorem 5. *A necessary and sufficient condition that there exist an orthogonal matrix T with the property that*

$$T'AT = \begin{bmatrix} \lambda_1 & & & & 0 \\ & \lambda_2 & & & \\ & & \cdot & & \\ & & & \cdot & \\ 0 & & & & \lambda_N \end{bmatrix} \qquad T'BT = \begin{bmatrix} \mu_1 & & & & 0 \\ & \mu_2 & & & \\ & & \cdot & & \\ & & & \cdot & \\ 0 & & & & \mu_N \end{bmatrix} \tag{1}$$

is that A and B commute.

Proof. The proof of the sufficiency is quite simple if either A or B has distinct characteristic roots. Assume that A has distinct characteristic roots. Then from

$$Ax = \lambda x \tag{2}$$

we obtain

$$A(Bx) = B(Ax) = B(\lambda x) = \lambda(Bx) \tag{3}$$

From this we see that Bx is a characteristic vector associated with λ whenever x is. Since the characteristic roots are assumed distinct, any two characteristic vectors associated with the same characteristic root must be proportional. Hence we have

$$Bx^i = \mu_i x^i \qquad i = 1, 2, \ldots, N \tag{4}$$

where the μ_i are scalars—which must then be the characteristic roots of the matrix B. We see then that A and B have the same characteristic vectors, $x^1, x^2, \ldots, x^N$.

T can be taken to be

$$T = (x^1,x^2, \ldots ,x^N) \tag{5}$$

Consider now the general case where λ_1 is a characteristic root of multiplicity k with associated characteristic vectors $x^1, x^2, \ldots, x^k$. Then from (3) we can only conclude that

$$Bx^i = \sum_{j=1}^{k} c_{ij}x^j \qquad i = 1, 2, \ldots, k \tag{6}$$

Let us, however, see whether we can form suitable linear combinations of the x^i which will be characteristic vectors of B. We note, to begin with, that the matrix $C = (c_{ij})$ is symmetric for the orthonormality of the x^i yields

$$(x^j,Bx^i) = c_{ij} = (Bx^j,x^i) = c_{ji} \tag{7}$$

Consider the linear combination $\sum_{i=1}^{k} a_ix^i$. We have

$$B\left(\sum_{i=1}^{k} a_ix^i\right) = \sum_{i=1}^{k} a_i\left(\sum_{j=1}^{k} c_{ij}x^j\right) = \sum_{j=1}^{k}\left(\sum_{i=1}^{k} c_{ij}a_i\right)x^j \tag{8}$$

Hence if the a_i are chosen so that

$$\sum_{i=1}^{k} c_{ij}a_i = r_1a_j \qquad j = 1, 2, \ldots, k \tag{9}$$

we will have

$$B\left(\sum_{i=1}^{k} a_ix^i\right) = r_1\left(\sum_{i=1}^{k} a_ix^i\right) \tag{10}$$

which means that r_1 is a characteristic root of B and $\sum_{i=1}^{k} a_ix^i$ is an associated characteristic vector.

The relation in (9) shows that r_1 is a characteristic root of C and the a_i components of an associated characteristic vector. Thus, if T_k is a k-dimensional orthogonal transformation which reduces C to diagonal form, the set of vectors z^i furnished by the relation

$$\begin{bmatrix} z^1 \\ z^2 \\ \cdot \\ \cdot \\ \cdot \\ z^k \end{bmatrix} = T'_k \begin{bmatrix} x^1 \\ x^2 \\ \cdot \\ \cdot \\ \cdot \\ x^k \end{bmatrix} \tag{11}$$

will be an orthonormal set which are simultaneously characteristic vectors of A and B.

Performing similar transformations for the characteristic vectors associated with each multiple characteristic root, we obtain the desired matrix T.

The necessity of the condition follows from the fact that two matrices of the form

$$A = T \begin{bmatrix} \lambda_1 & & & 0 \\ & \lambda_2 & & \\ & & \ddots & \\ 0 & & & \lambda_N \end{bmatrix} T' \qquad B = T \begin{bmatrix} \mu_1 & & & 0 \\ & \mu_2 & & \\ & & \ddots & \\ 0 & & & \mu_N \end{bmatrix} T' \tag{12}$$

always commute if T is orthogonal.

12. Simultaneous Reduction to Sum of Squares. As was pointed out in the previous section, the simultaneous reduction of two symmetric matrices A and B to diagonal form by means of an orthogonal transformation is possible if and only if A and B commute. For many purposes, however, it is sufficient to reduce A and B simultaneously to diagonal form by means of a nonsingular matrix. We wish to demonstrate Theorem 6.

Theorem 6. *Given two real symmetric matrices, A and B, with A positive definite, there exists a nonsingular matrix T such that*

$$T'AT = I \tag{1}$$

$$T'BT = \begin{bmatrix} \mu_1 & & & 0 \\ & \mu_2 & & \\ & & \ddots & \\ 0 & & & \mu_N \end{bmatrix}$$

Proof. Let S be an orthogonal matrix such that

$$A = S \begin{bmatrix} \lambda_1 & & & 0 \\ & \lambda_2 & & \\ & & \ddots & \\ 0 & & & \lambda_N \end{bmatrix} S' \tag{2}$$

Then, if $x = Sy$, we have

$$(x,Ax) = \sum_{i=1}^{N} \lambda_i y_i^2 \tag{3}$$
$$(x,Bx) = (y,S'BSy)$$

Now perform a "stretching" transformation

$$y_i = \frac{z_i}{\lambda_i^{1/2}} \qquad i = 1, 2, \ldots, N \tag{4}$$

that is $\qquad y = S_2 z$

Then

$$\sum_{i=1}^{N} \lambda_i y_i^2 = (z,z) \tag{5}$$
$$(y,S'BSy) = (z,S_2'S'BSS_2z)$$

Denote by C the matrix $S_2'S'BSS_2$, and let S_3 be an orthogonal matrix which reduces C to diagonal form. Then, if we set $z = S_3w$, we have

$$(z,Cz) = (w,S_3'CS_3w) = \sum_{i=1}^{N} \mu_i w_i^2 \tag{6}$$

while $(z,z) = (w,w)$. It follows that

$$T = SS_2S_3 \tag{7}$$

is the desired nonsingular matrix.

EXERCISES

1. What is the corresponding result in case A is non-negative definite?

2. Let the notation $A > B$ for two symmetric matrices denote the fact that $A - B$ is positive definite. Use the foregoing result to show that $A > B > 0$ implies that $B^{-1} > A^{-1}$.

13. Hermitian Matrices. It is clear that precisely the same techniques as used in establishing Theorem 2 enable one to establish Theorem 7.

Theorem 7. *If H is a Hermitian matrix, there exists a unitary matrix U such that*

$$H = U \begin{bmatrix} \lambda_1 & & & 0 \\ & \lambda_2 & & \\ & & \ddots & \\ 0 & & & \lambda_N \end{bmatrix} U^* \tag{1}$$

14. The Original Maximization Problem. We are now in a position to resolve the problem we used to motivate this study of the positivity and negativity of quadratic forms, namely, that of deciding when a stationary point of a function $f(x_1,x_2, \ldots ,x_N)$ is actually a local maximum or a local minimum.

Let $c = (c_1,c_2, \ldots ,c_N)$ be a stationary point, and suppose that f possesses continuous mixed derivatives of the second order. It is consequently sufficient to consider the nature of the quadratic form

$$Q = \left(\frac{\partial^2 f}{\partial c_i \partial c_j}\right) = (f_{c_i c_j}) \tag{1}$$

where, as before,

$$\frac{\partial^2 f}{\partial c_i \partial c_j} = \frac{\partial^2 f}{\partial x_i \partial x_j} \tag{2}$$

evaluated at the point $x_1 = c_1, x_2 = c_2, \ldots , x_N = c_N$.

We see then that a sufficient condition that c be a local minimum is that Q be positive definite, and a necessary condition that c be a local minimum is that Q be positive indefinite. The criterion given in Sec. 8 furnishes a method for determining whether Q is positive or negative definite. If Q vanishes identically, higher order terms must be examined.

If N is large, it is, however, not a useful method. Subsequently, in Chap. 6, we shall derive the most useful criterion for the positive definiteness of a matrix.

EXERCISE

1. Show that a set of sufficient conditions for $f(x_1,x_2)$ to have a local maximum is

$$f_{c_1c_1} < 0 \qquad \begin{vmatrix} f_{c_1c_1} & f_{c_1c_2} \\ f_{c_1c_2} & f_{c_2c_2} \end{vmatrix} > 0$$

15. Perturbation Theory—I. We can now discuss a problem of great theoretical and practical interest. Let A be a symmetric matrix possessing the known characteristic values $\lambda_1, \lambda_2, \ldots , \lambda_N$ and corresponding characteristic vectors $x^1, x^2, \ldots , x^N$. What can we say about the characteristic roots and vectors of $A + \epsilon B$, where B is a symmetric matrix and ϵ is a "small" quantity? How small ϵ has to be in order to be called so will not be discussed here, since we are interested only in the formal theory.

If A and B commute, both may be reduced to diagonal form by the same orthogonal transformation. Consequently, with a suitable reordering of the characteristic roots of B, the characteristic roots of $A + \epsilon B$ will be $\lambda_i + \epsilon\mu_i$, $i = 1, 2, \ldots , N$, while the characteristic vectors will be as before.

Let us then consider the interesting case where $AB \neq BA$. For the sake of simplicity, we shall consider only the case where the characteristic roots are distinct.

It is to be expected that the characteristic roots of $A + \epsilon B$ will be distinct, for ϵ sufficiently small, and that they will be close to the characteristic roots of A. It will follow from this, that the characteristic vectors of $A + \epsilon B$ will be close to those of A.

One way to proceed is the following. In place of the characteristic equation $|A - \lambda I| = 0$, we now have the characteristic equation $|A + \epsilon B - \lambda I| = 0$. The problem thus reduces to finding approximate expressions for the roots of this equation, given the roots of the original equation and the fact that ϵ is small.

To do this, let T be an orthogonal matrix reducing A to diagonal form. Then the determinantal equation may be written

$$\begin{vmatrix} \lambda_1 + \epsilon c_{11} - \lambda & \epsilon c_{12} & \cdots & \epsilon c_{1N} \\ \epsilon c_{21} & \lambda_2 + \epsilon c_{22} - \lambda & \cdots & \epsilon c_{2N} \\ \cdot & & & \cdot \\ \cdot & & & \cdot \\ \cdot & & & \cdot \\ \epsilon c_{N1} & \cdots & & \lambda_N + \epsilon c_{NN} - \lambda \end{vmatrix} = 0$$

where $T'BT = C = (c_{ij})$.

We now look for a set of solutions of this equation of the form

$$\lambda = \lambda_i + d_{1i}\epsilon + d_{2i}\epsilon^2 + \cdots$$

We leave it as a set of exercises for the reader versed in determinantal expansions to obtain the coefficients d_{1i}, d_{2i}, and so on, in terms of the elements c_{ij}.

We do not wish to pursue it in any detail since the method we present in the following section is more powerful, and, in addition, can be used to treat perturbation problems arising from more general operators.

16. Perturbation Theory—II. Let us now suppose not only that the characteristic roots and vectors of $A + \epsilon B$ are close to those of A, but that we actually have power series in ϵ for these new quantities.

Write μ_i, y^i as an associated pair of characteristic root and characteristic vector of $A + \epsilon B$ and λ_i, x^i as a similar set for A. Then we set

$$\begin{aligned} \mu_i &= \lambda_i + \epsilon\lambda_{1i} + \epsilon^2\lambda_{2i} + \cdots \\ y^i &= x^i + \epsilon x^{i1} + \epsilon^2 x^{i2} + \cdots \end{aligned} \tag{1}$$

To determine the unknown coefficients λ_{1i}, λ_{2i}, . . . and the unknown vectors x^{i1}, x^{i2}, . . . , we substitute these expressions in the equation

$$(A + \epsilon B)y^i = \mu_i y^i \tag{2}$$

and equate coefficients.

We thus obtain

$$\begin{aligned}(A + \epsilon B)(x^i + \epsilon x^{i1} + \epsilon^2 x^{i2} + \cdots) \\ = (\lambda_i + \epsilon\lambda_{i1} + \epsilon^2\lambda_{i2} + \cdots)(x^i + \epsilon x^{i1} + \epsilon^2 x^{i2} + \cdots)\end{aligned} \tag{3}$$

and the series of relations

$$\begin{aligned} Ax^i &= \lambda_i x^i \\ Ax^{i1} + Bx^i &= \lambda_i x^{i1} + \lambda_{i1} x^i \\ Ax^{i2} + Bx^{i1} &= \lambda_i x^{i2} + \lambda_{i1} x^{i1} + \lambda_{i2} x^i \\ &\ \ \vdots \end{aligned} \tag{4}$$

Examining these equations, we see that the first is identically satisfied, but that the second equation introduces two unknown quantities, the scalar λ_{i1} and the vector x^{i1}. Similarly, the third equation introduces two further unknowns, the scalar λ_{i2} and the vector x^{i2}.

At first sight this appears to invalidate the method. However, examining the second equation, we see that it has the form

$$(A - \lambda_i I)x^{i1} = (\lambda_{i1} I - B)x^i \tag{5}$$

where the coefficient matrix $A - \lambda_i I$ is singular. Consequently, there will be a solution of this equation only if the right-hand-side vector $(\lambda_{i1} I - B)x^i$ possesses special properties.

When we express the fact that $(\lambda_{i1} I - B)x^i$ possesses these properties, we will determine the unknown scalar λ_{i1}.

Let us simplify the notation, writing

$$\begin{aligned} x^{i1} &= y \\ (\lambda_{i1} I - B)x^i &= z \end{aligned} \tag{6}$$

Then we wish to determine under what conditions upon z the equation

$$(A - \lambda_i I)y = z \tag{7}$$

has a solution, when λ_i is a characteristic root of A, and what this solution is.

Let $x^1, x^2, \ldots, x^N$ be the normalized characteristic vectors of A, and consider the expressions for y and z as linear combinations of these vectors,

$$\begin{aligned} y &= \sum_{j=1}^{N} b_j x^j \\ z &= \sum_{j=1}^{N} c_j x^j \end{aligned} \tag{8}$$

where

$$b_j = (y,x^j) \qquad c_j = (z,x^j) \qquad i = 1, 2, \ldots, N \tag{9}$$

Substituting in (7), we obtain the equation

$$(A - \lambda_i I) \sum_{j=1}^{N} b_j x^j = \sum_{j=1}^{N} c_j x^j \tag{10}$$

or, since $Ax^j = \lambda_j x^j$, $j = 1, 2, \ldots, N$,

$$\sum_{i=1}^{N} b_j(\lambda_j - \lambda_i)x^j = \sum_{j=1}^{N} c_j x^j \tag{11}$$

Equating the coefficients of x^j, we obtain the relations

$$b_j(\lambda_j - \lambda_i) = c_j \qquad j = 1, 2, \ldots, N \tag{12}$$

We see that these equations are consistent if and only if

$$c_i = 0 \tag{13}$$

Then b_i is arbitrary, but the remaining b_j are given by

$$b_j = c_j/(\lambda_j - \lambda_i) \qquad j = 1, 2, \ldots, N, j \neq i \tag{14}$$

What the condition in (13) asserts is that (7) has a solution if and only if z is orthogonal to x^i, the characteristic vector associated with λ_i. If so, there is a one-parameter family of solutions of (7), having the form

$$y = b_i x^i + \sum_{j \neq i} \frac{c_j x^j}{\lambda_j - \lambda_i} \tag{15}$$

where b_i is arbitrary.

Returning to (6), we see that we can take $b_i = 0$, since a different choice of b_i merely affects the normalization of the new characteristic vector y^i.

The orthogonality condition yields the relation

$$(x^i, (\lambda_{i1} I - B)x^i) = 0 \tag{16}$$

or, the simple result,

$$\lambda_{i1} = (x^i, Bx^i) \tag{17}$$

EXERCISES

1. Find the value of λ_{i2}.
2. Consider the case of multiple roots, first for the 2×2 case, and then, in general.
3. Let λ_i denote the characteristic roots of A and $\lambda_i(z)$ those of $A + zB$. Show

that

$$\lambda_j(A + zB) = \sum_{m=0}^{\infty} \lambda_j^{(m)} z^m \qquad \lambda_j^{(0)} = \lambda_j$$

where
$$\lambda_j^{(m)} = (-1)^m m^{-1} tr \left(\sum_{\substack{k_1 + \cdots + k_m = m-1 \\ k_i \geq 0}} BS_j^{k_1} BS_j^{k_2} \cdots BS_j^{k_m} \right)$$

$$S_j = -E_j$$

The E_j are defined by the relation $A = \sum_j \lambda_j E_j$, and $S_j = \sum_{k \neq j} E_k/(\lambda_k - \lambda_j)$ (*T. Kato*).

4. Let A and B be 10×10 matrices of the form

$$A = \begin{bmatrix} 0 & 1 & \cdot & \cdot & \cdot & 0 \\ & 0 & 1 & & & \cdot \\ & & & \cdot & & \cdot \\ & & & & \cdot & \cdot \\ & & & & \cdot & 1 \\ 0 & & & & & 0 \end{bmatrix} \qquad B = \begin{bmatrix} 0 & 1 & \cdot & \cdot & \cdot & 0 \\ & 0 & 1 & & & \cdot \\ & & & \cdot & & \cdot \\ & & & & \cdot & \cdot \\ & & & & \cdot & 1 \\ 10^{-10} & & & & & 0 \end{bmatrix}$$

A and B are identical except for the element in the (10,1) place, which is 10^{-10} in the case of B. Show that $|\lambda I - A| = \lambda^{10}$ and that $|\lambda I - B| = \lambda^{10} - 10^{-10}$. Hence, the characteristic roots of B are $10^{-1}, 10^{-1}\omega, \ldots, 10^{-1}\omega^9$, where ω is an irreducible tenth root of unity. What is the explanation of this phenomenon? (*Forsythe.*)

MISCELLANEOUS EXERCISES

1. Show that a real symmetric matrix may be written in the form

$$A = \sum_{i=1}^{N} \lambda_i E_i$$

where the E_i are non-negative definite matrices satisfying the conditions

$$E_i E_j = 0, \qquad i \neq j, \; E_i^2 = E_i$$

and the λ_i are the characteristic roots of A. This is called the *spectral decomposition* of A.

2. Let A be a real skew-symmetric matrix; $a_{ij} = -a_{ji}$. Then the characteristic roots are pure imaginary or zero. Let $x + iy$ be a characteristic vector associated with $i\mu$, where μ is a real nonzero quantity and where x and y are real. Show that x and y are orthogonal.

3. Referring to the above problem, let T_1 be an orthogonal matrix whose first two columns are x and y,

$$T_1 = (x,y,x^3,x^4, \ldots ,x^N)$$

Show that

$$T_1'AT = \begin{bmatrix} \begin{pmatrix} 0 & \mu \\ -\mu & 0 \end{pmatrix} & 0 \\ 0 & A_{N-2} \end{bmatrix}$$

where A_{N-2} is again skew-symmetric.

4. Prove inductively that if A is a real skew-symmetric matrix of even dimension we can find an orthogonal matrix T such that

$$T'AT = \begin{bmatrix} \begin{pmatrix} 0 & \mu_1 \\ -\mu_1 & 0 \end{pmatrix} & & & & 0 \\ & \begin{pmatrix} 0 & \mu_2 \\ -\mu_2 & 0 \end{pmatrix} & & & \\ & & \cdot & & \\ & & & \cdot & \\ 0 & & & \cdot & \begin{pmatrix} 0 & \mu_N \\ -\mu_N & 0 \end{pmatrix} \end{bmatrix}$$

where some of the μ_i may be zero.

5. If A is of odd dimension, show that the canonical form is

$$T'AT = \begin{bmatrix} \begin{pmatrix} 0 & \mu_1 \\ -\mu_1 & 0 \end{pmatrix} & & & 0 \\ & \cdot & & \\ & & \begin{pmatrix} 0 & \mu_N \\ -\mu_N & 0 \end{pmatrix} & \\ 0 & & & 0 \end{bmatrix}$$

where again some of the μ_i may be zero.

6. Show that the determinant of a skew-symmetric matrix of odd dimension is zero.

7. The determinant of a skew-symmetric matrix of even dimension is the square of a polynomial in the elements of the matrix.

8. Let A be an orthogonal matrix, and let λ be a characteristic root of absolute value 1 but not equal to ± 1, with $x + iy$ an associated characteristic vector, x and y real. Show that x and y are orthogonal.

9. Proceeding inductively as before, show that every orthogonal matrix A can be reduced to the form

$$A = T \begin{bmatrix} \begin{pmatrix} \cos\lambda_1 & -\sin\lambda_1 \\ \sin\lambda_1 & \cos\lambda_1 \end{pmatrix} & & & & & 0 \\ & \cdot & & & & \\ & & \cdot & & & \\ & & & \begin{pmatrix} \cos\lambda_k & -\sin\lambda_k \\ \sin\lambda_k & \cos\lambda_k \end{pmatrix} & & \\ & & & & \pm 1 & \\ & & & & & \cdot \\ 0 & & & & & \pm 1 \end{bmatrix} T'$$

10. Prove that $T\Lambda T'$ is a positive definite matrix whenever T is an orthogonal matrix and Λ is a diagonal matrix with positive elements down the main diagonal.

11. Prove Theorem 5 by means of an inductive argument, along the lines of the proof given in Sec. 6.

12. Establish the analogue of the result in Exercise 9 for unitary matrices.

13. Write down a sufficient condition that the function $f(x_1,x_2, \ldots ,x_N)$ possess a local maximum at $x_i = c_i$, $i = 1, 2, \ldots , N$.

14. Given that A is a positive definite matrix, find all solutions of $XX' = A$.

15. Define the Gramian of N real functions $f_1, f_2, \ldots , f_N$ over (a,b) by means of the expression

$$G(f_1,f_2, \ldots ,f_N) = \left| \int_a^b f_i(t)f_j(t)\,dt \right|$$

Prove that if

$$Q(x_1,x_2, \ldots ,x_N) = \int_a^b \left[g(t) - \sum_{i=1}^{N} x_i f_i(t) \right]^2 dt$$

then

$$\min_x Q = \frac{G(g,f_1,f_2, \ldots ,f_N)}{G(f_1,f_2, \ldots ,f_N)}$$

For some further results, see J. Geronimus.[1] For many further results, see G. Szego.[2]

16. If we consider the $(N - 1)$-dimensional matrix

$$H = \begin{bmatrix} 2 & -1 & & & & \\ -1 & 2 & -1 & & & \\ & & \cdot & & & \\ & & & \cdot & & \\ & & & \cdot & & \\ & & & -1 & 2 & -1 \\ & & & & -1 & 1 \end{bmatrix}$$

where all undesignated elements are zero, show that

$$|\lambda I - H| = \prod_{k=1}^{N-1} \left(\lambda - 2 - 2\cos\frac{2k\pi}{2N - 1}\right)$$

and determine the characteristic vectors.

17. Hence, show that if $x_1 = 0$ and the x_i are real, then

$$\sum_{i=1}^{N-1} (x_i - x_{i+1})^2 \geq 4 \sin^2 \frac{\pi}{2(2N - 1)} \sum_{i=2}^{N} x_i^2$$

Determine the case of equality.

18. Show that if the x_i are all real and $x_0 = x_{N+1} = 0$ that

$$\sum_{i=0}^{N} (x_i - x_{i+1})^2 \geq 4 \sin^2 \frac{\pi}{2(N + 1)} \sum_{i=0}^{N} x_i^2$$

[1] J. Geronimus, On Some Persymmetric Determinants Formed by the Polynomials of M. Appell, *J. London Math. Soc.*, vol. 6, pp. 55–58, 1931.

[2] G. Szego, *Orthogonal Polynomials*, *Am. Math. Soc. Colloq. Publ.*, vol. 23, 1939.

and that if $x_1 = x_{N+1}$, $\sum_{i=1}^{N} x_i = 0$, then

$$\sum_{i=1}^{N} (x_i - x_{i+1})^2 \geq 4 \sin^2 \frac{\pi}{N} \sum_{i=1}^{N} x_i^2$$

(*Ky Fan, O. Taussky, and J. Todd*)

19. Find the characteristic roots and vectors of the matrix associated with the quadratic form $\sum_{i,j=1}^{N} (x_i - x_j)^2$.

20. Let A and B be real symmetric matrices such that A is non-negative definite. Then if $|A + iB| = 0$, there exists a nontrivial real vector such that $(A + iB)x = 0$ (*Peremans-Duparc-Lekkerkerker*).

21. If A and B are symmetric, the characteristic roots of AB are real, provided that at least one is non-negative definite.

22. If A and B are symmetric, the characteristic roots of $AB - BA$ are pure complex.

23. Reduce $Q_N(x) = x_1x_2 + x_2x_3 + \cdots + x_{N-1}x_N$ to diagonal form, determining the characteristic vectors and characteristic values. Similarly, reduce $P_N(x) = x_1x_2 + x_2x_3 + \cdots + x_{N-1}x_N + x_Nx_1$.

24. If A is a complex square matrix, then A is HU, where H is non-negative definite Hermitian and U is unitary.

Further results and references to earlier results of Autonne, Wintner, and Murnaghan may be found in J. Williamson.[1]

25. If A, B, and C are symmetric and positive definite, the roots of

$$|\lambda^2 A + \lambda B + C| = 0$$

have negative real parts (*Parodi*).

26. Let A be a square matrix. Set $|A| = |A|_+ - |A|_-$, where $|A|_+$ denotes the sum of the terms which are given a positive sign in the determinantal expansion. Show that if A is a positive definite Hermitian matrix, then $|A|_- \geq 0$ (*Schur-Leng*).

27. Let A be a real matrix with the property that there exists a positive definite matrix M such that $A'M = MA$. A matrix A with this property is called *symmetrizable*. Show that this matrix A possesses the following properties:

(*a*) All characteristic roots are real.

(*b*) In terms of a generalized inner product $[x,y] = (x,My)$, characteristic vectors associated with distinct characteristic roots are orthogonal.

(*c*) Let T be called "orthogonal" if it is real and its columns are orthogonal in the extended sense. Then there exists an "orthogonal" matrix T such that TAT' is diagonal.[2]

28. Extend in a similar fashion the concept of Hermitian matrix.

29. Call a matrix A idempotent if $A^2 = A$. Show that a necessary and sufficient condition that a symmetric matrix A of rank k be idempotent is that k of the characteristic roots are equal to one and the remaining $N - k$ are equal to zero. (The notion of *rank* is defined in Appendix A.)

30. If A is a symmetric idempotent matrix, then the rank of A is equal to the trace of A $\left(\text{The trace of } A \text{ is equal to } \sum_{i=1}^{N} a_{ii}.\right)$

[1] J. Williamson, A Generalization of the Polar Representation of Nonsingular Matrices, *Bull. Am. Math. Soc.*, vol. 48, pp. 856–863, 1942.

[2] See A. Kolmogoroff, *Math. Ann.*, vol. 112, pp. 155–160, 1936.

31. The only nonsingular symmetric idempotent matrix is the identity matrix.

32. If A is an $N \times N$ symmetric idempotent matrix of rank k, then A is positive definite if $k = N$, positive indefinite if $k < N$.

33. If A is a symmetric idempotent matrix with $a_{ii} = 0$, for a particular value of i, then every element in the ith row and ith column is equal to zero.

34. If each A_i is symmetric, and $\sum_{i=1}^{m} A_i = I$, then the three following conditions are equivalent:

(a) Each A_i is idempotent.

(b) $A_iA_j = 0 \qquad i \neq j$.

(c) $\sum_i n_i = N$, where n_i is the rank of A_i and N the dimension of the A_i.

35. Let A_i be a collection of $N \times N$ symmetric matrices where the rank of A_i is p_i. Let $A = \sum_i A_i$ have rank p. Consider the four conditions

$C1$. Each A_i is idempotent.

$C2$. $A_iA_j = 0 \qquad i \neq j$.

$C3$. A is idempotent.

$C4$. $p = \sum_{i=1}^{m} p_i$.

Then

1. Any two of the three conditions $C1$, $C2$, $C3$ imply all four of the conditions $C1$, $C2$, $C3$, $C4$.

2. $C3$ and $C4$ imply $C1$ and $C2$.

For a proof of the foregoing and some applications, see F. A. Graybill and G. Marsaglia,[1] D. J. Djoković.[2]

36. Using the fact that $2x_1x_2 < x_1^2 + x_2^2$ for any two real quantities x_1 and x_2, show that the quadratic form

$$\lambda \sum_{i=1}^{k} b_i x_i^2 + \sum_{i=k+1}^{N} b_i x_i^2 + \sum_{j=1}^{k} \sum_{i=k+1}^{N} a_{ij}x_ix_j + \sum_{i,j=1}^{k} a_{ij}x_ix_j$$

is positive definite if $b_i > 0$ and λ is sufficiently large.

37. Show that $\sum_{i,j=1}^{N} a_{ij}x_ix_j$ is positive definite if $\sum_i a_{ii}x_i^2 + \sum_{i \neq j} |a_{ij}|x_ix_j$ is positive definite.

38. A matrix $A = (a_{ij})$, $i, j = 1, 2, 3, 4$, is called a Lorentz matrix if the transformation $x = Ay$ leaves the quadratic form $Q(x) = x_1^2 - x_2^2 - x_3^2 - x_4^2$ unchanged; that is, $Q(x) = Q(y)$. Show that the product of two Lorentz matrices is again a Lorentz matrix.

39. Show that

$$\begin{aligned} y_1 &= (x_1 + \beta x_2)/(1 - \beta^2)^{1/2} \\ y_2 &= (\beta x_1 + x_2)/(1 - \beta^2)^{1/2} \\ y_3 &= x_3 \\ y_4 &= x_4 \end{aligned}$$

where $0 < \beta^2 < 1$ is a Lorentz transformation.

[1] F. A. Graybill and G. Marsaglia, Idempotent Matrices and Quadratic Forms in the General Linear Hypothesis, *Ann. Math. Stat.*, vol. 28, pp. 678–686, 1957.

[2] D. J. Djoković, Note on Two Problems on Matrices, *Amer. Math. Monthly*, to appear.

40. Show that any Lorentz transformation is a combination of an orthogonal transformation of the variables x_2, x_3, x_4 which leaves x_1 fixed, a transformation of the type appearing above, and a possible change of sign of one of the variables (a reflection).[1] Physical implications of Lorentz transformations may be found in J. L. Synge.[2]

41. Let H be a non-negative Hermitian matrix. Show that given any t interval $[a,b]$ we can find a sequence of complex functions $\{f_i(t)\}$, $i = 1, 2, \ldots$, such that

$$h_{ij} = \int_a^b f_i(t)\overline{f_j(t)}\,dt \qquad i, j = 1, 2, \ldots\dagger$$

42. Let $A = (a_{ij})$ be an $N \times N$ symmetric matrix, and A_{N-1} the symmetric matrix obtained by taking the $(N-1) \times (N-1)$ matrix whose elements are a_{ij}, $i, j = 1, 2, \ldots, N-1$. By means of an orthogonal transformation, reduce A to the form

$$A = \begin{bmatrix} \mu_1 & & & & & z_1 \\ & \mu_2 & 0 & & & z_2 \\ & & \cdot & & & \cdot \\ & & & \cdot & & \cdot \\ & & & & \cdot & \cdot \\ & 0 & & & \mu_{N-1} & z_{N-1} \\ z_1 & z_2 & & & z_{N-1} & a_{NN} \end{bmatrix}$$

where μ_i are the characteristic roots of A_{N-1}. Using this representation, determine the relations, if any, between multiple characteristic roots of A and the characteristic roots of A_{N-1}.

43. Using these results, show that the rank of a symmetric matrix can be defined as the order, N, minus the number of zero characteristic roots.

44. If H is a positive indefinite Hermitian matrix, then $H = TT^*$, where T is triangular and has real and non-negative diagonal elements. See H. C. Lee.[3]

45. Necessary and sufficient conditions for the numbers $d_1, d_2, \ldots, d_N$ to be the diagonal elements of a proper orthogonal matrix are $|d_j| \le 1$, $j = 1, 2, \ldots, N$, and $\sum_{k=1}^{N} |d_k| \le n - 2 - 2\lambda \min_{1\le j\le N} |d_j|$, where $\lambda = 1$ if the number of negative d_i is even and 0 otherwise (*L. Mirsky, Amer. Math. Monthly, vol.* 66, *pp.* 19–22, 1959).

46. If A is a symmetric matrix with no characteristic root in the interval $[a,b]$, then $(A - aI)(A - bI)$ is positive definite (*Kato's lemma*). Can we use this result to obtain estimates for the location of intervals which are free of characteristic roots of A?

47. Show that

$$\min_{a_i} \int_0^\infty e^{-t}\,(1 + a_1t + \cdots + a_nt^n)^2\,dt = 1/(n+1)$$

$$\min_{a_i} \int_0^1 (1 + a_1t + \cdots + a_nt^n)^2\,dt = 1/(n+1)^2$$

(*L. J. Mordell, Equationes Mathematicae, to appear*)

[1] Cf. I. G. Petrovsky, *Lectures on Partial Differential Equations*, Interscience Publishers, Inc., New York, 1954.

[2] J. L. Synge, *Relativity, the Special Theory*, Interscience Publishers, Inc., New York, 1956.

† I. Schur, *Math. Z.*, vol. 1, p. 206, 1918.

[3] H. C. Lee, Canonical Factorization of Non-negative Hermitian Matrices, *J. London Math. Soc.*, vol. 23, pp. 100–110, 1948.

48. Let z be an N-dimensional vector such that $(z,\bar{z}) = 1$. Determine the minimum of $(\bar{z} - \bar{b}, A(z - b))$ over this z-region assuming that A is Hermitian. See G. E. Forsythe and G. H. Golub, On the Stationary Values of a Second Degree Polynomial on the Unit Sphere, *J. Soc. Indus. Appl. Math.*, vol. 13, pp. 1050–1068, 1965.

49. Let λ_i, μ_i, $i = 1, 2, \ldots, N$, be respectively the characteristic roots of A and B. Set $M = \max\,(|a_{ij}|,|b_{ij}|)$, $d = \sum_{i,j} |b_{ij} - a_{ij}|/M$. To each root λ_i there is a root μ_j such that $|\lambda_i - \mu_j| \leq (N + 2)Md^{1/N}$. Furthermore, the λ_i and μ_i can be put into one-to-one correspondence in such a way that $|\lambda_i - \mu_i| \leq 2(N + 1)^2Md^{1/N}$ (*A. Ostrowski, Mathematical Miscellanea XXVI: On the Continuity of Characteristic Roots in Their Dependence on the Matrix Elements, Stanford University,* 1959).

50. Let $\{A_n\}$ be a bounded, monotone-increasing sequence of positive definite matrices in the sense that there exists a positive definite matrix B such that $B - A_n$ is non-negative definite for any n and such that $A_n - A_{n-1}$ is non-negative definite for any n. Show that A_n converges to a matrix A as $n \rightarrow \infty$ (*Riesz*).

Bibliography and Discussion

§2. We suppose that the reader has been exposed to the rudiments of the theory of linear systems of equations. For the occasional few who may have missed this or wish a review of some of the basic results, we have collected in Appendix A a statement and proof of the results required for the discussion in this chapter.

The reader who wishes may accept on faith the few results needed and at his leisure, at some subsequent time, fill in the proofs.

§5. The result in this section is a particular example of a fundamental principle of analysis which states that whenever a quantity is positive, there exists a formula for this quantity which makes this positivity apparent. Many times, it is not a trivial matter to find formulas of this type, nor to prove that they exist. See the discussion in

G. H. Hardy, J. E. Littlewood, and G. Polya, *Inequalities*, Cambridge University Press, New York, pp. 57–60, 1934.

and a related comment at the end of Chap. 5.

§8. The concept of a positive definite quadratic form, as a natural extension of the positivity of a scalar, is one of the most powerful and fruitful in all of mathematics. The paper by Ky Fan indicates a few of the many ways this concept can be used in analysis.

Ky Fan, On Positive Definite Sequences, *Ann. Math.*, vol. 47, pp. 593–607, 1946.

For an elegant and detailed discussion of other applications, see the monograph

Ky Fan, *Les Fonctions définies-positives et les Fonctions complètement monotones*, fascicule CXIV, *Mem. sci. math.*, 1950.

In the appendices at the end of the volume, we also indicate some of the ingenious ways in which quadratic forms may be used in various parts of analysis.

A discussion of the diagonalization of complex non-Hermitian symmetric matrices may be found in

C. L. Dolph, J. E. McLaughlin, and I. Marx, Symmetric Linear Transformations and Complex Quadratic Forms, *Comm. Pure and Appl. Math.*, vol. 7, 621–632, 1954.

Questions of this nature arise as special cases of more general problems dealing with the theory of characteristic values and functions of Sturm-Liouville equations with complex coefficients.

§15. For a further discussion of these problems, and additional references, see

R. D. Brown and I. M. Bassett, A Method for Calculating the First Order Perturbation of an Eigenvalue of a Finite Matrix, *Proc. Phys. Soc.*, vol. 71, pp. 724–732, 1958.

Y. W. Chen, On Series Expansion of a Determinant and Solution of the Secular Equation, *J. Math. Phys.*, vol. 7, pp. 27–34, 1966.

H. S. Green, *Matrix Mechanics*, Erven P. Noordhoff, Ltd., Groningen, Netherlands, 1965.

The book by Green contains a discussion of the matrix version of the factorization techniques of Infeld-Hull.

For some interesting extensions of the concept of positive definiteness, see

M. G. Krein, On an Application of the Fixed Point Principle in the Theory of Linear Transformations of Spaces with an Indefinite Metric, *Am. Math. Soc. Transl.*, (2), vol. 1, pp. 27–35, 1955,

where further references may be found.

§16. In a paper devoted to physical applications,

M. Lax, Localized Perturbations, *Phys. Rev.*, vol. 94, p. 1391, 1954,

the problem of obtaining the solution of $(A + B)x = \lambda x$ when only a few elements of B are nonzero is discussed.

For some interesting results concerning skew-symmetric matrices, see

M. P. Drazin, A Note on Skew-symmetric Matrices, *Math. Gaz.*, vol. XXXVI, pp. 253–255, 1952.

N. Jacobson, *Bull. Amer. Math. Soc.*, vol. 45, pp. 745–748, 1939.

With regard to Exercise 24, see also

H. Stenzel, Über die Darstellbarkeit einer Matrix als Product . . . , *Math. Zeit.*, vol. 15, pp. 1–25.

Finally, for some interesting connections between orthogonality and quadratic forms, see

W. Groebner, Über die Konstruktion von Systemen orthogonaler Polynome in ein- und zwei-dimensionaler Bereich, *Monatsh. Math.*, vol. 52, pp. 38–54, 1948.

H. Larcher, *Proc. Amer. Math. Soc.*, vol. 10, pp. 417–423, 1959.

For an extensive generalization of the results of the two foregoing chapters, see

F. V. Atkinson, Multiparametric Spectral Theory, *Bull. Am. Math. Soc.*, vol. 74, pp. 1–27, 1968.

5

Constrained Maxima

1. Introduction. In the previous chapters, we have discussed the problem of determining the set of values assumed by (x,Ax) as x ranged over the region $(x,x) = 1$. In particular, we were interested in determining whether (x,Ax) could assume both positive and negative values. In this chapter we shall investigate this question under the condition that in addition to the relation $(x,x) = 1$, x satisfies certain additional linear constraints of the form

$$(x,b^i) = c_i \qquad i = 1, 2, \ldots, k \tag{1}$$

In geometric terms, we were examining the set of values assumed by (x,Ax) as x roamed over the unit sphere. We now add the condition that x simultaneously lie on a set of planes, or alternatively, is a point on a given $N - k$ dimensional plane.

Constraints of this nature arise very naturally in various algebraic, analytic, and geometric investigations.

As a first step in this direction, we shall carry out the generalization of the algebraic method used in Chap. 1, obtaining in this way a more useful set of necessary and sufficient conditions for positive definiteness. These results, in turn, will be extended in the course of the chapter.

2. Determinantal Criteria for Positive Definiteness. In Chap. 4, it was demonstrated that a necessary and sufficient condition that a real symmetric matrix be positive definite is that all of its characteristic roots be positive. Although this is a result of theoretical value, it is relatively difficult to verify. For analytic and computational purposes, it is important to derive more usable criteria.

The reason why a criterion in terms of characteristic roots is not useful in applications is that the numerical determination of the characteristic roots of a matrix of large dimension is a very difficult matter. Any direct attempt based upon a straightforward expansion of the determinant $|A - \lambda I|$ is surely destined for failure because of the extraordinarily large number of terms appearing in the expansion of a determinant. A determinant of order N has $N!$ terms in its complete expansion. Since

$10! = 3{,}628{,}000$ and $20! \cong 2{,}433 \times 10^{15}$, it is clear that direct methods cannot be applied, even with the most powerful computers at one's disposal.[1]

The numerical determination of the characteristic roots and vectors constitutes one of the most significant domains of matrix theory. As we have previously indicated, we will not discuss any aspects of the problem here.

For the form in two variables,

$$Q(x_1,x_2) = a_{11}x_1^2 + 2a_{12}x_1x_2 + a_{22}x_2^2 \tag{1}$$

we obtained the representation

$$Q = a_{11}\left(x_1 + \frac{a_{12}}{a_{11}}x_2\right)^2 + \left(a_{22} - \frac{a_{12}^2}{a_{11}}\right)x_2^2 \tag{2}$$

under the assumption that $a_{11} \neq 0$.

From this, we concluded that the relations

$$a_{11} > 0 \qquad \begin{vmatrix} a_{11} & a_{12} \\ a_{12} & a_{22} \end{vmatrix} > 0 \tag{3}$$

were necessary and sufficient for A to be positive definite.

Let us now continue inductively from this point. Consider the form in three variables,

$$Q(x_1,x_2,x_3) = a_{11}x_1^2 + 2a_{12}x_1x_2 + a_{22}x_2^2 + 2a_{23}x_2x_3 + 2a_{13}x_1x_3 + a_{33}x_3^2 \tag{4}$$

Since $a_{11} > 0$ is clearly a necessary condition for positive definiteness, as we see upon setting $x_2 = x_3 = 0$, we can write

$$Q(x_1,x_2,x_3) = a_{11}\left(x_1 + \frac{a_{12}x_2}{a_{11}} + \frac{a_{13}x_3}{a_{11}}\right)^2 + \left(a_{22} - \frac{a_{12}^2}{a_{11}}\right)x_2^2 + 2\left(a_{23} - \frac{a_{12}a_{13}}{a_{11}}\right)x_2x_3 + \left(a_{33} - \frac{a_{13}^2}{a_{11}}\right)x_3^2 \tag{5}$$

If $Q(x_1,x_2,x_3)$ is to be positive definite, we see upon taking

$$x_1 + \frac{(a_{12}x_2 + a_{13}x_3)}{a_{11}} = 0$$

that the quadratic form in x_2 and x_3 must be positive definite. It follows, upon applying the result for 2×2 matrices given in (3), that a set of

[1] At the rate of one operation per microsecond, 20! operations would require over 50,000 years!

necessary and sufficient conditions that $Q(x_1,x_2,x_3)$ be positive definite are

$$a_{11} > 0 \qquad a_{22} - \frac{a_{12}^2}{a_{11}} > 0 \qquad \begin{vmatrix} a_{22} - \dfrac{a_{12}^2}{a_{11}} & a_{23} - \dfrac{a_{12}a_{13}}{a_{11}} \\ a_{23} - \dfrac{a_{12}a_{13}}{a_{11}} & a_{33} - \dfrac{a_{13}^2}{a_{11}} \end{vmatrix} > 0 \tag{6}$$

The first two conditions we recognize. Let us now see if we can persuade the third to assume a more tractable appearance. Consider the determinant

$$D_3 = \begin{vmatrix} a_{11} & a_{12} & a_{13} \\ a_{21} & a_{22} & a_{23} \\ a_{31} & a_{32} & a_{33} \end{vmatrix} = \begin{vmatrix} a_{11} & a_{12} & a_{13} \\ 0 & a_{22} - \dfrac{a_{12}^2}{a_{11}} & a_{23} - \dfrac{a_{12}a_{13}}{a_{11}} \\ 0 & a_{23} - \dfrac{a_{13}a_{12}}{a_{11}} & a_{33} - \dfrac{a_{13}^2}{a_{11}} \end{vmatrix} \tag{7}$$

The last equation follows upon subtracting the first row multiplied by a_{12}/a_{11} from the second, and the first row multiplied by a_{13}/a_{11} from the third (a method previously applied in Sec. 2 of Chap. 3).

It follows then that the conditions in (6) may be written in the suggestive form

$$a_{11} > 0 \qquad \begin{vmatrix} a_{11} & a_{12} \\ a_{21} & a_{22} \end{vmatrix} > 0 \qquad \begin{vmatrix} a_{11} & a_{12} & a_{13} \\ a_{21} & a_{22} & a_{23} \\ a_{31} & a_{32} & a_{33} \end{vmatrix} > 0 \tag{8}$$

Consequently we have all the ingredients of an inductive proof of

Theorem 1. *A necessary and sufficient set of conditions that A be positive definite is that the following relations hold:*

$$D_k > 0 \qquad k = 1, 2, \ldots, N \tag{9}$$

where

$$D_k = |a_{ij}| \qquad i, j = 1, 2, \ldots, k \tag{10}$$

We leave the details of the complete proof as an exercise for the reader.

3. Representation as Sum of Squares. Pursuing the analysis a bit further, we see that we can state the following result.

Theorem 2. *Provided that no D_k is equal to zero, we may write*

$$Q(x_1,x_2, \ldots ,x_N) = \sum_{k=1}^{N} (D_k/D_{k-1})y_k^2 \qquad D_0 = 1 \tag{1}$$

where

$$\begin{aligned} y_k &= x_k + \sum_{j=k+1}^{N} c_{kj}x_j \qquad k = 1, 2, \ldots, N-1 \\ y_N &= x_N \end{aligned} \tag{2}$$

The c_{ij} are rational functions of the a_{ij}.

This is the general form of the representation given in (2.5).

EXERCISE

1. What happens if one or more of the D_k are zero?

4. Constrained Variation and Finsler's Theorem. Let us now consider the problem of determining the positivity or negativity of the quadratic form $Q(x_1,x_2, \ldots ,x_N)$ when the variables are allowed to vary over all x_i satisfying the constraints

$$\sum_{j=1}^{N} b_{ij}x_j = 0 \qquad i = 1, 2, \ldots, k, k < N \tag{1}$$

As usual, all quantities occurring are taken to be real. Without loss of generality, we can suppose that these k equations are independent. Hence, using these equations, we can solve for k of the x_i as linear functions of the remaining $N - k$, substitute these relations in Q and use the criteria obtained in the preceding sections to treat the resulting $(N - k)$-dimensional problem.

Although this method can be carried through successfully, a good deal of determinantal manipulation is involved. We shall consequently pursue a different tack. However, we urge the reader to attempt an investigation along the lines just sketched at least for the case of one constraint, in order to appreciate what follows.

Let us begin by demonstrating the following result of Finsler.

Theorem 3. *If $(x,Ax) > 0$ whenever $(x,Bx) = 0$, where B is a positive indefinite matrix, then there exists a scalar constant λ such that $(x,Ax) + \lambda(x,Bx)$ is positive definite.*

Proof. Write $x = Ty$, where T is an orthogonal matrix chosen so that

$$(x,Bx) = \sum_{i=1}^{k} \mu_i y_i^2 \qquad \mu_i > 0,\ 1 \leq k < N \tag{2}$$

Since B is by assumption positive indefinite, we know that (x,Bx) can be put in this form. Under this transformation, we have

$$(x,Ax) = \sum_{i,j=1}^{N} c_{ij}y_iy_j \tag{3}$$

If (x,Ax) is to be positive whenever $(x,Bx) = \sum_{i=1}^{k} \mu_i y_i^2 = 0$, we must have the relation

$$\sum_{i,j=k+1}^{N} c_{ij}y_iy_j > 0 \tag{4}$$

for all nontrivial $y_{k+1}, y_{k+2}, \ldots, y_N$.

In the $(N - k)$-dimensional space $(y_{k+1},y_{k+2}, \ldots ,y_N)$, make an orthogonal transformation which converts the quadratic form in (4) into a sum of squares. Let the variables $y_1, y_2, \ldots , y_k$ remain unchanged. In N-dimensional y space, write $y = Sw$ to represent this transformation.

In the w_i variables, we have

$$(x,Ax) = Q(w) = \sum_{i,j=1}^{N} c_{ij}y_iy_j = \sum_{i=k+1}^{N} \mu_i w_i^2 + 2\sum_{i=k+1}^{N}\sum_{j=1}^{k} d_{ij}w_iw_j + \sum_{i,j=1}^{k} d_{ij}w_iw_j \quad (5)$$

Our problem then reduces to demonstrating that a quadratic form of the type

$$\lambda \sum_{i=1}^{k} \mu_i w_i^2 + Q(w) \quad (6)$$

is positive definite whenever $\mu_i > 0$, $i = 1, 2, \ldots , N$, and λ is sufficiently large. Our proof is complete upon using the result indicated in Exercise 36 at the end of Chap. 4.

Returning to the original variational problem, we now make the observation that the k equations in (1) can be combined into one equation,

$$\sum_{i=1}^{k} \Big(\sum_{j=1}^{N} b_{ij}x_j\Big)^2 = 0 \quad (7)$$

Regarding $\sum_{i=1}^{k} \Big(\sum_{j=1}^{N} b_{ij}x_j\Big)^2$ as a positive indefinite form, we see that Theorem 3 yields the following result:

A necessary and sufficient condition that $Q(x_1,x_2, \ldots ,x_N)$ be positive for all nontrivial values satisfying the linear equations in (1) *is that the quadratic form*

$$P(x_1,x_2, \ldots ,x_N) = Q(x_1,x_2, \ldots ,x_N) + \lambda \sum_{i=1}^{k} \Big(\sum_{j=1}^{N} b_{ij}x_j\Big)^2 \quad (8)$$

be positive definite for all sufficiently large positive λ.

EXERCISE

1. Establish the foregoing result by considering the maximum over all x_i of the function

$$f(x) = -\frac{Q(x_1,x_2, \ldots ,x_N)}{\sum_{i=1}^{k} \Big(\sum_{j=1}^{N} b_{ij}x_j\Big)^2}$$

(*I. Herstein*)

5. The Case $k = 1$. Let us begin the task of obtaining a usable criterion from the foregoing result with a case of frequent occurrence, $k = 1$. The quadratic form P is then

$$P = \sum_{i,j=1}^{N} (a_{ij} + \lambda b_{1i} b_{1j}) x_i x_j \tag{1}$$

For this to be positive definite for all large positive λ, we must have, in accord with Theorem 1, the determinantal inequalities

$$|a_{ij} + \lambda b_{1i} b_{1j}| > 0 \qquad i, j = 1, 2, \ldots, k \tag{2}$$

for $k = 1, 2, \ldots, N$.

In order to obtain simpler equivalents of these inequalities, we employ the following matrix relation.

$$\begin{bmatrix} a_{11} & a_{12} & \cdots & a_{1k} & \lambda b_1 \\ a_{21} & a_{22} & \cdots & a_{2k} & \lambda b_2 \\ \cdot & & & & \\ \cdot & & & & \\ \cdot & & & & \\ a_{k1} & a_{k2} & \cdots & a_{kk} & \lambda b_k \\ b_1 & b_2 & \cdots & b_k & -1 \end{bmatrix} \begin{bmatrix} 1 & & & & 0 \\ & 1 & & 0 & 0 \\ & & \cdot & & \cdot \\ & & & \cdot & \cdot \\ & & & \cdot & \cdot \\ 0 & & & 1 & 0 \\ b_1 & b_2 & \cdots & b_k & 1 \end{bmatrix}$$
$$= \begin{bmatrix} a_{11} + \lambda b_1^2 & a_{12} + \lambda b_1 b_2 & \cdots & a_{1k} + \lambda b_1 b_k & \lambda b_1 \\ a_{21} + \lambda b_1 b_2 & a_{22} + \lambda b_2^2 & \cdots & a_{2k} + \lambda b_2 b_k & \lambda b_2 \\ \cdot & & & & \\ \cdot & & & & \\ \cdot & & & & \\ a_{k1} + \lambda b_k b_1 & a_{k2} + \lambda b_k b_2 & \cdots & a_{kk} + \lambda b_k^2 & \lambda b_k \\ 0 & 0 & \cdots & 0 & -1 \end{bmatrix} \tag{3}$$

Taking determinants of both sides, we see that positivity of the determinant $|a_{ij} + \lambda b_{1i} b_{1j}|$, $i, j = 1, 2, \ldots, k$, for all large λ is equivalent to negativity of the determinant

$$\begin{vmatrix} a_{11} & a_{12} & \cdots & a_{1k} & \lambda b_1 \\ a_{21} & a_{22} & \cdots & a_{2k} & \lambda b_2 \\ \cdot & & & & \\ \cdot & & & & \\ \cdot & & & & \\ a_{k1} & a_{k2} & \cdots & a_{kk} & \lambda b_k \\ b_1 & b_2 & \cdots & b_k & -1 \end{vmatrix} \tag{4}$$

for large positive λ and $k = 1, 2, \ldots, N$.

Hence, we must have the relation

$$\lambda \begin{vmatrix} a_{11} & a_{12} & \cdots & a_{1k} & b_1 \\ a_{21} & a_{22} & \cdots & a_{2k} & b_2 \\ \cdot & & & & \\ \cdot & & & & \\ \cdot & & & & \\ a_{k1} & a_{k2} & \cdots & a_{kk} & b_k \\ b_1 & b_2 & \cdots & b_k & 0 \end{vmatrix} - \begin{vmatrix} a_{11} & a_{12} & \cdots & a_{1k} \\ a_{21} & a_{22} & \cdots & a_{2k} \\ \cdot & & & \\ \cdot & & & \\ \cdot & & & \\ a_{k1} & a_{k2} & \cdots & a_{kk} \end{vmatrix} < 0 \tag{5}$$

for all large positive λ and $k = 1, 2, \ldots, N$.

Consequently, we see that a *sufficient* condition that Q be positive for all nontrivial x_i satisfying the linear relation in

$$b_1x_1 + b_2x_2 + \cdots + b_Nx_N = 0$$

is that the bordered determinants satisfy the relations

$$\begin{vmatrix} a_{11} & a_{12} & \cdots & a_{1k} & b_1 \\ a_{21} & a_{22} & \cdots & a_{2k} & b_2 \\ \cdot & & & & \\ \cdot & & & & \\ \cdot & & & & \\ a_{k1} & a_{k2} & \cdots & a_{kk} & b_k \\ b_1 & b_2 & \cdots & b_k & 0 \end{vmatrix} < 0 \tag{6}$$

for $k = 1, 2, \ldots, N$. It is clear that the conditions

$$\begin{vmatrix} a_{11} & a_{12} & \cdots & a_{1k} & b_1 \\ a_{21} & a_{22} & \cdots & a_{2k} & b_2 \\ \cdot & & & & \\ \cdot & & & & \\ \cdot & & & & \\ a_{k1} & a_{k2} & \cdots & a_{kk} & b_k \\ b_1 & b_2 & \cdots & b_k & 0 \end{vmatrix} \leq 0 \tag{7}$$

for $k = 1, 2, \ldots, N$ are necessary, but that some of these determinants can be zero without disturbing the positivity property of Q. The simplest way in which these can be zero is for some of the b_i to be zero.

Referring to (5), we see that if some element of the sequence of bordered determinants is zero, say the kth, then a necessary condition for positivity is that the relation

$$\begin{vmatrix} a_{11} & \cdots & a_{1k} \\ a_{21} & \cdots & a_{2k} \\ \cdot & & \\ \cdot & & \\ \cdot & & \\ a_{k1} & \cdots & a_{kk} \end{vmatrix} > 0 \tag{8}$$

be valid.

A further discussion of special cases is contained in the exercises following.

EXERCISES

1. Assuming that $b_N \neq 0$, obtain the foregoing results by solving for x_N, $x_N = -(b_1x_1 + b_2x_2 + \cdots + b_{N-1}x_{N-1})/b_N$, and considering $Q(x)$ as a quadratic form in $x_1, x_2, \ldots, x_{N-1}$. Does this yield an inductive approach?

2. If $b_1 = 0$, the condition in (7) yields $-a_{11}b_2^2 \leq 0$. Hence, if $b_2 \neq 0$, we must have $a_{11} \geq 0$. What form do the relations in (6) and (7) take if $b_1 = b_2 = \cdots = b_r = 0$, $1 \leq r < N$?

3. Consider in homogeneous coordinates, the quadric surface $\sum_{i,j=1}^{4} a_{ij}x_ix_j = 0$ and the plane $\sum_{i=1}^{4} u_ix_i = 0$. Prove that a necessary and sufficient condition that the plane be tangent to the surface is that[1]

$$\begin{vmatrix} a_{11} & a_{12} & a_{13} & a_{14} & u_1 \\ a_{21} & a_{22} & a_{23} & a_{24} & u_2 \\ a_{31} & a_{32} & a_{33} & a_{34} & u_3 \\ a_{41} & a_{42} & a_{43} & a_{44} & u_4 \\ u_1 & u_2 & u_3 & u_4 & 0 \end{vmatrix} = 0$$

4. Determine the minimum distance from the quadric surface $(x,Ax) = 0$ to the plane $(u,x) = 0$.

5. Prove that a necessary and sufficient condition that the line determined by the planes $(u,x) = (v,x) = 0$ be either tangent to $(x,Ax) = 0$ or a generator, is that the bordered determinant

$$\begin{vmatrix} & & & & u_1 & v_1 \\ & & & & u_2 & v_2 \\ & & A & & u_3 & v_3 \\ & & & & u_4 & v_4 \\ u_1 & u_2 & u_3 & u_4 & 0 & 0 \\ v_1 & v_2 & v_3 & v_4 & 0 & 0 \end{vmatrix} = 0$$

[1] Cf. M. Bocher, *Introduction to Higher Algebra*, Chap. 12, The Macmillan Company, New York, 1947.

6. Determine the minimum distance from the quadric surface $(x,Ax) = 0$ to the line determined by the planes $(u,x) = (v,x) = 0$.

6. A Minimization Problem. Closely related to the foregoing is the problem of determining the minimum of (x,x) over all x satisfying the constraints

$$(x,a^i) = b_i \qquad i = 1, 2, \ldots, k \tag{1}$$

As usual, the vectors and scalars appearing are real. Without loss of generality, we can assume that the vectors a^i are linearly independent.

For any x we know that the inequality

$$\begin{vmatrix} (x,x) & (x,a^1) & \cdots & (x,a^k) \\ (x,a^1) & (a^1,a^1) & \cdots & (a^1,a^k) \\ \vdots & & & \\ (x,a^k) & (a^1,a^k) & \cdots & (a^k,a^k) \end{vmatrix} \geq 0 \tag{2}$$

is valid. This is equivalent to the relation

$$(x,x) \geq - \frac{\begin{vmatrix} 0 & b_1 & b_2 & \cdots & b_k \\ b_1 & & & & \\ \vdots & & (a^i,a^j) & & \\ b_k & & & & \end{vmatrix}}{|(a^i,a^j)|} \tag{3}$$

since the determinant $|(a^i,a^j)|$ is positive in view of the linear independence of the a^i.

The right side in (3) is the actual minimum value, since equality is attained in (2) if x is chosen to be a linear combination of the a^i,

$$x = \sum_{j=1}^{k} c_j a^j \tag{4}$$

In order to satisfy (1), the c_j are taken to be the solutions of the equations

$$\sum_{j=1}^{k} c_j(a^j,a^i) = b_i \qquad i = 1, 2, \ldots, k \tag{5}$$

Since $|(a^i,a^j)| \neq 0$, there is a unique solution, which yields the unique minimizing vector x.

EXERCISES

1. Deduce from this result that the quadratic form

$$Q(x) = \begin{vmatrix} 0 & x_1 & x_2 & \cdots & x_N \\ x_1 & & & & \\ \cdot & & A & & \\ \cdot & & & & \\ \cdot & & & & \\ x_N & & & & \end{vmatrix}$$

is negative definite whenever A is positive definite.

2. Give an independent proof of this result by showing that the associated quadratic form

$$P(z) = 2z_1(x_1z_2 + x_2z_3 + \cdots + x_Nz_{N+1}) + (z,Az)$$

cannot be positive definite, or even positive indefinite if A is positive definite.

3. Assuming that (x,Ax) possesses a positive minimum on the intersection of the planes $(x,a^i) = b_i$, $i = 1, 2, \ldots, k$, determine this minimum and thus derive the results of the preceding section in an alternate fashion.

7. General Values of k. Let us now return to the problem posed in Sec. 4. It is not difficult to utilize the same techniques in the treatment of the general problem. Since, however, the notation becomes quite cumbersome, it is worthwhile at this point to introduce the concept of a rectangular matrix and to show how this new notation greatly facilitates the handling of the general problem.

Once having introduced rectangular matrices, which it would be well to distinguish by a name such as "array," as Cayley himself wished to do, we are led to discuss matrices whose elements are themselves matrices. We have hinted at this in some of our previous notation, and we shall meet it again in the study of Kronecker products.

8. Rectangular Arrays. A set of complex quantities a_{ij} arranged in the form

$$A = \begin{bmatrix} a_{11} & a_{12} & \cdots & a_{1M} \\ a_{21} & a_{22} & \cdots & a_{2M} \\ \cdot & & & \\ \cdot & & & \\ \cdot & & & \\ a_{N1} & a_{N2} & \cdots & a_{NM} \end{bmatrix} \tag{1}$$

will be called a *rectangular matrix*. There is now no restriction that $M = N$. We will call this array an $M \times N$ matrix. Observe the order of M and N.

There is no difficulty in adding two $M \times N$ matrices, but the concept of multiplication is perhaps not so clear. We shall allow multiplication of a $K \times M$ matrix B by an $M \times N$ matrix A. This arises as before from the iteration of linear transformations. Thus

$$AB = C = (c_{ij}) \tag{2}$$

Schematically,

$$\overset{M}{\boxed{A}}\;N\quad\overset{K}{\boxed{B}}\;M = \overset{K}{\boxed{C}}\;N$$

a $K \times N$ matrix, where

$$c_{ij} = \sum_{k=1}^{M} a_{ik}b_{kj} \tag{3}$$

The transpose A' of an $M \times N$ matrix is an $N \times M$ matrix obtained by interchanging rows and columns.

In many cases, these rectangular matrices can be used to unify a presentation in the sense that the distinction between vectors and matrices disappears. However, since there *is* a great conceptual difference between vectors and square matrices, we feel that particularly in analysis this blending is not always desirable. Consequently, rectangular matrices, although of great significance in many domains of mathematics, will play a small role in the parts of matrix theory of interest to us here.

Hence, although in the following section we wish to give an example of the simplification that occasionally ensues when this notation is used, we urge the reader to proceed with caution. Since the underlying mathematical ideas are the important quantities, no notation should be adhered to slavishly. It is all a question of who is master.

EXERCISES

1. Let x and y be N-dimensional column vectors. Show that

$$xy' = (x_i y_j) \qquad i, j = 1, 2, \ldots, N$$

2. If x is an N-dimensional column vector, then

$$|\lambda I - xx'| = \lambda^N - (x',x)\lambda^{N-1}$$

9. Composite Matrices. We have previously defined matrices whose elements were complex quantities. Let us now define matrices whose elements are complex matrices,

$$A = (A_{ij}) \tag{1}$$

This notation will be particularly useful if it turns out that, for example, a 4×4 matrix may be written in the form

$$\begin{bmatrix} a_{11} & a_{12} & a_{13} & a_{14} \\ a_{21} & a_{22} & a_{23} & a_{24} \\ a_{31} & a_{32} & a_{33} & a_{34} \\ a_{41} & a_{42} & a_{43} & a_{44} \end{bmatrix} = \begin{bmatrix} A_{11} & A_{12} \\ A_{21} & A_{22} \end{bmatrix} \tag{2}$$

with the A_{ij} defined by

$$\begin{aligned} A_{11} &= \begin{bmatrix} a_{11} & a_{12} \\ a_{21} & a_{22} \end{bmatrix} & A_{12} &= \begin{bmatrix} a_{13} & a_{14} \\ a_{23} & a_{24} \end{bmatrix} \\ A_{21} &= \begin{bmatrix} a_{31} & a_{32} \\ a_{41} & a_{42} \end{bmatrix} & A_{22} &= \begin{bmatrix} a_{33} & a_{34} \\ a_{43} & a_{44} \end{bmatrix} \end{aligned} \tag{3}$$

or, alternatively, with the definitions

$$\begin{aligned} A_{11} &= \begin{bmatrix} a_{11} & a_{12} & a_{13} \\ a_{21} & a_{22} & a_{23} \\ a_{31} & a_{32} & a_{33} \end{bmatrix} & A_{12} &= \begin{bmatrix} a_{14} \\ a_{24} \\ a_{34} \end{bmatrix} \\ A_{21} &= (a_{41} \quad a_{42} \quad a_{43}) & A_{22} &= (a_{44}) \end{aligned} \tag{4}$$

For various purposes, as we shall see, this new notation possesses certain advantages. Naturally, we wish to preserve the customary rules for addition and multiplication.

It is easy to verify that addition is carried out in the expected fashion, and we leave it as an exercise for the reader to verify that the product of (A_{ij}) and (B_{ij}) may safely be defined as follows:

$$(A_{ij})(B_{ij}) = (C_{ij}) \tag{5}$$

where

$$C_{ij} = \sum_{k=1}^{N} A_{ik}B_{kj} \tag{6}$$

provided that all the products $A_{ik}B_{kj}$ are defined. Furthermore, the associativity of multiplication is also preserved.

EXERCISES

1. From the relation

$$\begin{bmatrix} A & B \\ C & D \end{bmatrix} \begin{bmatrix} D & -B \\ -C & A \end{bmatrix} = \begin{bmatrix} AD - BC & BA - AB \\ CD - DC & DA - CB \end{bmatrix}$$

deduce that

$$\begin{vmatrix} A & B \\ C & D \end{vmatrix} = |AD - BC|$$

if $BA = AB$ or $CD = DC$, and the required products exist.

2. Let $M = A + iB$, A and B real, and let

$$M_1 = \begin{bmatrix} A & -B \\ B & A \end{bmatrix}$$

Show that

$$|M_1 - \lambda I| = |M - \lambda I|\,|\bar{M} - \lambda I| \qquad \textit{(Szaraki and Wazewski)}$$

3. Hence, show that $|M_1| = |M|\,|\bar{M}|$, and that the characteristic roots of M_1 are those of M and $\bar{M}$.

4. Show that if M is Hermitian, then M_1 is symmetric.

10. The Result for General k. Let us now discuss the general problem posed in Sec. 4. As Theorem 3 asserts, the quadratic form

$$P = \sum_{i,j=1}^{N} \left(a_{ij} + \lambda \sum_{r=1}^{k} b_{ri} b_{rj} \right) x_i x_j \tag{1}$$

must be positive definite for large λ.

Hence we must have the determinantal relations

$$\left| a_{ij} + \lambda \sum_{r=1}^{k} b_{ri} b_{rj} \right| > 0 \qquad i, j = 1, 2, \ldots, n \tag{2}$$

for $n = 1, 2, \ldots, N$ and all sufficiently large positive λ.

In order to simplify this relation, we use the same device as employed in Sec. 5, together with our new notation. Let B_N denote the rectangular array

$$B_N = \begin{bmatrix} b_{11} & b_{21} & \cdots & b_{k1} \\ b_{12} & b_{22} & & b_{k2} \\ \cdot & & & \cdot \\ \cdot & & & \cdot \\ \cdot & & & \cdot \\ b_{1N} & b_{2N} & \cdots & b_{kN} \end{bmatrix} \tag{3}$$

Consider the following product of $(N + k)$-dimensional matrices,

$$\begin{bmatrix} A_N & \lambda B_N \\ B'_N & -I_k \end{bmatrix} \begin{bmatrix} I_N & 0 \\ B'_N & I_k \end{bmatrix} = \begin{bmatrix} A_N + \lambda B_N B'_N & \lambda B_N \\ 0 & -I_k \end{bmatrix} \tag{4}$$

where I_k is the k dimensional unit matrix and

$$A_N = (a_{ij}) \qquad i, j = 1, 2, \ldots, N$$

Taking determinants of both sides, we have

$$(-1)^k \begin{vmatrix} A_N & \lambda B_N \\ B'_N & -I_k \end{vmatrix} = |A_N + \lambda B_N B'_N| \tag{5}$$

Hence, the conditions of (2) are replaced by

$$(-1)^k \begin{vmatrix} A_N & \lambda B_N \\ B_N' & -I_k \end{vmatrix} > 0 \tag{6}$$

for all large λ.

It follows as before that sufficient conditions for positivity are readily obtained. The necessary conditions require a more detailed discussion because of the possible occurrence of zero values, both on the part of the coefficients b_{ij} and the bordered determinants.

MISCELLANEOUS EXERCISES

1. If $\{A_i\}$ is an arbitrary set of $r \times r$ matrices, there are unitary matrices U and V of order $r \times r$, such that $UA_iV = D_i$, D_i diagonal and real, if and only if $A_i\bar{A}_j = A_j\bar{A}_i$ and $\bar{A}_jA_i = \bar{A}_iA_j$ for all i and j (*N. A. Wiegmann*).

2. Let $\sum_{i,j=1}^{N} a_{ij}x_ix_j$ be a positive definite form. Show that

$$\sum_{i,j\neq k} \begin{vmatrix} a_{kk} & a_{ik} \\ a_{jk} & a_{ij} \end{vmatrix} x_ix_j$$

is a positive definite form in the $N - 1$ variables $x_1, x_2, \ldots, x_{k-1}, x_{k+1}, \ldots, x_N$ What are the characteristic roots and vectors of the matrix of this quadratic form? Generalize this result.

3. Introduce the following correspondence between complex numbers and matrices:

$$z = x + iy \sim Z = \begin{bmatrix} x & y \\ -y & x \end{bmatrix}$$

Show that

(*a*) $ZW = WZ$

(*b*) $\bar{z} \sim Z'$

(*c*) $z \sim Z, \quad w \sim W$ implies that $zw \sim ZW$

4. Use this correspondence and de Moivre's theorem to determine the form of Z^n for $n = 1, 2, \ldots$.

5. From Exercise 3, we see that we have the correspondence

$$1 \sim \begin{bmatrix} 1 & 0 \\ 0 & 1 \end{bmatrix} \qquad i \sim \begin{bmatrix} 0 & 1 \\ -1 & 0 \end{bmatrix}$$

Using this in the matrix

$$Q = \begin{bmatrix} x_1 + ix_2 & x_3 + ix_4 \\ -x_3 + ix_4 & x_1 - ix_2 \end{bmatrix}$$

we are led to the supermatrix

$$Q_s = \begin{bmatrix} \begin{pmatrix} x_1 & x_2 \\ -x_2 & x_1 \end{pmatrix} & \begin{pmatrix} x_3 & x_4 \\ -x_4 & x_3 \end{pmatrix} \\ \begin{pmatrix} -x_3 & x_4 \\ -x_4 & x_3 \end{pmatrix} & \begin{pmatrix} x_1 & -x_2 \\ +x_2 & x_1 \end{pmatrix} \end{bmatrix}$$

or, dropping parentheses,

$$Q_s = \begin{bmatrix} x_1 & x_2 & x_3 & x_4 \\ -x_2 & x_1 & -x_4 & x_3 \\ -x_3 & x_4 & x_1 & -x_2 \\ -x_4 & x_3 & -x_2 & x_1 \end{bmatrix}$$

Consider the equivalence $Q \sim Q_s$. Show that $Q \sim Q_s$, $W \sim W_s$ results in $QW \sim Q_sW_s$. Hence, determine $|Q_s|$, and the characteristic values of Q_s.

6. How does one determine the maximum value of (x,Ax) subject to the constraints $(x,x) = 1$, $(x,c^i) = 0$, $i = 1, 2, \ldots, k$?

7. Write, for two symmetric matrices, A and B, $A \geq B$ to indicate that $A - B$ is non-negative definite. Show that $A \geq B$ does not necessarily imply that $A^2 \geq B^2$.

8. Let $g(t) \geq 0$ and consider the quadratic form

$$Q_N(x) = \int_0^1 g(t) \left(\sum_{k=0}^{N} x_k t^k \right)^2 dt$$

Show that $Q_N(x)$ is positive indefinite, and, hence, derive various determinantal inequalities for the quantities $m_k = \int_0^1 g(t)t^k\,dt$ (*Stieltjes*).

9. Let $g(t) \geq 0$ and consider the Hermitian form

$$H_N(x) = \int_0^1 g(t) \left| \sum_{k=0}^{N} x_k e^{2\pi ikt} \right|^2 dt$$

Show that $H_N(x)$ is positive definite, and, hence, derive determinantal inequalities for the quantities $m_k = \int_0^1 g(t)e^{2\pi ikt}\,dt$ (*O. Toeplitz*).

10. Show that

$$\sum_{m,n=0}^{N} \frac{x_m x_n}{m + n + 1} < \pi \sum_{n=0}^{N} x_n^2 \qquad \textit{(Hilbert's inequality)}$$

11. Establish Theorem 1 inductively. See C. J. Seelye.[1]

12. Determine the minimum value over all x_i of the expression

$$Q_N(x) = \int_0^\pi \left| e^{-i\theta} - \sum_{k=0}^{N} x_k e^{ik\theta} \right|^2 d\theta$$

13. What is the limit of the minimum value as $N \to \infty$? (This result plays an important role in prediction theory. See U. Grenander and G. Szego, *Toeplitz Forms and their Applications*, University of California Press, 1958.)

14. Determine the minimum value over all x_i of the expression

$$Q_N(x) = \int_0^1 \left| t^k - \sum_{i=0}^{N} x_i t^{\lambda_i} \right|^2 dt$$

where $\{\lambda_i\}$ is a sequence of real numbers, $\lambda_0 < \lambda_1 < \cdots$.

[1] C. J. Seelye, *Am. Math. Monthly*, vol. 65, pp. 355–356, 1958.

15. Consider the problem of determining the minimum and maximum values of $(x,Ax) + 2(b,x)$ on the sphere $(x,x) = 1$. How many stationary values are there, and are they all real?

16. How many normals, real and complex, can one draw from a given point in the plane (x_1,y_1) to the ellipse $x^2/a^2 + y^2/b^2 = 1$?

17. If $A = (a_{ij})$, $C = (c_{ij})$, and $\sum_j a_{ij} = 0$, $\sum_i a_{ij} = 0$, $c_{ij} = c_i + c_j$, then AB and $A(B + C)$ have the same characteristic equation (*A. Brauer*).

18. If A, C_1, and C_2 are such that $C_1A = AC_2 = 0$, $C = C_1 + C_2$, then AB and $A(B + C)$ have the same characteristic equation (*Parker*).

19. If $ACA = 0$, then AB and $A(B + C)$ have the same characteristic equation (*Parker*).

20. Let H be a Hermitian matrix with the characteristic roots $\lambda_1 \leq \lambda_2 \leq \cdots \leq \lambda_N$, and let the characteristic roots of $S'HS$ be $\mu_1 \leq \mu_2 \leq \cdots \leq \mu_N$, where S is an arbitrary real nonsingular matrix. Then the number of positive, negative, and zero terms in the two lists of characteristic roots is the same. A quantitative sharpening of this result (due to Sylvester and Jacobi), often called the Law of Inertia, may be found in A. M. Ostrowski, A Quantitative Formulation of Sylvester's Law of Inertia, *Proc. Nat. Acad. Sci. U.S.*, vol. 45, pp. 740–743, 1959.

21. Let A and B be real symmetric matrices. A necessary and sufficient condition that the pencil of matrices $\lambda A + \mu B$, λ, μ real scalars, contains a positive definite matrix is that $(x,Ax) = 0$ and $(x,Bx) = 0$ imply that $x = 0$, provided that the dimension N is greater than 2 (*Calabi*).

22. Any pencil $\lambda A + \mu B$ for which the foregoing holds can be transformed into diagonal form; see O. Taussky, Positive Definite Matrices, *Inequalities*, Academic Press, Inc., New York, 1967.

Bibliography

§2. These determinantal criteria can also be obtained by use of a Sturm's sequence derived from the characteristic polynomial $|A - \lambda I|$ and this technique can be used to prove the reality of the characteristic roots. See

W. S. Burnside and A. W. Panton, *Theory of Equations*, vol. II, p. 65, Exercise 39, etc., Longmans, Green & Co., Inc., New York, 1928.

A remarkable extension of these inequalities is contained in

I. Schur, Über endliche Gruppen und Hermitesche Formen, *Math. Z.*, vol. 1, pp. 184–207, 1918.

Some interesting results concerning positivity may also be found in

I. J. Good, A Note on Positive Determinants, *J. London Math. Soc.*, vol. 22, pp. 92–95, 1947.

§4. The question treated here is part of a group of investigations, reference to which may be found in

L. L. Dines, On the Mapping of Quadratic Forms, *Bull. Am. Math. Soc.*, vol. 47, pp. 494–498, 1941.

L. L. Dines, On Linear Combinations of Quadratic Forms, *Bull. Am. Math. Soc.*, vol. 49, 1943.

The particular result we employ is due to Finsler; cf.

P. Finsler, Über das Vorkommen definiter und semidefiniter Formen in Scharen quadratischen Formen, *Commentarii Mathematicii Helveticii*, vol. 9, pp. 188–192, 1937.

The result was derived independently by Herstein, using the argument given in the exercise at the end of the section. We follow his derivation of the conditions here. A number of other derivations appear in later literature, e.g.,

S. N. Afriat, The Quadratic Form Positive Definite on a Linear Manifold, *Proc. Cambridge Phil. Soc.*, vol. 47, pp. 1–6, 1951.

§5. The exercises at the end of this section, 3 to 6, illustrate an important heuristic concept which can be made precise by means of the theory of invariants: Whenever we obtain a positivity condition in the form of an inequality, there is a more precise result expressing distances, areas, volumes, etc., from which the inequality becomes obvious.

§6. The problem of determining when $(x,Ax) \geq 0$ for $x \geq 0$ is treated in

J. W. Gaddum, Linear Inequalities and Quadratic Forms, *Pacific J. Math.*, vol. 8, pp. 411–414, 1958.

6

Functions of Matrices

1. Introduction. In this chapter, we wish to concentrate on the concept of a function of a matrix. We shall first discuss two important matrix functions, the inverse function, defined generally, and the square root, of particular interest in connection with positive definite matrices. Following this, we shall consider the most important scalar functions of a matrix, the coefficients in the characteristic polynomial.

The problem of defining matrix functions of general matrices is a rather more difficult problem than it might seem at first glance, and we shall in consequence postpone any detailed discussion of various methods that have been proposed until a later chapter, Chap. 11. For the case of symmetric matrices, the existence of the diagonal canonical form removes most of the difficulty.

This diagonal representation, established in Chap. 4, will also be used to obtain a parametric representation for the elements of symmetric matrices which is often useful in demonstrating various results. As an example of this we shall prove an interesting result due to Schur concerning the composition of two positive definite matrices.

Finally, we shall derive an important relation between the determinant of a positive definite matrix and the associated quadratic form which will be made a cornerstone of a subsequent chapter devoted to inequalities, and obtain also an analogous result for Hermitian matrices.

2. Functions of Symmetric Matrices. As we have seen, every real symmetric matrix can be represented in the form

$$A = T \begin{bmatrix} \lambda_1 & & & & 0 \\ & \lambda_2 & & & \\ & & \cdot & & \\ & & & \cdot & \\ 0 & & & & \lambda_N \end{bmatrix} T' \tag{1}$$

where T is an orthogonal matrix. From this it follows that

$$A^k = T \begin{bmatrix} \lambda_1^k & & & & \\ & \lambda_2^k & & 0 & \\ & & \cdot & & \\ & & & \cdot & \\ & 0 & & & \cdot \\ & & & & \lambda_N^k \end{bmatrix} T' \tag{2}$$

for any integer k. Hence, for any analytic function $f(z)$ we can define $f(A)$ to be

$$f(A) = T \begin{bmatrix} f(\lambda_1) & & & & \\ & f(\lambda_2) & & 0 & \\ & & \cdot & & \\ & & & \cdot & \\ & 0 & & & \cdot \\ & & & & f(\lambda_N) \end{bmatrix} T' \tag{3}$$

whenever the scalar functions $f(\lambda_i)$ are well defined.

3. The Inverse Matrix. Following the path pursued above, we are led to define an inverse matrix as follows.

$$A^{-1} = T \begin{bmatrix} \lambda_1^{-1} & & & & \\ & \lambda_2^{-1} & & 0 & \\ & & \cdot & & \\ & & & \cdot & \\ & 0 & & & \cdot \\ & & & & \lambda_N^{-1} \end{bmatrix} T' \tag{1}$$

whenever the λ_i are all nonzero.

It is clear that A^{-1} has the appropriate properties of an inverse, namely,

$$AA^{-1} = A^{-1}A = I \tag{2}$$

4. Uniqueness of Inverse. It is natural to inquire whether or not the inverse matrix is unique, if it exists. In the first place, let us observe that the condition that the λ_i are all nonzero is equivalent to the condition that $|A| \neq 0$. For if $\lambda = 0$ is a root of $|A - \lambda I| = 0$, we see that $|A| = 0$, and if $|A| = 0$, then $\lambda = 0$ is a root.

If A is a matrix for which $|A| = 0$, it will be called *singular*. If $|A| \neq 0$, A is called *nonsingular*.

It is immediately seen that a singular matrix cannot have an inverse. For, from $AB = I$, we have $|AB| = 1$ or $|A|\,|B| = 1$. No such equation can hold if $|A| = 0$.

Let us now show, however, that any nonsingular matrix, symmetric or not, possesses a unique inverse. For the case where A is symmetric, the matrix found in this new way must coincide with that given by (3.1).

The equation $AB = I$ is equivalent to the N^2 equations

$$\sum_{k=1}^{N} a_{ik}b_{kj} = \delta_{ij} \qquad i, j = 1, 2, \ldots, N \tag{1}$$

Fixing the value of j, we obtain N linear equations for the quantities b_{kj}, $k = 1, 2, \ldots, N$. The determinant of this set of equations is $|A| = |a_{ij}|$, regardless of the value of j.

It follows then that if A is nonsingular, there is a unique solution of the equations in (1). Furthermore, we see that the solution is given by

$$A^{-1} = (\alpha_{ij}/|A|) \tag{2}$$

where α_{ij} is the cofactor of a_{ji} in the expansion of the determinant of A.

EXERCISES

1. Show that A^{-1}, if it exists, is symmetric if A is symmetric in three ways:
(*a*) Directly from (3.1)
(*b*) Using the expression in (4.2)
(*c*) Using the relations $AA^{-1} = A^{-1}A = I$ and the uniqueness of the inverse

2. If A is singular, show that we can find a matrix B whose elements are arbitrarily small and such that $A + B$ is nonsingular.

3. Let $|\lambda I - A| = \lambda^N + c_1\lambda^{N-1} + \cdots + c_N$, the characteristic polynomial of A. Show that $A^{-1} = -(A^{N-1} + c_1A^{N-2} + \cdots + c_{N-1})/c_N$, whenever $c_N \neq 0$.

4. If A_1, A_2, B_1, B_2 are nonsingular square matrices of the same order, then

$$\begin{bmatrix} A_1 & B_1 \\ A_2 & B_2 \end{bmatrix}^{-1} = \begin{bmatrix} (A_1 - B_1B_2^{-1}A_2)^{-1} & (A_2 - B_2B_1^{-1}A_1)^{-1} \\ (B_1 - A_1A_2^{-1}B_2)^{-1} & (B_2 - A_2A_1^{-1}B_1)^{-1} \end{bmatrix}$$

5. Let A be an $M \times N$ matrix and consider the set of inconsistent linear equations $x = Ay$, where x is an N-dimensional vector and y an M-dimensional vector. To obtain an approximate solution to these equations, we determine y as the minimum of the quadratic function $(x - Ay, x - Ay)$. Show that, under the above assumption of inconsistency, there is a unique solution given by $y = (A'A)^{-1}A'x$, and determine the minimum value of $(x - Ay, x - Ay)$.

6. Let $A \geq B$, for two symmetric matrices A and B, denote as previously the fact that $A - B$ is non-negative definite. Show that $A \geq B$ implies that $B^{-1} \geq A^{-1}$, provided that A^{-1} and B^{-1} exist.

7. If S is a real skew-symmetric matrix, then $I + S$ is nonsingular.

8. If S is a real skew symmetric, then

$$T = (I - S)(I + S)^{-1}$$

is orthogonal (*the Cayley transform*).

9. If A is an orthogonal matrix such that $A + I$ is nonsingular, then we can write

$$A = (I - S)(I + S)^{-1}$$

where S is a real skew-symmetric matrix.

10. Given a matrix A, we can find a matrix J, having ± 1s along the main diagonal and zeroes elsewhere, such that $JA + I$ is nonsingular.

11. Using this result, show that every orthogonal matrix A can be written in the form

$$A = J(I - S)(I + S)^{-1}$$

where J is as above.

12. Show that

(a) $A_1 \geq B_1$, $A_2 \geq B_2$, implies $A_1 + A_2 \geq B_1 + B_2$

(b) $A_1 \geq B_1$, $B_1 \geq C_1$, implies $A_1 \geq C_1$

(c) $A_1 \geq B_1$, $A_2 \geq B_2$, does not necessarily imply $A_1A_2 \geq B_1B_2$, even if A_1A_2 and B_1B_2 are both symmetric. Thus, $A \geq B$ does not necessarily imply that $A^2 \geq B^2$.

(d) $A_1 \geq B_1$ implies that $T'A_1T \geq T'B_1T$

13. Show that A^{-1} can be defined by the relation

$$(x,A^{-1}x) = \max_y \, [2(x,y) - (y,Ay)]$$

if A is positive definite.

14. From this, show that $B^{-1} \geq A^{-1}$ if A and B are symmetric and $A \geq B > 0$.

15. If A is a matrix whose elements are rational integers, and if $|A| = \pm 1$, then A^{-1} is a matrix of the same type.

16. If A is a matrix whose elements are complex integers, i.e., numbers of the form $x_1 + iy_1$ where x_1 and y_1 are rational integers, and if $|A| = \pm 1$ or $\pm i$, then A^{-1} is a matrix of the same type.

17. What is a necessary and sufficient condition that the solutions of $Ax = y$ be integers whenever the components of y are integers, given that the elements of A are integers?

18. Show that $|A + iB|^2 = |A|^2|I + A^{-1}BA^{-1}B|$ if $H = A + iB$ is Hermitian, and A is nonsingular.

19. Show that $(\bar{z},Hz) = (x,Ax) + (y,Ay) + 2(Bx,y)$ if $H = A + iB$ is Hermitian and $z = x + iy$.

20. Show that $(x,Ax) = (x,A_sx)$ where $A_s = (A + A')/2$.

21. If A and B are alike except for one column, show how to deduce the elements of A^{-1} from those of B^{-1}.

22. If A has the form

$$A = \begin{bmatrix} 0 & x_1 & 0 & \cdots & & \\ y_1 & 0 & x_2 & 0 & \cdots & \\ 0 & y_2 & 0 & x_3 & \cdots & \\ \cdot & 0 & y_3 & & & \\ \cdot & \cdot & \cdot & & & \\ \cdot & \cdot & \cdot & & & \\ & \cdot & \cdot & & & \end{bmatrix}$$

then, provided that $y_i \neq 0$, in A^{-1} the element b_{1k} is given by

$$\begin{aligned} b_{1k} &= 0 \qquad k \text{ odd} \\ &= (-1)^{k/2-1} \sum_{i=0}^{k/2-1} \frac{x_{2i}}{y_{2i+1}}, \qquad k \text{ even} \end{aligned}$$

where $x_0 = 1$. *(Clement)*

23. From this deduce the form of A^{-1}.

5. Square Roots. Since a positive definite matrix represents a natural generalization of a positive number, it is interesting to inquire whether or not a positive definite matrix possesses a positive definite square root.

Proceeding as in Sec. 2, we can define $A^{1/2}$ by means of the relation

$$A^{1/2} = T \begin{bmatrix} \lambda_1^{1/2} & & & 0 \\ & \lambda_2^{1/2} & & \\ & & \ddots & \\ 0 & & & \lambda_N^{1/2} \end{bmatrix} T' \tag{1}$$

obtaining in this way a matrix satisfying the relation $B^2 = A$, which is positive definite if A is positive definite.

There still remains the question of uniqueness. To settle it, we can proceed as follows. Since B is symmetric, it possesses a representation

$$B = S \begin{bmatrix} \mu_1 & & & 0 \\ & \mu_2 & & \\ & & \ddots & \\ 0 & & & \mu_N \end{bmatrix} S' \tag{2}$$

where S is orthogonal. It follows that B and $B^2 = A$ commute. Hence, both can be reduced to diagonal form by the same orthogonal matrix T. The relation $B^2 = A$ shows that B has the form given in (1) where the positive square roots are taken.

EXERCISES

1. How many symmetric square roots does an $N \times N$ positive definite matrix possess?

2. Can a symmetric matrix possess nonsymmetric square roots?

3. If B is Hermitian and $B \geq 0$, and if $A^2 = B$ and $(A + A^*) \geq 0$, then A is the unique non-negative Hermitian square root of B (*Putnam, On Square Roots and Logarithms of Operators, Purdue University, PRF*-1421, 1958).

4. If $e^A = B$, B positive definite, and $A \leq (2 \log 2)I$, then A is the unique positive definite logarithm of B (*Putnam*).

6. Parametric Representation. The representation

$$A = T \begin{bmatrix} \lambda_1 & & & 0 \\ & \lambda_2 & & \\ & & \ddots & \\ 0 & & & \lambda_N \end{bmatrix} T' \tag{1}$$

furnishes us a parametric representation for the elements of a symmetric matrix in terms of the elements t_{ij} of an orthogonal matrix T, namely,

$$a_{ij} = \sum_{k=1}^{N} \lambda_k t_{ik} t_{jk} \tag{2}$$

which can be used as we shall see in a moment to derive certain properties of the a_{ij}.

EXERCISE

1. Obtain a parametric representation in terms of $\cos\theta$ and $\sin\theta$ for the elements of a general 2×2 symmetric matrix.

7. A Result of I. Schur. Using this representation, we can readily prove Theorem 1.

Theorem 1. *If $A = (a_{ij})$, $B = (b_{ij})$ are positive definite, then*

$$C = (a_{ij}b_{ij})$$

is positive definite.

Proof. We have

$$\sum_{i,j=1}^{N} a_{ij}b_{ij}x_ix_j = \sum_{i,j=1}^{N} b_{ij}x_ix_j \left(\sum_{k=1}^{N} \lambda_k t_{ik}t_{jk}\right)$$
$$= \sum_{k=1}^{N} \lambda_k \left(\sum_{i,j=1}^{N} b_{ij}x_it_{ik}x_jt_{jk}\right) \qquad \lambda_k > 0 \tag{1}$$

As a quadratic form in the variables x_it_{ik} the expression $\sum_{i,j=1}^{N} b_{ij}x_it_{ik}x_jt_{jk}$ is positive unless the quantities x_it_{ik} are all zero. Since

$$\sum_{i,k} x_i^2 t_{ik}^2 = \sum_i x_i^2 \sum_k t_{ik}^2 = \sum_i x_i^2 \tag{2}$$

we see that this cannot be true for all i, unless all the x_i are zero.

This establishes the required positive definite character.

EXERCISE

1. Use Exercise 2 of the miscellaneous exercises of Chap. 5 to obtain an inductive proof of this result, starting with the easily established result for $N = 2$.

8. The Fundamental Scalar Functions. Let us now consider some properties of the scalar functions of A determined by the characteristic polynomial,

$$|\lambda I - A| = \lambda^N - \phi_1(A)\lambda^{N-1} + \phi_2(A)\lambda^{N-2} + \cdots + (-1)^N\phi_N(A) \tag{1}$$

In this section the matrices which occur are general square matrices, not necessarily symmetric.

It follows from the relations between the coefficients and the roots of a polynomial equation that

$$\begin{aligned}
\phi_1(A) &= \lambda_1 + \lambda_2 + \cdots + \lambda_N \\
\phi_2(A) &= \sum_{i \neq j} \lambda_i \lambda_j \\
&\vdots \\
\phi_N(A) &= \lambda_1 \lambda_2 \cdots \lambda_N
\end{aligned} \tag{2}$$

On the other hand, writing out the term in λ^{N-1} of the characteristic polynomial as it arises from the expansion of the determinant $|A - \lambda I|$, we see that

$$\phi_1(A) = a_{11} + a_{22} + \cdots + a_{NN} \tag{3}$$

while setting $\lambda = 0$ yields

$$\phi_N(A) = |A| \tag{4}$$

The linear function $\phi_1(A)$ is of paramount importance in matrix theory. It is called the *trace* of A, and usually written tr (A).

It follows from (3) that

$$\operatorname{tr}(A + B) = \operatorname{tr}(A) + \operatorname{tr}(B) \tag{5}$$

and a little calculation shows that

$$\operatorname{tr}(AB) = \operatorname{tr}(BA) \tag{6}$$

for any two matrices A and B. These relations hold despite the fact that there is no simple relation connecting the characteristic roots of A, B and $A + B$ or AB.

This last result is a special case of Theorem 2.

Theorem 2. *For any two matrices A and B, we have*

$$\phi_k(AB) = \phi_k(BA) \qquad k = 1, 2, \ldots, N \tag{7}$$

The proof will depend upon a fact that we already know, namely, that the result is valid for $k = N$, that is, $|AB| = |BA|$ for any two matrices.

If A is nonsingular, we have the relations

$$|\lambda I - AB| = |A(\lambda A^{-1} - B)| = |(\lambda A^{-1} - B)A| = |\lambda I - BA| \tag{8}$$

which yield the desired result.

If A is singular, we can obtain (8) by way of a limiting relation, starting with $A + \epsilon I$, where ϵ is small and $A + \epsilon I$ is nonsingular.

In the exercises immediately below, we indicate another approach which does not use any nonalgebraic results.

EXERCISES

1. Show that $|I + tA| = 1 + t \operatorname{tr}(A) + \cdots$.

2. Show that $\phi_k(TAT^{-1}) = \phi_k(A)$; $\phi_k(TAT') = \phi_k(A)$ if T is orthogonal.

3. Show that any polynomial $p(a_{11},a_{12}, \ldots ,a_{NN})$ in the elements of A which has the property that $p(AB) = p(BA)$ for all A and B must be a polynomial in the functions $\phi_k(A)$.

4. Show that $\operatorname{tr}((AB)^k) = \operatorname{tr}((BA)^k)$ for $k = 1, 2, \ldots$, and hence that $\phi_k(AB) = \phi_k(BA)$ for $k = 1, 2, \ldots, N$.

5. Show that $\operatorname{tr}(AX) = 0$ for all X implies that $A = 0$.

6. Exercise 2 and Theorem 2 above are both special cases of the result that for any m matrices $A_1, A_2, \ldots, A_m$, of order N, the quantity $\phi_k(A_1A_2 \cdots A_m)$ remains unchanged after a cyclic permutation of the factors $A_1, A_2, \ldots, A_m$.

9. The Infinite Integral $\int_{-\infty}^{\infty} e^{-(x,Ax)}\,dx$. One of the most important integrals in analysis is the following:

$$I_N = \int_{-\infty}^{\infty} \cdots \int_{-\infty}^{\infty} e^{-(x,Ax)}\,dx \tag{1}$$

where $dx = dx_1\,dx_2 \cdots dx_N$. Although the indefinite integral cannot be evaluated in finite terms, it turns out that the definite integral has a quite simple value.

Theorem 3. *If A is positive definite, we have*

$$I_N = \frac{\pi^{N/2}}{|A|^{1/2}} \tag{2}$$

We shall give two proofs.

First Proof. Let T be an orthogonal matrix reducing A to diagonal form, and make the change of variable $x = Ty$. Then

$$(x,Ax) = (y,\Lambda y) \tag{3}$$

Furthermore, $\prod_{i=1}^{N} dx_i = \prod_{i=1}^{N} dy_i$, since the Jacobian of the transformation $x = Ty$ is $|T|$, which can be taken to be $+1$. Since $x = Ty$ is a one-to-one transformation, it follows that

$$\begin{aligned} I_N &= \int_{-\infty}^{\infty} \cdots \int_{-\infty}^{\infty} e^{-\lambda_1 y_1^2 - \lambda_2 y_2^2 - \cdots - \lambda_N y_N^2}\,dy \\ &= \prod_{i=1}^{N} \int_{-\infty}^{\infty} e^{-\lambda_i y_i^2}\,dy_i = \frac{\pi^{N/2}}{(\lambda_1\lambda_2 \cdots \lambda_N)^{1/2}} \end{aligned} \tag{4}$$

From the fact that $\prod_{i=1}^{N} \lambda_i = |A|$, and the known evaluation

$$\int_{-\infty}^{\infty} e^{-x^2}\, dx = \pi^{1/2},$$

we obtain the result stated in (2).

Let us now give a second proof. In Sec. 3 of Chap. 5, it was shown that we may write

$$(x,Ax) = \sum_{k=1}^{N} \frac{D_k}{D_{k-1}} y_k^2 \qquad D_0 = 1 \tag{5}$$

where

$$y_k = x_k + \sum_{j=k+1}^{N} b_{kj} x_j \qquad k = 1, 2, \ldots, N \tag{6}$$

provided that A is positive definite. Let us then make a change of variable from the x_k to the y_k. The Jacobian is again equal to 1, and the transformation is one-to-one. It follows that

$$I_N = \left(\int_{-\infty}^{\infty} e^{-D_1 y_1^2}\, dy_1\right)\left(\int_{-\infty}^{\infty} e^{-D_2 y_2^2/D_1}\, dy_2\right) \cdots \left(\int_{-\infty}^{\infty} e^{-D_N y_N^2/D_{N-1}}\, dy_N\right) = \frac{\pi^{N/2}}{D_N^{1/2}} = \frac{\pi^{N/2}}{|A|^{1/2}} \tag{7}$$

EXERCISES

1. Show that if A is positive definite and B symmetric, we have

$$\int_{-\infty}^{\infty} \cdots \int_{-\infty}^{\infty} e^{-(x,Ax)-i(x,Bx)}\, dx = \frac{\pi^{N/2}}{|A + iB|^{1/2}}$$

where the principal value of $|A + iB|^{1/2}$ is taken, in the following steps:

(a) Set $x = Ty$, where $T'AT = \Lambda$, a diagonal matrix

(b) Make the further change of variable, $y_k = z_k/\lambda_k^{1/2}$

(c) The integrand is now

$$e^{-\sum_{k=1}^{N} z_k^2 - i(z,Cz)}$$

Reduce C to diagonal form by means of an orthogonal transformation $z = Sw$.

(d) Evaluate the remaining integral.

2. Evaluate integrals of the form

$$I_{mn} = \int_0^{\infty}\int_0^{\infty} e^{-a_{11}x_1^2 - 2a_{12}x_1x_2 - a_{22}x_2^2} x_1^m x_2^n\, dx_1\, dx_2$$

m, n positive integers or zero, in the following fashion. Write

$$I_{0,0} = \frac{\pi}{(a_{11}a_{22} - a_{12}^2)^{1/2}}$$

and thus

$$I_{2m,2n} = \left(\frac{\partial}{\partial a_{11}}\right)^m \left(\frac{\partial}{\partial a_{22}}\right)^n I_{0,0}$$

How does one treat the general case where either m or n may be odd?

3. If A and B are positive definite, evaluate

$$I_N = \int_{-\infty}^{\infty} e^{-(x,ABx)}\,dx$$

4. Evaluate

$$I_N = \int_{-\infty}^{\infty} e^{-(x,Ax)+2i(x,y)}\,dx$$

10. An Analogue for Hermitian Matrices. By analogy with (9.1), let us consider the integral

$$J(H) = \int_{-\infty}^{\infty} e^{-(\bar{z},Hz)}\,dx\,dy \tag{1}$$

where $dx = dx_1\,dx_2 \cdots dx_N$, $dy = dy_1\,dy_2 \cdots dy_N$ and H is a positive definite Hermitian matrix.

Write $H = A + iB$, where A is real and positive definite and B is real and skew-symmetric. Then

$$J(H) = \int_{-\infty}^{\infty} e^{-(x,Ax)-2(Bx,y)-(y,Ay)}\,dx\,dy \tag{2}$$

Since the integral is absolutely convergent, it may be evaluated by integration first with respect to x and then with respect to y.

Using the relation

$$\begin{aligned}(x,Ax) + 2(Bx,y) &= (x,Ax) - 2(x,By) \\ &= (A(x - A^{-1}By),x - A^{-1}By) - (By,A^{-1}By)\end{aligned} \tag{3}$$

we see that

$$\int_{-\infty}^{\infty} e^{-(x,Ax)-2(Bx,y)}\,dx = \frac{\pi^{N/2}e^{(By,A^{-1}By)}}{|A|^{1/2}} \tag{4}$$

Hence,

$$\begin{aligned}J(H) &= \frac{\pi^{N/2}}{|A|^{1/2}} \int_{-\infty}^{\infty} e^{-[(y,Ay)+(y,BA^{-1}By)]}\,dy \\ &= \frac{\pi^{N/2}}{|A|^{1/2}} \cdot \frac{\pi^{N/2}}{|A + BA^{-1}B|^{1/2}} \\ &= \frac{\pi^N}{|A|\,|I + A^{-1}BA^{-1}B|^{1/2}}\end{aligned} \tag{5}$$

11. Relation between $J(H)$ and $|H|$. It remains to express $J(H)$ in terms of $|H|$. We have

$$\begin{aligned}|H| &= |A + iB| = |A|\,|I + iA^{-1}B| \\ |H'| &= |A - iB| = |A|\,|I - iA^{-1}B|\end{aligned} \tag{1}$$

Since $|H| = |H'|$, we have

$$|H|^2 = |A|^2|(I + iA^{-1}B)(I - iA^{-1}B)| \\ = |A|^2|I + A^{-1}BA^{-1}B| \tag{2}$$

Combining the foregoing results, we see that

$$J(H) = \pi^N/|H| \tag{3}$$

MISCELLANEOUS EXERCISES

1. Show that the characteristic roots of $p(A)$, where $p(\lambda)$ is a polynomial, are $p(\lambda_i)$, and thus that $|p(A)| = \prod_i p(\lambda_i)$.

2. Under what conditions does the corresponding result hold for rational functions?

3. Determine the inverses of the following matrices:

$$C = \begin{vmatrix} x & -y \\ y & x \end{vmatrix}$$

$$Q = \begin{bmatrix} x_1 & x_2 & x_3 & x_4 \\ -x_2 & x_1 & -x_4 & x_3 \\ -x_3 & x_4 & x_1 & x_2 \\ -x_4 & x_3 & -x_2 & x_1 \end{bmatrix}$$

4. Let $A_0 = I$, and the sequence of scalars and matrices be determined by $c_k = \text{tr}\,(AA_{k-1})/k$, $k = 1, 2, \ldots, A_k = AA_{k-1} - c_kI$. Then $A^{-1} = A_{n-1}/c_n$ (*Frame*).

5. Show that $|A + tB| = |A|(1 + t\,\text{tr}\,(A^{-1}B) + \cdots)$.

6. Using the representation

$$|A + \epsilon B + \lambda I| = \phi_N(A + \epsilon B) + \lambda\phi_{N-1}(A + \epsilon B) + \cdots \\ = |A + \lambda I|(1 + \epsilon\,\text{tr}\,((A + \lambda I)^{-1}B) + \cdots)$$

we obtain the relation

$$\phi_{N-1}(A + \epsilon B) = \phi_{N-1}(A) - \epsilon\phi_N(A)\,\text{tr}\,(A^{-2}B) + \cdots$$

and corresponding results for each $\phi_k(A)$.

7. Show that a necessary and sufficient condition that A be positive definite is that $\text{tr}\,(AB) > 0$ for all positive definite B.

8. Show that

$$|I + \lambda A| = e^{\sum_{k=1}^{\infty} \frac{(-1)^{k-1}}{k} \lambda^k \text{tr}\,(A^k)}$$

and hence, obtain relations connecting $\phi_k(A)$ and $\text{tr}\,(A^k)$.

9. Show that $AB + B$ and $BA + B$ have the same determinant by showing that $\text{tr}\,((AB + B)^n) = \text{tr}\,((BA + B)^n)$ for $n = 1, 2, \ldots$.

10. Let $B_0 = I$, $B_i = AB_{i-1} + k_iI$, where $k_i = \text{tr}\,(AB_{i-1})/i$; then

$$(-1)^n|A - \lambda I| = \lambda^n + k_1\lambda^{n-1} + \cdots \qquad (\textit{Leverrier})$$

11. Let A, B, . . . , be commutative matrices, and let $f(x_1,x_2, \ldots)$ be any rational function. The characteristic roots $a_1, a_2, \ldots, a_n$ of A, $b_1, b_2, \ldots, b_n$ of B, . . . , can be ordered in such a way that the characteristic roots of $f(A,B, \ldots)$ are $f(a_1,b_1, \ldots)$, $f(a_2,b_2, \ldots)$, and so on.

12. Show that every nonsingular matrix may be represented as a product of two not necessarily real symmetric matrices in an infinite number of ways (*Voss*).

13. If H is non-negative definite Hermitian, there exists a triangular matrix T such that $H = TT^*$ (*Toeplitz*).

14. For every A there is a triangular matrix such that TA is unitary (*E. Schmidt*).

15. Let P, Q, R, X be matrices of the second order. Then every characteristic root of a solution X of $PX^2 + QX + R = 0$ is a root of $|P\lambda^2 + Q\lambda + R| = 0$ (*Sylvester*).

16. If A, B, C are positive definite, then the roots of $|A\lambda^2 + B\lambda + C| = 0$ have negative real parts.

17. Consider matrices A and B such that $AB = r_1BA$, where r_1 is a qth root of unity. Show that the characteristic roots of A and B, λ_i and μ_i, may be arranged so that those of $A + B$ are $(\lambda_i^q + \mu_i^q)^{1/q}$, and those of AB are $r^{(q-1)/2}\lambda_i\mu_i$. Different branches of the qth root may have to be chosen for different values of i (*Potter*).

18. Using the representation

$$\frac{\pi^{N/2}}{|A + \epsilon B|^{1/2}} = \int_{-\infty}^{\infty} e^{-(x,Ax)-\epsilon(x,Bx)} \prod_i dx_i$$

and expanding both sides in powers of ϵ, obtain representations for the coefficients in the expansion of $|A + \epsilon B|$ in powers of ϵ, and thus of the fundamental functions $\phi_k(A)$.

19. Let A and B be Hermitian matrices with the property that $c_1A + c_2B$ has the characteristic roots $c_1\lambda_i + c_2\mu_i$ for all scalars c_1 and c_2, where λ_i, μ_i are, respectively, the characteristic roots of A and B. Then $AB = BA$ (*Motzkin*).

20. Show that A is nonsingular if $|a_{ii}| > \sum_{k \neq i} |a_{ik}|$, $i = 1, 2, \ldots, N$.

21. The characteristic vectors of A are characteristic vectors of $p(A)$, for any polynomial p, but not necessarily conversely.

22. If A and B are symmetric matrices such that

$$|I - \lambda A|\,|I - \mu B| = |I - \lambda A - \mu B|$$

for all λ and μ, then $AB = 0$ (*Craig-Hotelling*).

23. Let $AB = 0$, and let $p(A,B)$ be a polynomial having no constant term. Then $|\lambda I - p(A,B)| = \lambda^{-N}|\lambda I - p(A,0)|\,|\lambda I - p(B,0)|$ (*H. Schneider*).

24. Generalize this result to $p(A_1,A_2, \ldots, A_n)$, where $A_iA_j = 0$, $i < j$ (*H. Schneider*).

25. Let A and B be positive definite. By writing $|AB| = |A^{1/2}|\,|B|\,|A^{1/2}|$, show that

$$\frac{\sqrt{\pi}}{|AB|} = \int_{-\infty}^{\infty} e^{-(A^{1/2}x,BA^{1/2}x)} \prod_{i=1}^{N} dx_i$$

26. Show that

$$\int_{-\infty}^{\infty} e^{-(x,Ax)/2+(y,x)} \prod dx_i = (2\pi)^{N/2}|A|^{-1/2}e^{-(y,A^{-1}y)/2}$$

27. Let Y be a positive definite 2×2 matrix and X a 2×2 symmetric matrix. Consider the integral

$$J(Y) = \int_{X>0} e^{-tr\,(XY)}|X|^{s-3/2}\,dx_{11}\,dx_{12}\,dx_{22}$$

where the region of integration is determined by the inequalities $x_{11} > 0$, $x_{11}x_{22} - x_{12}^2 > 0$. In other words, this is the region where X is positive definite. Then

$$J(Y) = \frac{\pi^{1/2}\Gamma(s)\Gamma(s-1/2)}{|Y|^s}$$

for Re $(s) > 3/2$ (*Ingham-Siegel*). (See the discussion of Sec. 9 for further information concerning results of this nature.)

28. Let A be a complex matrix of order N, $A = (a_{rs}) = (b_{rs} + ic_{rs})$. Let B be the real matrix of order $2N$ obtained by replacing $b_{rs} + ic_{rs}$ by its matrix equivalent

$$b_{rs} + ic_{rs} \sim \begin{bmatrix} b_{rs} & c_{rs} \\ -c_{rs} & b_{rs} \end{bmatrix}$$

Show that A^{-1} can be obtained from B^{-1} by reversing this process after B^{-1} has been obtained.

29. If A is symmetric and $|A| = 0$, then the bordered determinant

$$B(x) = \begin{vmatrix} a_{11} & \cdots & a_{1N} & x_1 \\ a_{21} & \cdots & a_{2N} & x_2 \\ \cdot & & & \\ \cdot & & & \\ \cdot & & & \\ a_{N1} & \cdots & a_{NN} & x_N \\ x_1 & \cdots & x_N & 0 \end{vmatrix}$$

multiplied by the leading second minor of A is the square of a linear function of the x_i (*Burnside-Panton*).

30. Consequently, if A is symmetric and the leading first minor vanishes, the determinant and its leading second minor have opposite signs.

31. Let X and A be 2×2 matrices. Find all solutions of the equation $X^2 = A$ which have the form $X = c_1I + c_2A$, where c_1 and c_2 are scalars.

32. Similarly, treat the case where X and A are 3×3 matrices, with $X = c_1I + c_2A + c_3A^2$.

33. Let $f(t)$ be a polynomial of degree less than or equal to $N - 1$, and let $\lambda_1, \lambda_2, \ldots, \lambda_N$ be the N distinct characteristic roots of A. Then

$$f(A) = \sum_{i=1}^{N} f(\lambda_i) \prod_{\substack{1 \le j \le N \\ j \ne i}} \left[\frac{A - \lambda_i I}{\lambda_i - \lambda_j}\right]$$

(*Sylvester interpolation formula*)

What is the right-hand side equal to when $f(\lambda)$ is a polynomial of degree greater than N?

34. Is the formula valid for $f(A) = A^{-1}$?

35. Show that

$$\int_{(x,Ax)\le 1} dx_1\,dx_2 \cdots dx_N = \frac{\pi^{N/2}|A|^{-1/2}}{\Gamma(N/2+1)}$$

36. If the elements a_{ij} of the $N \times N$ matrix A are real and satisfy the conditions $a_{ii} > \sum_{j \neq i} |a_{ij}|$, $i = 1, 2, \ldots, N$, then the absolute term in the Laurent expansion of

$$f(z_1,z_2, \ldots ,z_N) = \prod_{j=1}^{N} \left(\sum_{k=1}^{N} a_{jk}z_k \Big/ z_j \right)^{-1}$$

on $|z_1| = |z_2| = \cdots = |z_N|$ is equal to $|A|^{-1}$.

Alternatively,

$$\frac{1}{|A|} = \frac{1}{(2\pi i)^N} \oint \cdots \oint \prod_{j=1}^{N} \left[\frac{dz_j}{\sum_k a_{jk}z_k} \right]$$

where the contour of integration is $|z_j| = 1$.

This result remains valid if A is complex and $|a_{ii}| > \sum_{j \neq i} |a_{ij}|$. A number of further results are given in P. Whittle.[1]

The original result was first given by Jacobi and applications of it may be found in H. Weyl[2] and in A. R. Forsyth.[3]

37. Show that if A and B are non-negative definite, then $\sum_{i,j} a_{ij}b_{ij} \geq 0$, and thus establish Schur's theorem, given in Sec. 7 (*Fejer*).

38. Let $\{A_i\}$ be a set of $N \times N$ symmetric matrices, and introduce the inner product

$$(A_i,A_j) = \text{tr } (A_iA_j)$$

Given a set of $M = N(N + 1)/2$ linearly independent symmetric matrices $\{A_i\}$, construct an orthonormal linearly independent set $\{Y_i\}$ using the foregoing inner product and show that every $N \times N$ symmetric matrix X can be written in the form $X = \sum_{i=1}^{M} c_iY_i$, where $c_i = (X,Y_i)$.

39. For the case of 2×2 matrices, take

$$A_1 = \begin{bmatrix} 1 & 0 \\ 0 & 0 \end{bmatrix} \qquad A_2 = \begin{bmatrix} 0 & 0 \\ 0 & 1 \end{bmatrix} \qquad A_3 = \begin{bmatrix} 0 & 1 \\ 1 & 0 \end{bmatrix}$$

What is the corresponding orthonormal set?

40. Similarly, in the general case, take for the A_i the matrices obtained in the analogous fashion, and construct the Y_i.

41. The Cayley-Hamilton theorem asserts that A satisfies a scalar polynomial of degree N. Call the *minimum* polynomial associated with A the scalar polynomial of minimum degree, $q(\lambda)$, with the property that $q(A) = 0$. Show that $q(\lambda)$ divides $|A - \lambda I|$.

[1] P. Whittle, Some Combinatorial Results for Matrix Powers, *Quart. J. Math.*, vol. 7, pp. 316–320, 1956.

[2] H. Weyl, *The Classical Groups . . .* , Princeton University Press, Princeton, N.J., 1946.

[3] A. R. Forsyth, *Lectures Introductory to the Theory of Functions of Two Complex Variables*, Cambridge University Press, New York, 1914.

42. Find the Jacobian of the transformation $Y = X^{-1}$.*

43. Evaluate the matrix function

$$\frac{1}{2\pi i}\int_C f(z)(zI - A)^{-1}\,dz$$

under various assumptions concerning A and the contour C.

The first use of a contour integral $\frac{1}{2\pi i}\int_C (\lambda I - T)^{-1} f(\lambda)\,d\lambda$ to represent $f(T)$ is found in Poincaré (*H. Poincaré, Sur les groupes continus, Trans. Cambridge Phil. Soc., vol.* 18, *pp.* 220–225, 1899, *and Oeuvres, vol.* 3, *pp.* 173–212).

44. If $Ax = b$ has a solution for *all* b, then A^{-1} exists.

45. For any two positive constants k_1 and k_2, there exist matrices A and B such that every element of $AB - I$ is less than k_1 in absolute value and such that there exist elements of $BA - I$ greater than k_2 (*Householder*).

46. If A_1 is nonsingular, then

$$\begin{vmatrix} A_1 & A_2 \\ A_3 & A_4 \end{vmatrix} = |A_1|\,|A_4 - A_3A_1^{-1}A_2|$$

47. If $BC = CB$, and

$$A = \begin{bmatrix} I & 0 & \cdots & & & \\ -B & I & 0 & \cdots & & \\ C & -B & I & 0 & \cdots & \\ 0 & C & -B & I & 0 & \cdots \\ \cdot & & & & & \\ \cdot & & & & & \\ \cdot & & & & & \end{bmatrix}$$

then

$$A^{-1} = \begin{bmatrix} I & 0 & \cdots & & & \\ D_1 & I & 0 & \cdots & & \\ D_2 & D_1 & I & 0 & \cdots & \\ D_3 & D_2 & D_1 & I & 0 & \cdots \end{bmatrix}$$

Determine recurrence relations for the D_i.

48. If A is an $N \times M$ matrix of rank r, $r < N$, C an $M \times N$ matrix such that $ACA = kA$, where k is a scalar, B an $M \times N$ matrix, then the characteristic equation of AB is $\lambda^{N-r}\phi(\lambda) = 0$ and that of $A(B + C)$ is $\lambda^{N-r}\phi(\lambda - k) = 0$ (*Parker*).

49. If $Y = (AX + B)(CX + D)^{-1}$, express X in terms of Y.

50. We have previously encountered the correspondence

$$a + bi \sim \begin{vmatrix} a & b \\ -b & a \end{vmatrix}$$

Prove that a necessary and sufficient condition for

$$f(a + bi) \sim f\begin{vmatrix} a & b \\ -b & a \end{vmatrix}$$

where $f(z)$ is a power series in z, is that $f(\bar{z}) = \overline{f(z)}$. For a generalization, see D. W. Robinson, *Mathematics Magazine*, vol. 32, pp. 213–215, 1959.

* I. Olkin, Note on the Jacobians of Certain Matrix Transformations Useful in Multivariate Analysis, *Biometrika*, vol. 40, pp. 43–6, 1953.

51. If $F(z) = \sum_{n=0}^{\infty} c_n z^n$ with $c_n \geq 0$, then $F(A)$ is positive definite whenever A is positive definite.

52. If $F(A)$ is positive definite for every A of the form (a_{i-j}), then $F(z) = \sum_{n=0}^{\infty} c_n z^n$ with $c_n \geq 0$ (*W. Rudin*). An earlier result is due to Schoenberg, *Duke Math. J.*, 1942.

53. A matrix $A = (a_{ij})$ is called a Jacobi matrix if $a_{ij} = 0$, for $|i - j| \geq 2$. If A is a symmetric Jacobi matrix, to what extent are the elements a_{ij} determined, if the characteristic roots are given? This is a finite version of a problem of great importance in mathematical physics, namely that of determining the coefficient function $\phi(x)$, given the spectrum of $u'' + \lambda\phi(x)u = 0$, subject to various boundary conditions.[1]

54. If A has the form

$$A = \begin{bmatrix} a_1 & 1 & 0 & \cdots & & & \\ 1 & a_2 & 1 & 0 & \cdots & & \\ 0 & 1 & a_3 & 1 & 0 & \cdots & \end{bmatrix}$$

how does one determine the a_i given the characteristic roots?

55. Show that for any A, there is a generalized inverse X such that $AXA = A$, $XAX = X$, $XA = (XA)'$, $AX = (AX)'$ (*Penrose*[2]).

56. X is called a generalized inverse for A if and only if $AXA = A$. Show that X is a generalized inverse if and only if there exist matrices U, V, and W such that

$$X = Q\begin{bmatrix} I_R & U \\ V & W \end{bmatrix} P$$

where

$$PAQ = \begin{bmatrix} I_R & 0 \\ 0 & 0 \end{bmatrix}$$

and I_R is the identity matrix of dimension n.

57. X is called a reflexive generalized inverse for A if and only if $AXA = A$ and $XAX = X$. Show that

$$X = Q\begin{bmatrix} I_r & U \\ V & VU \end{bmatrix} P \qquad (\textit{C. A. Rhode})$$

58. Show that if $Ax = y$ is consistent, the general solution is given by

$$x = A^{\dagger}y + (I - A^{\dagger}A)z$$

where $A^{\dagger}$ is the Penrose-Moore inverse and z is arbitrary (*Penrose*). For computational aspects see Urquehart.[3]

[1] G. Borg, Eine Umkehrung der Sturm-Liouvilleschen Eigenwerte Aufgabe, *Acta Math.*, vol. 78, pp. 1–96, 1946. I. M. Gelfand and B. M. Leviton, On the Determination of a Differential Equation from Its Spectral Function, *Trans. Am. Math. Soc.*, vol. 2, pp. 253–304, 1955.

[2] See the reference in the Bibliography and Discussion, Sec. 3.

[3] N. S. Urquehart, Computation of Generalized Inverse Matrices which Satisfy Specified Conditions, *SIAM Review*, vol. 10, pp. 216–218, 1968.

See also Langenhop,[1] Radhakrishna Rao,[2] and Schwerdtfeger.[3]

59. If A^{-1} exists and if more than N of the elements of A are positive with the others non-negative, then A^{-1} has at least one negative element. A has the form PD with P a permutation matrix and D a diagonal matrix with positive elements if and only if A and A^{-1} both have non-negative entries (*Spira*).

60. Consider the matrix $A + \epsilon B$ where A is positive definite, B is real and symmetric and ϵ is small. Show that we can write $(A + \epsilon B)^{-1} = A^{-1/2}(1 + \epsilon S)^{-1}A^{-1/2}$, where S is symmetric, and thus obtain a symmetric perturbation formula,

$$(A + \epsilon B)^{-1} = A^{-1} - \epsilon A^{-1/2}SA^{-1/2} + \cdots.$$

61. Let $f(z)$ be analytic for $|z| < 1$, $f(z) = a_0 + a_1z + \cdots$. Write $f(z)^k = a_{0k} + a_{1k}z + a_{2k}z^2 + \cdots$, $k = 1, 2, \ldots$. Can one evaluate $|a_{ij}|$, $i, j = 0, 1, \ldots, N$, in simple terms?

62. Write $f(z) = b_1z + b_2z^2 + \cdots$, and $f^{(k)}(z) = b_{1k}z + b_{2k}z^2 + \cdots$, where $f^{(k)}$ is the kth iterate, that is, $f^{(2)} = f(f), \ldots, f^{(k)} = f(f^{(k-1)}), \ldots$. Can one evaluate $|b_{ij}|$, $i, j = 0, 1, \ldots, N$, in simple terms?

Bibliography and Discussion

§1. There is an extensive discussion of functions of matrices in the book by MacDuffee,

C. C. MacDuffee, *The Theory of Matrices*, Ergebnisse der Mathematik, Reprint, Chelsea Publishing Co., New York, 1946.

More recent papers are

R. F. Rinehart, The Derivative of a Matrix Function, *Proc. Am. Math. Soc.*, vol. 7, pp. 2–5, 1956.

H. Richter, Über Matrixfunktionen, *Math. Ann.*, vol. 122, pp. 16–34, 1950.

S. N. Afriat, Analytic Functions of Finite Dimensional Linear Transformations, *Proc. Cambridge Phil. Soc.*, vol. 55, pp. 51–56, 1959.

and an excellent survey of the subject is given in

R. F. Rinehart, The Equivalence of Definitions of a Matric Function, *Am. Math. Monthly*, vol. 62, pp. 395–414, 1955.

[1] C. E. Langenhop, On Generalized Inverses of Matrices, *SIAM J. Appl. Math.*, vol. 15, pp. 1239–1246, 1967.

[2] C. Radhakrishna Rao, Calculus of Generalized Inverses of Matrices, I: General Theory, *Sankhya*, ser. A, vol. 29, pp. 317–342, 1967.

[3] H. Schwerdtfeger, Remarks on the Generalized Inverse of a Matrix, *Lin. Algebra Appl.*, vol. 1, pp. 325–328, 1968.

See also

P. S. Dwyer and M. S. Macphail, Symbolic Matrix Derivatives, *Ann. Math. Stat.*, vol. 19, pp. 517–534, 1948.

The subject is part of the broader field of functions of hypercomplex quantities. For the case of quaternions, which is to say matrices having the special form

$$Q = \begin{bmatrix} x_1 & x_2 & x_3 & x_4 \\ -x_2 & x_1 & -x_4 & x_3 \\ -x_3 & x_4 & x_1 & -x_2 \\ -x_4 & -x_3 & x_2 & x_1 \end{bmatrix}$$

there is an extensive theory quite analogous to the usual theory of functions of a complex variable (which may, after all, be considered to be the theory of matrices of the special form

$$z = \begin{bmatrix} x & y \\ -y & x \end{bmatrix}$$

with x and y real); see

R. Fucter, *Functions of a Hyper Complex Variable,* University of Zurich, 1948–1949, reprinted by Argonne National Laboratory, 1959.

R. Fueter, *Commentarii Math. Helveticii,* vol. 20, pp. 419–20, where many other references can be found.

R. F. Rinehart, Elements of a Theory of Intrinsic Functions on Algebras, *Duke Math. J.*, vol. 27, pp. 1–20, 1960.

Finally, let us mention a paper by W. E. Roth,

W. E. Roth, A Solution of the Matrix Equation $P(X) = A$, *Trans. Am. Math. Soc.*, vol. 30, pp. 597–599, 1928.

which contains a thorough discussion of the early results of Sylvester, Cayley, and Frobenius on the problem of solving polynomial equations of this form.

Further references to works dealing with functions of matrices will be given at the end of Chap. 10.

§3. The concept of a generalized inverse for singular matrices was discussed first by

E. H. Moore, General Analysis, Part I, *Mem. Am. Phil. Soc.*, vol. 1, p. 197, 1935.

and then independently discovered by

R. Penrose, A Generalized Inverse for Matrices, *Proc. Cambridge Phil. Soc.*, vol. 51, pp. 406–413, 1955.

These and other matters are discussed in

M. R. Hestenes, Inversion of Matrices by Biorthogonalization, *J. Soc. Indust. Appl. Math.*, vol. 6, pp. 51–90, 1958.

An exposition of the computational aspects of matrix inversion which contains a detailed account of many of the associated analytic questions may be found in

J. von Neumann and H. Goldstine, Numerical Inverting of Matrices of High Order, *Bull. Am. Math. Soc.*, vol. 53, pp. 1021–1099, 1947.

§4. An excellent account of various methods of matrix inversion, together with many references, is given in

D. Greenspan, Methods of Matrix Inversion, *Am. Math. Monthly*, vol. 62, pp. 303–319, 1955.

The problem of deducing feasible, as distinguished from theoretical, methods of matrix inversion is one of the fundamental problems of numerical analysis. It will be extensively discussed in a succeeding volume by G. Forsythe.

The problem of determining when a matrix is not singular is one of great importance, and usually one that cannot be answered by any direct calculation. It is consequently quite convenient to have various simple tests that can be applied readily to guarantee the nonvanishing of the determinant of a matrix.

Perhaps the most useful is the following (Exercise 20 of Miscellaneous Exercises):

If A is a real matrix and $|a_{ii}| > \sum_{j \neq i} |a_{ij}| \qquad i = 1, 2, \ldots, N$, then $|A| \neq 0$.

For the history of this result and a very elegant discussion of extensions, see

O. Taussky, A Recurring Theorem on Determinants, *Am. Math. Monthly*, vol. 56, pp. 672–676, 1949.

This topic is intimately related to the problem of determining simple estimates for the location of the characteristic values of A in terms of the elements of A, a topic we shall encounter in part in Chap. 16. Generally, any extensive discussion will be postponed until a later volume.

A useful bibliography of works in this field is

O. Taussky, *Bibliography on Bounds for Characteristic Roots of Finite Matrices*, National Bureau of Standards Report, September, 1951.

It should be noted, however, that the foregoing result immediately yields the useful result of Gersgorin that the characteristic roots of A must lie inside the circles of center a_{ii} and radius $\sum_{j \neq i} |a_{ij}|$, for $i = 1, 2, \ldots, N$. The original result is contained in

S. Gersgorin, Über die Abgrenzung der Eigenwerte einer Matrix, *Izv. Akad. Nauk SSSR*, vol. 7, pp. 749–754, 1931.

§6. As mentioned previously, a parametric representation of different type can be obtained for 3×3 matrices in terms of elliptic functions; see

F. Caspary, Zur Theorie der Thetafunktionen mit zwei Argumenten, *Kronecker J.*, pp. 74–86, vol. 94.

F. Caspary, Sur les systèmes orthogonaux formés par les fonctions théta, *Comptes Rendus de Paris*, pp. 490–493, vol. 104.

as well as a number of other papers by the same author over the same period.

§7. See

I. Schur, Bemerkungen zur Theorie der beschränkten Bilinearformen mit unendlich vielen Veränderlichen, *J. Math.*, vol. 140, pp. 1–28, 1911.

See also

L. Fejer, Über die Eindeutigkeit der Lösung der linearen partiellen Differentialgleichung zweiter Ordnung, *Math. Z.*, vol. 1, pp. 70–79, 1918.

where there is reference to an earlier work by T. Moutard. See also

H. Lewy, Composition of Solutions of Linear Partial Differential Equations in Two Independent Variables, *J. Math. and Mech.*, vol. 8, pp. 185–192, 1959.

The inner product $(A,B) = \Sigma a_{ij}b_{ij}$ is used by R. Oldenburger for another purpose in

R. Oldenburger, Expansions of Quadratic Forms, *Bull. Am. Math. Soc.*, vol. 49, pp. 136–141, 1943.

§9. This infinite integral plays a fundamental role in many areas of analysis. We shall base our discussion in the chapter on inequalities upon it, and an extension.

An extensive generalization of this representation may be obtained from integrals first introduced by Ingham and Siegel. See, for the original results,

A. E. Ingham, An Integral which Occurs in Statistics, *Proc. Cambridge Phil. Soc.*, vol. 29, pp. 271–276, 1933.

C. L. Siegel, Über die analytische Theorie der quadratischen Formen, *Ann. Math.*, vol. 36, pp. 527–606, 1935.

For extensions and related results, see

R. Bellman, A Generalization of Some Integral Identities Due to Ingham and Siegel, *Duke Math. J.*, vol. 24, pp. 571–578, 1956.

C. S. Herz, Bessel Functions of Matrix Argument, *Ann. Math.*, vol. 61, pp. 474–523, 1955.

I. Olkin, A Class of Integral Identities with Matrix Argument, *Duke Math. J.*, vol. 26, pp. 207–213, 1959.

S. Bochner, "Group Invariance of Cauchy's Formula in Several Variables," *Ann. Math.*, vol. 45, pp. 686–707, 1944.

R. Bellman, Generalized Eisenstein Series and Nonanalytic Automorphic Functions, *Proc. Natl. Acad. Sci. U.S.*, vol. 36, pp. 356–359, 1950.

H. Maass, Zur Theorie der Kugelfunktionen einer Matrixvariablen, *Math. Z.*, vol. 135, pp. 391–416, 1958.

H. S. A. Potter, The Volume of a Certain Matrix Domain, *Duke Math. J.*, vol. 18, pp. 391–397, 1951.

Integrals of this type, analogues and extensions, arise naturally in the field of statistics in the domain of multivariate analysis. See, for example,

T. W. Anderson and M. A. Girshick, Some Extensions of the Wishart Distribution, *Ann. Math. Stat.*, vol. 15, pp. 345–357, 1944.

G. Rasch, A Functional Equation for Wishart's Distribution, *Ann. Math. Stat.*, vol. 19, pp. 262–266, 1948.

R. Sitgreaves, On the Distribution of Two Random Matrices Used in Classification Procedures, *Ann. Math. Stat.*, vol. 23, pp. 263–270, 1952.

In connection with the concepts of functions of matrices, it is appropriate to mention the researches of Loewner, which have remarkable connections with various parts of mathematical physics and applied mathematics. As we have noted in various exercises, $A \geq B$, in the sense that A, B are symmetric and $A - B$ is non-negative definite, does not necessarily imply that $A^2 \geq B^2$, although it is true that $B^{-1} \geq A^{-1}$ if $A \geq B > 0$. The general question was first discussed in

C. Loewner, *Math. Z.*, vol. 38, pp. 177–216, 1934.

See also

R. Dobsch, *Math. Z.*, vol. 43, pp. 353–388, 1937.

In his paper Loewner studies the problem of determining functions $f(x)$ for which $A \geq B$ implies that $f(A) \geq f(B)$. This leads to a class of functions called *positive real*, possessing the property that Re $[f(z)] \geq 0$ whenever Re $(z) \geq 0$. Not only are these functions of great importance in modern network theory, cf.

L. Weinberg and P. Slepian, *Positive Real Matrices*, Hughes Research Laboratories, Culver City, Calif., 1958.

F. H. Effertz, On the Synthesis of Networks Containing Two Kinds of Elements, *Proc. Symposium on Modern Network Synthesis*, Polytechnic Institute of Brooklyn, 1955,

where an excellent survey of the field is given, and

R. J. Duffin, Elementary Operations Which Generate Network Matrices, *Proc. Am. Math. Soc.*, vol. 6, pp. 335–339, 1955;

but they also play a paramount role in certain parts of modern physics; see the expository papers

A. M. Lane and R. G. Thomas, *R*-matrix Theory of Nuclear Reactions, *Revs. Mod. Physics*, vol. 30, pp. 257–352, 1958.

E. Wigner, On a Class of Analytic Functions from the Quantum Theory of Collisions, *Ann. Math.*, vol. 53, pp. 36–67, 1951.

Further discussion may be found in a more recent paper,

C. Loewner, Some Classes of Functions Defined by Difference or Differential Inequalities, *Bull. Am. Math. Soc.*, vol. 56, pp. 308–319, 1950.

and in

J. Bendat and S. Sherman, Monotone and Convex Operator Functions, *Trans. Am. Math. Soc.*, vol. 79, pp. 58–71, 1955.

The concept of convex matrix functions is also discussed here, treated first by

F. Kraus, *Math. Z.*, vol. 41, pp. 18–42, 1936.

§10. This result is contained in

R. Bellman, Representation Theorems and Inequalities for Hermitian Matrices, *Duke Math. J.*, 1959.

7

Variational Description of Characteristic Roots

1. Introduction. In earlier chapters, we saw that both the largest and smallest characteristic roots had a quite simple geometric significance which led to variational problems for their determination.

As we shall see, this same geometric setting leads to corresponding variational problems for the other characteristic values. However, this variational formulation is far surpassed in usefulness and elegance by another approach due to R. Courant and E. Fischer. Both depend upon the canonical form derived in Chap. 4.

We shall then obtain some interesting consequences of this second representation, which is, in many ways, the link between the elementary and advanced analytic theory of symmetric matrices.

2. The Rayleigh Quotient. As we know, the set of values assumed by the quadratic form (x,Ax) on the sphere $(x,x) = 1$ is precisely the same set taken by the quadratic form $(y,\Lambda y) = \lambda_1 y_1^2 + \lambda_2 y_2^2 + \cdots + \lambda_N y_N^2$ on $(y,y) = 1$, $\Lambda = T'AT$, $y = Tx$, with T orthogonal. Let us henceforth suppose that

$$\lambda_1 \geq \lambda_2 \geq \cdots \geq \lambda_N \tag{1}$$

Since, with this convention,

$$\begin{aligned} (y,\Lambda y) &\geq \lambda_N(y_1^2 + y_2^2 + \cdots + y_N^2) \\ (y,\Lambda y) &\leq \lambda_1(y_1^2 + y_2^2 + \cdots + y_N^2) \end{aligned} \tag{2}$$

we readily obtain the representations

$$\begin{aligned} \lambda_1 &= \max_y \frac{(y,\Lambda y)}{(y,y)} = \max_x \frac{(x,Ax)}{(x,x)} \\ \lambda_N &= \min_y \frac{(y,\Lambda y)}{(y,y)} = \min_x \frac{(x,Ax)}{(x,x)} \end{aligned} \tag{3}$$

The quotient

$$q(x) = \frac{(x,Ax)}{(x,x)} \tag{4}$$

is often called the *Rayleigh quotient.*

From the relation in (3), we see that for all x we have

$$\lambda_1 \geq \frac{(x,Ax)}{(x,x)} \geq \lambda_N \tag{5}$$

Relations of this type are of great importance in applications where one is interested in obtaining quick estimates of the characteristic values. Remarkably, rudimentary choices of the x_i often yield quite accurate approximations for λ_1 and λ_N. Much of the success of theoretical physics depends upon this phenomenon.

The reason for this interest in the characteristic roots, and not so much the characteristic vectors, lies in the fact that in many physical applications the λ_i represent characteristic frequencies. Generally speaking, in applications the λ_i are more often the observables which allow a comparison of theory with experiment.

Matters of this type will be discussed in some slight detail in Sec. 12 below.

3. Variational Description of Characteristic Roots. Let us begin our investigation of variational properties by proving Theorem 1.

Theorem 1. *Let x^i be a set of N characteristic vectors associated with the λ_i. Then, for $k = 1, 2, \ldots, N$,*

$$\lambda_k = \max_{R_k} (x,Ax)/(x,x) \tag{1}$$

where R_k is the region of x space determined by the orthogonality relations

$$(x,x^i) = 0 \qquad i = 1, 2, \ldots, k-1, x \neq 0 \tag{2}$$

Geometrically, the result is clear. To determine, for example, the second largest semiaxis of an ellipsoid, we determine the maximum distance from the origin in a plane perpendicular to the largest semiaxis.

To demonstrate the result analytically, write

$$x = \sum_{k=1}^{N} u_k x^k \tag{3}$$

Then

$$(x,Ax) = \sum_{k=1}^{N} \lambda_k u_k^2 \tag{4}$$

$$(x,x) = \sum_{k=1}^{N} u_k^2$$

The result for λ_1 is precisely that given in Sec. 2. Consider the expression for λ_2. The condition $(x,x^1) = 0$ is the same as the assertion that

$u_1 = 0$. It is clear then that the maximum of $\sum_{k=2}^{N} \lambda_k u_k^2$ subject to $\sum_{k=2}^{N} u_k^2 = 1$ is equal to λ_2.

In exactly the same fashion, we see that λ_k has the stated value for the remaining values of k.

4. Discussion. It is known from physical considerations that stiffening of a bar or plate results in an increase in *all* the natural frequencies. The analytic equivalent of this fact is the statement that the characteristic roots of $A + B$ are uniformly larger than those of A if B is positive definite.

To demonstrate the validity of this statement for λ_1 or λ_N is easy. We have, using an obvious notation,

$$\begin{aligned} \lambda_1(A + B) &= \max_x \frac{[x,(A + B)x]}{(x,x)} \\ &= \max_x \left[\frac{(x,Ax)}{(x,x)} + \frac{(x,Bx)}{(x,x)} \right] \\ &\geq \max_x \frac{(x,Ax)}{(x,x)} + \min_x \frac{(x,Bx)}{(x,x)} \\ &\geq \lambda_1(A) + \lambda_N(B) \end{aligned} \tag{1}$$

Since by assumption $\lambda_N(B) > 0$, we see that $\lambda_1(A + B) > \lambda_1(A)$ if B is positive definite. The proof that $\lambda_N(A + B) > \lambda_N(A)$ proceeds similarly.

If, however, we attempt to carry through a proof of this type for the roots $\lambda_2, \lambda_3, \ldots, \lambda_{N-1}$, we are baffled by the fact that the variational description of Theorem 1 is inductive. The formula for λ_2 depends upon the characteristic vector x^1, and so on.

Furthermore, as noted above, in general, in numerous investigations of engineering and physical origin, we are primarily concerned with the characteristic values which represent resonant frequencies. The characteristic vectors are of only secondary interest. We wish then to derive a representation of λ_k which is independent of the other characteristic values and their associated characteristic vectors.

EXERCISES

1. Show that $\lambda_N(A + B) \geq \lambda_N(A) + \lambda_N(B)$ if A and B are positive definite.
2. Suppose that B is merely positive indefinite. Must $\lambda_N(A + B)$ actually be greater than $\lambda_N(A)$?

5. Geometric Preliminary. In order to see how to do this, let us take an ellipsoid of the form $x^2/a^2 + y^2/b^2 + z^2/c^2 = 1$, and pass a plane through the origin of coordinates. The cross section will be an ellipse which will have a major semiaxis and a minor semiaxis. If we now rotate this cross-sectional plane until the major semiaxis assumes its smallest value, we will have determined the semiaxis of the ellipsoid of second greatest length.

The analytic transliteration of this observation will furnish us the desired characterization. Like many results in analysis which can be demonstrated quite easily, the great merit of this contribution of Courant and Fischer lies in its discovery.

6. The Courant-Fischer min-max Theorem. Let us now demonstrate Theorem 2.

Theorem 2. *The characteristic roots λ_i, $i = 1, 2, \ldots, N$, may be defined as follows:*

$$\begin{aligned} \lambda_1 &= \max_x \, (x,Ax)/(x,x) \\ \lambda_2 &= \min_{(y,y)=1} \max_{(x,y)=0} (x,Ax)/(x,x) \\ &\;\;\vdots \\ \lambda_k &= \min_{\substack{(y^i,y^i)=1 \\ i=1,2,\ldots,k-1}} \max_{(x,y^i)=0} (x,Ax)/(x,x) \\ &\;\;\vdots \end{aligned} \tag{1}$$

Equivalently,

$$\begin{aligned} \lambda_N &= \min_x \, (x,Ax)/(x,x) \\ \lambda_{N-1} &= \max_{(y,y)=1} \min_{(x,y)=0} (x,Ax)/(x,x) \end{aligned} \tag{2}$$

and so on.

Proof. Let us consider the characteristic root λ_2. Define the quantity

$$\mu_2 = \min_{(y,y)=1} \max_{(x,y)=0} (x,Ax)/(x,x) \tag{3}$$

We shall begin by making an orthogonal transformation which converts A into diagonal form, $x = Tz$. Then

$$\begin{aligned} \mu_2 &= \min_{(y,y)=1} \max_{(Tz,y)=0} \left\{ \sum_{k=1}^{N} \lambda_k z_k^2 \Big/ \sum_{k=1}^{N} z_k^2 \right\} \\ &= \min_{(x,y)=1} \max_{(z,T'y)=0} \{\cdots\} \end{aligned} \tag{4}$$

Setting $T'y = y^1$, this becomes

$$\mu_2 = \min_{(Ty^1,Ty^1)=1} \max_{(z,y^1)=0} \{\cdots\} \tag{5}$$

Since $(Ty^1,Ty^1) = (y^1,y^1)$, we may write

$$\mu_2 = \min_{(y,y)=1} \max_{(z,y)=0} \{\cdot \cdot \cdot\} \tag{6}$$

In other words, it suffices to assume that A is a diagonal matrix.

In place of maximizing over all z, let us maximize over the subregion defined by

$$S: \quad z_3 = z_4 = \cdot \cdot \cdot = z_N = 0 \text{ and } (z,y) = 0 \tag{7}$$

Since this is a subregion of the z region defined by $(z,y) = 0$, we have

$$\mu_2 \geq \min_{(y,y)=1} \max_{S} \{\lambda_1 z_1^2 + \lambda_2 z_2^2 / z_1^2 + z_2^2\} \tag{8}$$

Since $(\lambda_1 z_1^2 + \lambda_2^1 z_2^2)/(z_1^2 + z_2^2) \geq \lambda_2$ for all z_1 and z_2, it follows that we have demonstrated that $\mu_2 \geq \lambda_2$. Now let us show that $\mu_2 \leq \lambda_2$. To do this, in the region defined by $(y,y) = 1$, consider the set consisting of the single value $y_1 = 1$. This choice of the vector y forces z to satisfy the condition that its first component z_1 must be zero. Since the minimum over a subset must always be greater than or equal to the minimum over the larger set, we see that

$$\begin{aligned} \mu_2 &\leq \max_{z_1=0} \{(\lambda_1 z_1^2 + \lambda_2 z_2^2 + \cdot \cdot \cdot + \lambda_N z_N^2)/(z_1^2 + z_2^2 + \cdot \cdot \cdot + z_N^2)\} \\ &\leq \max_{z} \{(\lambda_2 z_2^2 + \cdot \cdot \cdot + \lambda_N z_N^2)/(z_2^2 + \cdot \cdot \cdot + z_N^2)\} = \lambda_2 \end{aligned} \tag{9}$$

It follows that $\mu_2 = \lambda_2$.

Now let us pursue the same techniques to establish the result for general k. Consider the quantity

$$\begin{aligned} \mu_k &= \min_{\substack{(y^i,y^i)=1 \\ i=1,2,\ldots,k-1}} \max_{(x,y^i)=0} (x,Ax)/(x,x) \\ &= \min_{\substack{(y^i,y^i)=1 \\ i=1,2,\ldots,k-1}} \max_{(z,y^i)=0} \left\{ \sum_{k=1}^{N} \lambda_k z_k^2 \Big/ \sum_{k=1}^{N} z_k^2 \right\} \end{aligned} \tag{10}$$

Consider maximization over the subregion defined by

$$S: \quad z_{k+1} = z_{k+2} = \cdot \cdot \cdot = z_N = 0 \tag{11}$$

It follows that

$$\mu_k \geq \min_{\substack{(y^i,y^i)=1 \\ i=1,2,\ldots,k-1}} \max_{S} \{\cdot \cdot \cdot\} \geq \lambda_k \tag{12}$$

since $\sum_{i=1}^{k} \lambda_i z_i^2 \Big/ \sum_{i=1}^{k} z_i^2 \geq \lambda_k$ for all z.

Similarly, to show that $\lambda_k \leq \mu_k$, we consider minimization over the subregion of $(y^i,y^i) = 1$, $i = 1, 2, \ldots, k - 1$, consisting of the $(k - 1)$

vectors

$$y^1 = \begin{bmatrix} 1 \\ 0 \\ \cdot \\ \cdot \\ \cdot \\ 0 \end{bmatrix} \qquad y^2 = \begin{bmatrix} 0 \\ 1 \\ 0 \\ \cdot \\ \cdot \\ \cdot \\ 0 \end{bmatrix} \qquad \cdots \qquad y^{k-1} = \begin{bmatrix} 0 \\ \cdot \\ \cdot \\ \cdot \\ 1 \\ \cdot \\ \cdot \\ \cdot \\ 0 \end{bmatrix} \tag{13}$$

The orthogonality conditions $(y^i,z) = 0$ are then equivalent to

$$z_1 = z_2 = \cdots = z_{k-1} = 0 \tag{14}$$

As before, we see that $\mu_k \le \lambda_k$, and thus that $\mu_k = \lambda_k$.

7. Monotone Behavior of $\lambda_k(A)$. The representation for λ_k given in Theorem 2 permits us to conclude the following result.

Theorem 3. *Let A and B be symmetric matrices, with B non-negative definite. Then*

$$\lambda_k(A + B) \ge \lambda_k(A) \qquad k = 1, 2, \ldots, N \tag{1}$$

If B is positive definite, then

$$\lambda_k(A + B) > \lambda_k(A) \qquad k = 1, 2, \ldots, N \tag{2}$$

EXERCISE

1. Obtain a lower bound for the difference $\lambda_k(A + B) - \lambda_k(A)$.

8. A Sturmian Separation Theorem. Theorem 2 also permits us to demonstrate Theorem 4.

Theorem 4. *Consider the sequence of symmetric matrices*

$$A_r = (a_{ij}) \qquad i, j = 1, 2, \ldots, r \tag{1}$$

for $r = 1, 2, \ldots, N$.

Let $\lambda_k(A_r)$, $k = 1, 2, \ldots, r$, denote the kth characteristic root of A_r, where, consistent with the previous notation,

$$\lambda_1(A_r) \ge \lambda_2(A_r) \ge \cdots \ge \lambda_r(A_r) \tag{2}$$

Then

$$\lambda_{k+1}(A_{i+1}) \le \lambda_k(A_i) \le \lambda_k(A_{i+1}) \tag{3}$$

Proof. Let us merely sketch the proof, leaving it as a useful exercise for the reader to fill in the details. Let us prove

$$\lambda_2(A_{i+1}) \le \lambda_1(A_i) < \lambda_1(A_{i+1}) \tag{4}$$

We have

$$\begin{aligned} \lambda_1(A_i) &= \max_x (x,A_ix)/(x,x) \\ \lambda_1(A_{i+1}) &= \max_x (x,A_{i+1}x)/(x,x) \end{aligned} \tag{5}$$

where the x appearing in the first expression is i-dimensional, and the x appearing in the second expression is $(i+1)$-dimensional. Considering variation over the set of $(i+1)$-dimensional vectors x with $(i+1)$st component zero, we see that $\lambda_1(A_i) \le \lambda_1(A_{i+1})$.

To obtain the inequality $\lambda_2(A_{i+1}) \le \lambda_1(A_i)$, we use the definition of the λ_k in terms of max-min. Thus

$$\begin{aligned} \lambda_1(A_i) &= \max_{\substack{(y^k,y^k)=1 \\ k=1,2,\ldots,i-1}} \min_{\substack{(x,y^k)=0 \\ k=1,2,\ldots,i-1}} (x,Ax)/(x,x) \\ \lambda_2(A_{i+1}) &= \max_{\substack{(y^k,y^k)=1 \\ k=1,2,\ldots,i-1}} \min_{\substack{(x,y^k)=0 \\ k=1,2,\ldots,i-1}} (x,A_{i+1}x)/(x,x) \end{aligned} \tag{6}$$

The vectors appearing in the expression for $\lambda_1(A_i)$ are all i-dimensional, while those appearing in the expression for $\lambda_2(A_{i+1})$ are all $(i+1)$-dimensional. In this second expression, consider only vectors x for which the $(i+1)$st component is zero. Then it is clear that for vectors of this type we maximize over the y^k by taking vectors y^k whose $(i+1)$st components are zero. It follows that $\lambda_2(A_{i+1}) \le \lambda_1(A_i)$.

9. A Necessary and Sufficient Condition that A be Positive Definite. Using the foregoing result, we can obtain another proof of the fact that a necessary and sufficient condition that A be positive definite is that $|A_k| > 0$, $k = 1, 2, \ldots, N$.

As we know, a necessary and sufficient condition that A be positive definite is that $\lambda_k(A) > 0$, $k = 1, 2, \ldots, N$. If A is positive definite, and thus $\lambda_k(A) > 0$, we must have, by virtue of the above separation theorem, $\lambda_k(A_i) > 0$, $k = 1, 2, \ldots, i$, and thus $|A_i| = \prod_{k=1}^{i} \lambda_k(A) > 0$.

On the other hand, if the determinants $|A_k|$ are positive for all k, we must have as the particular case $k = 1$, $\lambda_1(A_1) > 0$. Since

$$\lambda_1(A_2) \ge \lambda_1(A_1) \ge \lambda_2(A_2) \tag{1}$$

the condition $|A_2| > 0$ ensures that $\lambda_2(A_2) > 0$. Proceeding inductively, we establish that $\lambda_k(A_i) > 0$, $k = 1, 2, \ldots, i$, for $i = 1, 2, \ldots, N$, and thus obtain the desired result. Again the details of the complete proof are left as an exercise for the reader.

10. The Poincaré Separation Theorem. Let us now establish the following result which is useful for analytic and computational purposes.

Theorem 5. *Let* $\{y^k\}$, $k = 1, 2, \ldots, K$, *be a set of* K *orthonormal vectors and set* $x = \sum_{k=1}^{K} u_k y^k$, *so that*

$$(x,Ax) = \sum_{k,l=1}^{K} u_k u_l (y^k, Ay^l) \tag{1}$$

Set

$$B_K = (y^k, Ay^l) \qquad k, l = 1, 2, \ldots, K \tag{2}$$

Then

$$\begin{aligned} \lambda_i(B_K) &\le \lambda_i(A) \qquad i = 1, 2, \ldots, K \\ \lambda_{K-j}(B_K) &\ge \lambda_{N-j}(A) \qquad j = 0, 1, 2, \ldots, K-1 \end{aligned} \tag{3}$$

The results follow immediately from Theorem 4.

11. A Representation Theorem. Let us introduce the notation

$$|A|_k = \lambda_N \lambda_{N-1} \cdots \lambda_{N-k+1} \tag{1}$$

We wish to demonstrate

Theorem 6. *If* A *is positive definite,*

$$\frac{\pi^{k/2}}{|A|_k^{1/2}} = \max_R \int_R e^{-(x,Ax)}\, dV_k \tag{2}$$

where the integration is over a k*-dimensional linear subspace of* N*-dimensional space* R*, whose volume element is* dV_k*, and the maximization is over all* R.

Proof. It is easy to see that it is sufficient to take (x,Ax) in the form $\lambda_1 x_1^2 + \lambda_2 x_2^2 + \cdots + \lambda_N x_N^2$. Hence, we must show that

$$\frac{\pi^{k/2}}{(\lambda_N \lambda_{N-1} \cdots \lambda_{N-k+1})^{1/2}} = \max_R \int_R e^{-(\lambda_1 x_1^2 + \lambda_2 x_2^2 + \cdots + \lambda_N x_N^2)}\, dV_k \tag{3}$$

Denote by $V_a(\rho)$ the volume of the region defined by

$$\begin{aligned} \lambda_1 x_1^2 + \lambda_2 x_2^2 + \cdots + \lambda_N x_N^2 &\le \rho \\ (x, a^i) = 0 \qquad i &= 1, 2, \ldots, N-k \end{aligned} \tag{4}$$

where the a^i are $N - k$ linearly independent vectors.

It is clear that

$$V_a(\rho) = \rho^{k/2} V_a(1) \tag{5}$$

Then

$$\begin{aligned} \int_R &= \int_{(x,a^i)=0} e^{-\lambda_1 x_1^2 - \lambda_2 x_2^2 - \cdots - \lambda_N x_N^2}\, dV_k \\ &= \int_{-\infty}^{\infty} e^{-\rho}\, dV_a(\rho) = \frac{k V_a(1)}{2} \int_{-\infty}^{\infty} e^{-\rho} \rho^{k/2-1}\, d\rho \end{aligned} \tag{6}$$

To complete the proof, we must show that the maximum of $V_a(1)$ is attained when the relations $(x,a^i) = 0$ are $x_1 = 0, x_2 = 0, \ldots, x_k = 0$.

This, however, is a consequence of the formula for the volume of the ellipsoid, $1 = \sum_{i=N-k+1}^{N} \lambda_k x_k^2$, and the min-max characterization of the characteristic roots given in Theorem 2.

12. Approximate Techniques. The problem of determining the minimum of $\int_0^1 u'^2\,dt$ over all functions satisfying the constraints

$$\int_0^1 q(t)u^2\,dt = 1 \tag{1a}$$

$$u(0) = u(1) = 0 \tag{1b}$$

is one that can be treated by means of the calculus of variations. Using standard variational techniques, we are led to consider the Sturm-Liouville problem of determining values of λ which yield nontrivial solutions of the equation

$$u'' + \lambda q(t)u = 0 \qquad u(0) = u(1) = 0 \tag{2}$$

Since, in general, the differential equation cannot be solved in terms of the elementary transcendents, various approximate techniques must be employed to resolve this problem.

In place of obtaining the *exact variational* equation of (2) and using an *approximate method* to solve it, we can always replace the original variational problem by an *approximate variational* problem, and then use an *exact method* to solve this new problem. One way of doing this is the following.

Let $\{u_i(t)\}$ be a sequence of linearly independent functions over $[0,1]$ satisfying the conditions of (1b), that is, $u_i(0) = u_i(1) = 0$. We attempt to find an approximate solution to the original variational problem having the form

$$u = \sum_{i=1}^{N} x_i u_i(t) \tag{3}$$

The problem that confronts us now is finite dimensional, involving only the unknowns $x_1, x_2, \ldots, x_N$. We wish to minimize the quadratic form

$$\sum_{i,j=1}^{N} x_i x_j \int_0^1 u_i'(t)u_j'(t)\,dt \tag{4}$$

subject to the constraint

$$\sum_{i,j=1}^{N} x_i x_j \int_0^1 q(t)u_i(t)u_j(t)\,dt = 1 \tag{5}$$

It is clear that important simplifications result if we choose the sequence $\{u_i(t)\}$ so that either

$$\int_0^1 u_i'(t)u_j'(t)\,dt = \delta_{ij} \qquad \text{or} \qquad \int_0^1 q(t)u_i(t)u_j(t)\,dt = \delta_{ij} \tag{6}$$

The first condition is usually easier to arrange, if $q(t)$ is not a function of particularly simple type. Thus, for example, we may take

$$u_k'(t) = \sin \pi kt \tag{7}$$

properly normalized, or

$$u_k'(t) = P_k(t) \tag{8}$$

the kth Legendre polynomial, again suitably normalized. This last would be an appropriate choice if $q(t)$ were a polynomial in t, since integrals of the form $\int_0^1 t^k u_i(t)u_j(t)\,dt$ are readily evaluated if the $u_i(t)$ are given by (8).

This procedure leads to a large number of interesting and significant problems. In the first place, we are concerned with the question of convergence of the solution of the finite-dimensional problem to the solution of the original problem.

Second, we wish to know something about the rate of convergence, a matter of great practical importance, and the mode of convergence, monotonic, oscillatory, and so forth. We shall, however, not pursue this path any further here.

EXERCISE

1. Let $\lambda_1^{(N)}, \lambda_2^{(N)}, \ldots, \lambda_N^{(N)}$ denote characteristic values associated with the problem posed in (4) and (5), for $N = 2, 3, \ldots$. What inequalities exist connecting $\lambda_i^{(N)}$ and $\lambda_i^{(N-1)}$?

MISCELLANEOUS EXERCISES

1. Let A and B be Hermitian matrices with respective characteristic values $\lambda_1 \geq \lambda_2 \geq \cdots$, $\mu_1 \geq \mu_2 \geq \cdots$, and let the characteristic values of $A + B$ be $\nu_1 \geq \nu_2 \geq \cdots$, then $\lambda_i + \mu_j \geq \nu_{i+j-1}$, for $i + j \leq N + 1$ (*Weyl*).

2. By considering A as already reduced to diagonal form, construct an inductive proof of the Poincaré separation theorem.

3. What are necessary and sufficient conditions that $a_{11}x_1^2 + 2a_{12}x_1x_2 + a_{22}x_2^2 \geq 0$ for all $x_1, x_2 \geq 0$?

4. What are necessary and sufficient conditions that $\sum_{i,j=1}^{3} a_{ij}x_ix_j \geq 0$ for $x_1, x_2, x_3 \geq 0$?

5. Can one obtain corresponding results for $\sum_{i,j=1}^{N} a_{ij}x_ix_j$?

6. Prove that

$$A(r) = \begin{bmatrix} a_{11}^r & a_{12}^r \\ a_{21}^r & a_{22}^r \end{bmatrix}$$

is positive definite for $r > 0$ if $A = (a_{ij})$ is positive definite and $a_{ij} \geq 0$.

7. Does a corresponding result hold for 3×3 matrices?

8. Prove that (a_{ij}) is positive definite if $a_{ii} > 0$ and $(|a_{ij}|)$ is positive definite.

9. Show that the characteristic roots of $A^{1/2}BA^{1/2}$ are less than those of A if B is a symmetric matrix all of whose roots are between 0 and 1 and A is positive definite.

10. If T is orthogonal, x real, is $(Tx,Ax) \leq (x,Ax)$ for all x if A is positive definite?

11. Define *rank* for a symmetric matrix as the order of A minus the number of zero characteristic roots. Show, using the results of this chapter, that this definition is equivalent to the definition given in Appendix A.

12. Suppose that we have a set of real symmetric matrices depending upon a parameter, scalar or vector, $q, \{A(q)\}$. We wish to determine the maximum characteristic root of each matrix, a quantity we shall call $f(q)$, and then to determine the maximum over q of this function. Let us proceed in the following way. Choose an initial value of q, say q_0, and let x^0 be a characteristic vector associated with the characteristic root $f(q_0)$. To determine our next choice q_1, we maximize the expression $(x^0,A(q)x^0)/(x^0,x^0)$ over all q values. Call one of the values yielding a maximum q_1, and let x^1 be a characteristic vector associated with the characteristic root $f(q_1)$. We continue in this way, determining a sequence of values $f(q_0), f(q_1), \ldots$. Show that

$$f(q_0) \leq f(q_1) \leq f(q_2) \leq \cdots$$

13. Show, however, by means of an example, that we may not reach the value $\max_q f(q)$ in this way.

14. Under what conditions on $A(q)$ can we guarantee reaching the absolute maximum in this way?

15. Let the *spectral radius* $r(A)$, of a square matrix A be defined to be the greatest of the absolute values of its characteristic roots. Let H be a positive definite Hermitian matrix and let

$$g(A,H) = \max_x [(Ax,HAx)/(x,Hx)]^{1/2}$$

Show that $r(A) = \min_H g(A,H)$.

16. If A is a real matrix, show that $r(A) = \min_S g(A,S)$, where now the minimum is taken over all real symmetric matrices (*H. Osborn, The Existence of Conservation Laws, Annals of Math., vol.* 69, *pp.* 105–118, 1959).

Bibliography and Discussion

§1. The extension by R. Courant of the min-max characterization from the finite case to the operators defined by partial differential equations is one of the foundations of the modern theory of the characteristic values of operators. For a discussion of these matters, see

R. Courant and O. Hilbert, *Methoden der mathematischen Physik*, Berlin, 1931, reprinted by Interscience Publishers, Inc., New York.

§2. The Rayleigh quotient affords a simple means for obtaining upper bounds for the smallest characteristic value. The problem of obtaining lower bounds is much more difficult. For the case of more general operators a method developed by A. Weinstein and systematically extended by N. Aronszajn and others is quite powerful. For an exposition and further references, see

J. B. Diaz, Upper and Lower Bounds for Eigenvalues, *Calculus of Variations and Its Applications*, L. M. Graves (ed.), McGraw-Hill Book Company, Inc., New York, 1958.

See also

G. Temple and W. G. Bickley, *Rayleigh's Principle and Its Application to Engineering*, Oxford, 1933.

An important problem is that of determining intervals within which characteristic values can or cannot lie. See

T. Kato, On the Upper and Lower Bounds of Eigenvalues, *J. Phys. Soc. Japan*, vol. 4, pp. 334–339, 1949.

H. Bateman, *Trans. Cambridge Phil. Soc.*, vol. 20, pp. 371–382, 1908.

G. Temple, An Elementary Proof of Kato's Lemma, *Matematika*, vol. 2, pp. 39–41, 1955.

§6. See the book by Courant and Hilbert cited above and

E. Fischer, Über quadratische Formen mit reellen Koeffizienten, *Monatsh. Math. u. Physik*, vol. 16, pp. 234–249, 1905.

For a discussion, and some applications, see

G. Polya, Estimates for Eigenvalues, *Studies in Mathematics and Mechanics*, presented to R. Von Mises, Academic Press, Inc., New York, 1954.

There are extensive generalizations of this result. Some are given in Chap. 7, Secs. 10, 11, and 13, results due to Ky Fan. See

Ky Fan, On a Theorem of Weyl Concerning Eigenvalues of Linear Transformations, I, *Proc. Natl. Acad. Sci. U.S.*, vol. 35, pp. 652–655, 1949.

An extension of this is due to Wielandt,

H. Wielandt, An Extremum Property of Sums of Eigenvalues, *Proc. Am. Math. Soc.*, vol. 6, pp. 106–110, 1955.

See also

Ali R. Amir-Moez, Extreme Properties of Eigenvalues of a Hermitian Transformation and Singular Values of the Sum and Product of a Linear Transformation, *Duke Math. J.*, vol. 23, pp. 463–477, 1956.

A further extension is due to

M. Marcus and R. Thompson, A Note on Symmetric Functions of Eigenvalues, *Duke Math. J.*, vol. 24, pp. 43–46, 1957.

M. Marcus, Convex Functions of Quadratic Forms, *Duke Math. J.*, vol. 24, pp. 321–325, 1957.

§8. Results of this type may be established by means of the classical techniques of Sturm. See

W. S. Burnside and A. W. Panton, *Theory of Equations*, vol. II, Longmans, Green & Co., Inc., New York, 1928.

§10. H. Poincaré, Sur les équations aux dérivées partielles de la physique mathématique, *Am. J. Math.*, vol. 12, pp. 211–294, 1890.

§11. R. Bellman, I. Glicksberg, and O. Gross, Notes on Matrix Theory—VI, *Am. Math. Monthly*, vol. 62, pp. 571–572, 1955.

§12. An extensive discussion of computational techniques of this type is contained in

L. Collatz, *Eigenwertproblemen und ihre numerische Behandlung*, New York, 1948.

Finally, let us mention the monograph

M. Parodi, *Sur quelques propriétés des valeurs charactéristiques des matrices carrées*, fascicule CXVII, *Mém. sci. mathé.*, 1952.

The problem of obtaining corresponding variational expressions for the characteristic values of complex matrices, which arise in the study of linear dissipative systems, is one of great difficulty. For some steps in this direction, see

R. J. Duffin, A Minimax Theory for Overdamped Networks, *J. Rat. Mech. Analysis*, vol. 4, pp. 221–233, 1955, and the results of Dolph *et al.*, previously referred to.

R. J. Duffin, The Rayleigh-Ritz Method for Dissipative or Gyroscopic Systems, *Q. Appl. Math.*, vol. 18, pp. 215–222, 1960.

R. J. Duffin and A. Schild, On the Change of Natural Frequencies Induced by Small Constraints, *J. Rat. Mech. Anal.*, vol. 6, pp. 731–758, 1957.

A problem of great interest in many parts of physics is that of determining $q(t)$ given the characteristic values corresponding to various boundary conditions. See

G. Borg, An Inversion of the Sturm-Liouville Eigenvalue Problem. Determination of the Differential Equation from the Eigenvalues, *Acta Math.*, vol. 78, pp. 1–96, 1946.

I. M. Gelfand and B. M. Levitan, On the Determination of a Differential Equation from Its Spectral Function, *Izv. Akad. Nauk SSSR, Ser. Math.*, vol. 15, pp. 309–360, 1951.

B. M. Levitan and M. G. Gasymov, Determination of a Differential Equation from Two Spectra, *Uspekhi Mat. Nauk*, vol. 19, pp. 3–64, 1964.

R. Bellman, A Note on an Inverse Problem in Mathematical Physics, *Q. Appl. Math.*, vol. 19, pp. 269–271, 1961.

L. E. Anderson, Summary of Some Results Concerning the Determination of the Operator from Given Spectral Data in the Case of a Difference Equation Corresponding to a Sturm-Liouville Differential Equation, to appear.

See also

R. Yarlagadda, An Application of Tridiagonal Matrices to Network Synthesis, *SIAM J. Appl. Math.*, vol. 16, pp. 1146–1162, 1968.

8

Inequalities

1. Introduction. In this chapter, we shall establish a number of interesting inequalities concerning characteristic values and determinants of symmetric matrices. Our fundamental tools will be the integral identity

$$\frac{\pi^{N/2}}{|A|^{1/2}} = \int_{-\infty}^{\infty} \cdots \int_{-\infty}^{\infty} e^{-(x,Ax)}\, dx \tag{1}$$

valid if A is positive definite, its extension given in Theorem 6 of Chap. 7, and some extensions of min-max characterizations of Courant-Fischer, due to Ky Fan.

Before deriving some inequalities pertaining to matrix theory, we shall establish the standard inequalities of Cauchy-Schwarz and Hölder. Subsequently, we shall also prove the arithmetic-geometric mean inequality, since we shall require it for one of the results of the chapter.

2. The Cauchy-Schwarz Inequality. The first result we obtain has already been noted in the exercises. However, it is well worth restating.

Theorem 1. *For any two real vectors x and y we have*

$$(x,y)^2 \leq (x,x)(y,y) \tag{1}$$

Proof. Consider the quadratic form in u and v,

$$\begin{aligned} Q(u,v) &= (ux + vy,\ ux + vy) \\ &= u^2(x,x) + 2uv(x,y) + v^2(y,y) \end{aligned} \tag{2}$$

Since $Q(u,v)$ is clearly non-negative definite, we must have the relation in (1). We see that (1) is a special, but most important case, of the non-negativity of the Gramian determinant established in Sec. 4 of Chap. 4.

3. Integral Version. In exactly the same way, we can establish the integral version of the preceding inequality.

Theorem 2. *Let $f(x)$, $g(x)$ be functions of x defined over some region R. Then*

$$\left(\int_R fg\, dV\right)^2 \leq \left(\int_R f^2\, dV\right)\left(\int_R g^2\, dV\right) \tag{1}$$

Proof. Since $(f - g)^2 \geq 0$, we have

$$2fg \leq f^2 + g^2 \tag{2}$$

Hence, fg is integrable if f^2 and g^2 are. Now consider the integral

$$\int_R (fu + gv)^2\, dV \tag{3}$$

which by virtue of the above inequality exists. As a quadratic form in u and v, the expression in (3) is non-negative definite. Consequently, as before, we see that (1) must be satisfied.

EXERCISES

1. We may also proceed as follows:

$$\begin{aligned}(x,x)(y,y) - (x,y)^2 &= (y,y)\left[(x,x) - 2\frac{(x,y)^2}{(y,y)} + \frac{(x,y)^2}{(y,y)}\right] \\ &= (y,y)\left(x - y\frac{(x,y)}{(y,y)}, x - y\frac{(x,y)}{(y,y)}\right) \geq 0\end{aligned}$$

2. Prove that $\left(\sum_{k=1}^N x_k y_k\right) \leq \left(\sum_{k=1}^N \lambda_k x_k^2\right)\left(\sum_{k=1}^N y_k^2/\lambda_k\right)$ if the λ_k are positive.

3. Hence, show that $(x,Ax)(y,A^{-1}y) \geq (x,y)^2$ if A is positive definite. Establish the result without employing the diagonal form.

4. Hölder Inequality. As an extension of (1), let us prove

Theorem 3. *Let $p > 1$, and $q = p/(p-1)$; then*

$$\sum_{k=1}^N x_k y_k \leq \left(\sum_{k=1}^N x_k^p\right)^{1/p}\left(\sum_{k=1}^N y_k^q\right)^{1/q} \tag{1}$$

if $x_k, y_k \geq 0$.

Proof. Consider the curve

$$v = u^{p-1} \tag{2}$$

where $p > 1$.

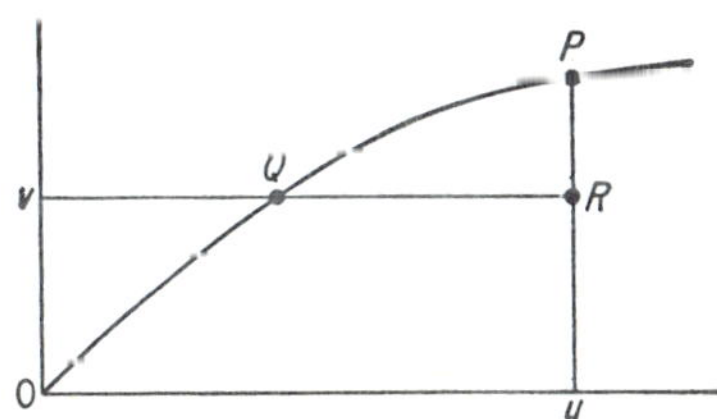

It is clear that the area of the rectangle $OvRu$ is less than or equal to the sum of the areas OPu and OQv,

$$uv \le \int_0^u u_1^{p-1}\,du_1 + \int_0^v v_1^{1/(p-1)}\,dv_1 \tag{3}$$

with equality only if $v = u^{p-1}$. Hence, if $u, v \ge 0$, $p > 1$, we have

$$uv \le \frac{u^p}{p} + \frac{v^q}{q} \tag{4}$$

Now set successively

$$\begin{aligned} u &= x_k \Big/ \Big(\sum_{k=1}^N x_k^p\Big)^{1/p} \\ v &= y_k \Big/ \Big(\sum_{k=1}^N y_k^q\Big)^{1/q} \end{aligned} \tag{5}$$

$k = 1, 2, \ldots, N$, and sum over k.

The result is

$$\frac{\sum_{k=1}^N x_k y_k}{\Big(\sum_{k=1}^N x_k^p\Big)^{1/p}\Big(\sum_{k=1}^N y_k^q\Big)^{1/q}} \le \frac{1}{p}\frac{\sum_{k=1}^N x_k^p}{\sum_{k=1}^N x_k^p} + \frac{1}{q}\frac{\sum_{k=1}^N y_k^q}{\sum_{k=1}^N y_k^q} \tag{6}$$

Since $1/p + 1/q = 1$, the result is as stated in (1).

Setting

$$u = \frac{f(x)}{\Big(\int_R f(x)^p\,dV\Big)^{1/p}} \qquad v = \frac{g(x)}{\Big(\int_R g(x)^q\,dV\Big)^{1/q}} \tag{7}$$

the inequality in (4) yields

$$\frac{f(x)g(x)}{\Big(\int_R f(x)^p\,dV\Big)^{1/p}\Big(\int_R g(x)^q\,dV\Big)^{1/q}} \le \frac{1}{p}\frac{f(x)^p}{\int_R f(x)^p\,dV} + \frac{1}{q}\frac{g(x)^q}{\int_R g(x)^q\,dV}. \tag{8}$$

Integrating over R, the result is

$$\int_R f(x)g(x)\,dV \le \Big(\int_R f(x)^p\,dV\Big)^{1/p}\Big(\int_R g(x)^q\,dV\Big)^{1/q} \tag{9}$$

valid when $f(x), g(x) \ge 0$ and the integrals on the right exist.

5. Concavity of $|A|$. Let us now derive some consequences of the integral identity of (1.1). The first is

Theorem 4. *If A and B are positive definite, then*

$$|\lambda A + (1-\lambda)B| \ge |A|^\lambda |B|^{1-\lambda} \tag{1}$$

for $0 \le \lambda \le 1$.

Proof. We have

$$\frac{\pi^{N/2}}{|\lambda A + (1-\lambda)B|^{1/2}} = \int_{-\infty}^{\infty} \cdots \int_{-\infty}^{\infty} e^{-\lambda(x,Ax)-(1-\lambda)(x,Bx)}\, dx \tag{2}$$

Let us now use the integral form of Hölder's inequality given in (4.9) with $p = 1/\lambda$, $q = 1/(1-\lambda)$. Then

$$\frac{\pi^{N/2}}{|\lambda A + (1-\lambda)B|^{1/2}} \leq \left(\int_{-\infty}^{\infty} \cdots \int_{-\infty}^{\infty} e^{-(x,Ax)}\, dx\right)^{\lambda} \cdot \left(\int_{-\infty}^{\infty} \cdots \int_{-\infty}^{\infty} e^{-(x,Bx)}\, dx\right)^{(1-\lambda)}$$

$$\leq \frac{\pi^{N\lambda/2}}{|A|^{\lambda/2}} \frac{\pi^{N(1-\lambda)/2}}{|B|^{(1-\lambda)/2}} \tag{3}$$

The result when simplified is precisely (1), which is a special case of a more general result we shall derive below.

EXERCISES

1. Prove that $|A + iB| \geq |A|$ if A is positive definite and B is real symmetric.
2. Show that $\lambda x + (1-\lambda)y \geq x^{\lambda}y^{1-\lambda}$ for $x, y \geq 0$, $0 \leq \lambda \leq 1$.

6. A Useful Inequality. A result we shall use below is given in Theorem 5.

Theorem 5. *If A is positive definite, then*

$$|A| \leq a_{11}a_{22} \cdots a_{NN} \tag{1}$$

Let us give two proofs.

First Proof. We have

$$|A| = a_{11} \begin{vmatrix} a_{22} & \cdots & a_{2N} \\ a_{32} & \cdots & a_{3N} \\ \vdots & & \\ a_{N2} & \cdots & a_{NN} \end{vmatrix} + \begin{vmatrix} 0 & a_{12} & \cdots & a_{1N} \\ a_{12} & a_{22} & \cdots & a_{2N} \\ \vdots & \vdots & & \vdots \\ a_{N1} & a_{N2} & \cdots & a_{NN} \end{vmatrix} \tag{2}$$

Since A is positive definite, the matrix (a_{ij}), $i, j = 2, \ldots, N$, is positive definite; hence the quadratic form in $a_{12}, a_{13}, \ldots, a_{1N}$ appearing as the second term on the right in (2) is negative definite. Thus

$$|A| \leq a_{11} \begin{vmatrix} a_{22} & \cdots & a_{2N} \\ a_{32} & \cdots & a_{3N} \\ \vdots & & \\ a_{N2} & \cdots & a_{NN} \end{vmatrix} \tag{3}$$

whence the result follows inductively.

Second Proof. Consider the integral in (1.1) for the case $N = 3$. Replacing x_1 by $-x_1$ and adding, we have

$$\frac{\pi^{3/2}}{|A|^{1/2}} = \int_{-\infty}^{\infty} e^{-a_{22}x_2^2 - 2a_{23}x_2x_3 - a_{33}x_3^2} e^{-a_{11}x_1^2} \left(\frac{z + z^{-1}}{2}\right) dx_1\, dx_2\, dx_3 \tag{4}$$

where

$$z = e^{-2a_{13}x_1x_3 - 2a_{12}x_1x_2} \tag{5}$$

Since $z + z^{-1} \geq 2$ for all $z \geq 0$, we have

$$\begin{aligned} \frac{\pi^{3/2}}{|A|^{1/2}} &\geq \left(\int_{-\infty}^{\infty} e^{-(a_{22}x_2^2 + 2a_{23}x_2x_3 + a_{33}x_3^2)}\, dx_2\, dx_3\right)\left(\int_{-\infty}^{\infty} e^{-a_{11}x_1^2}\, dx_1\right) \\ &\geq \frac{\pi^{1/2}}{|a_{11}|^{1/2}} \frac{\pi}{|A_2|^{1/2}} \end{aligned} \tag{6}$$

where $A_2 = (a_{ij})$, $i, j = 2, 3$. Hence,

$$|a_{11}|^{1/2} |A_2|^{1/2} \geq |A|^{1/2} \tag{7}$$

whence the result for $N = 3$ follows inductively. The inequality for general N is derived similarly.

EXERCISE

1. Let $D_1 = |a_{ij}|$, $i, j = 1, 2, \ldots, n_1$, $D_2 = |a_{ij}|$, $i, j = n_1 + 1, \ldots, n_2$, $D_r = |a_{ij}|$, $i, j = n_{r-1} + 1, \ldots, N$. Then

$$|A| \leq D_1 D_2 \cdots D_r$$

7. Hadamard's Inequality. The most famous determinantal inequality is due to Hadamard.

Theorem 6. *Let B be an arbitrary nonsingular real square matrix. Then*

$$|B|^2 \leq \prod_{i=1}^{N} \left(\sum_{k=1}^{N} b_{ik}^2\right) \tag{1}$$

Proof. Apply Theorem 5 to the positive definite square matrix $A = BB'$.

8. Concavity of $\lambda_N \lambda_{N-1} \cdots \lambda_k$. Now define the matrix function

$$|A|_k = \lambda_N \lambda_{N-1} \cdots \lambda_k \tag{1}$$

Then we may state Theorem 7.

Theorem 7. *If A and B are positive definite, we have*

$$|\lambda A + (1 - \lambda) B|_k \geq |A|_k^{\lambda} |B|_k^{1-\lambda} \tag{2}$$

for $0 \leq \lambda \leq 1$, $k = 1, 2, \ldots, N$.

The proof follows from Theorem 6 of Chap. 7 in exactly the same way that Theorem 4 followed from (1.1).

EXERCISES

1. Show that

$$\phi(A) = \frac{|A|}{|A_i|} = \min_x (x,Ax)$$

where x is constrained by the condition that $x_i = 1$, and A_i is the matrix obtained by deleting the ith row and ith column of A.

2. Hence, show that

$$\phi[\lambda A + (1-\lambda)B] \geq \phi(A)^{\lambda}\phi(B)^{(1-\lambda)}$$

for $0 \leq \lambda \leq 1$ (*Bergstrom's inequality*).

3. Let A and B be two positive definite matrices of order N and let $C = \lambda A + (1-\lambda)B$, $0 \leq \lambda \leq 1$. For each $j = 1, 2, \ldots, N$, let A^j denote the submatrix of A obtained by deleting the first $(j-1)$ rows and columns. If $k_1, k_2, \ldots, k_N$ are N real numbers such that $\sum_{i=1}^{j} k_i \geq 0$, then

$$\prod_{j=1}^{N} |C^j|^{k_j} \geq \prod_{j=1}^{N} |A^j|^{\lambda k_j}|B^j|^{(1-\lambda)k_j} \qquad (Ky\ Fan)$$

4. Establish Hadamard's inequality for Hermitian matrices.

5. If A is a positive definite $N \times N$ matrix and P_k denotes the product of all principal k-rowed minors of A, then

$$P_1 \geq P_2^{1/\binom{N-1}{1}} \geq P_3^{1/\binom{N-1}{2}} \geq P_{N-1}^{1/\binom{N-1}{N-2}} \geq P_N \qquad (Szasz)$$

See L. Mirsky.[1]

6. Let A_p denote the principal submatrix of A formed by the first p rows and p columns, and let B_p and C_p have similar meanings. Then

$$\left(\frac{|C|}{|C_p|}\right)^{1/(N-p)} \geq \left(\frac{|A|}{|A_p|}\right)^{1/(N-p)}\left(\frac{|B|}{|B_p|}\right)^{1/(N-p)} \qquad (Ky\ Fan)$$

9. Additive Inequalities from Multiplicative. Let us now give an example of a metamathematical principle which asserts that every multiplicative inequality has an associated additive analogue. Consider Theorem 7 applied to the matrices

$$A = I + \epsilon X \qquad B = I + \epsilon Y \qquad \epsilon > 0 \tag{1}$$

where X and Y are now merely restricted to be symmetric. If ϵ is sufficiently small, A and B will be positive definite.

Let $x_1 \geq x_2 \geq \cdots \geq x_N$, $y_1 \geq y_2 \geq \cdots \geq y_N$ be the characteristic roots of X and Y, respectively, and $z_1 \geq z_2 \geq \cdots \geq z_N$ the character-

[1] L. Mirsky, On a Generalization of Hadamard's Determinantal Inequality Due to Szasz, *Arch. Math.*, vol. VIII, pp. 274–275, 1957.

istic roots of $\lambda X + (1 - \lambda)Y$. Then Theorem 7 yields the result

$$\begin{aligned}(1 + \epsilon z_N)(1 + \epsilon z_{N-1}) \cdots (1 + \epsilon z_k) &\\ \geq [(1 + \epsilon x_N)(1 + \epsilon x_{N-1}) &\cdots (1 + \epsilon x_k)]^{\lambda} \\ \cdot [(1 + \epsilon y_N)(1 + \epsilon y_{N-1}) &\cdots (1 + \epsilon y_k)]^{1-\lambda} \end{aligned} \quad (2)$$

To first-order terms in ϵ, this relation yields

$$\begin{aligned}1 + \epsilon(z_N + z_{N-1} + \cdots + z_k) \geq 1 + \lambda\epsilon(x_N + x_{N-1} + \cdots + x_k) &\\ + (1 - \lambda)\epsilon(y_N + y_{N-1} + \cdots + y_k) + 0(\epsilon^2)&\end{aligned} \quad (3)$$

Letting $\epsilon \to 0$, we obtain

$$\begin{aligned}z_N + z_{N-1} + \cdots + z_k \geq (x_N + \cdots + x_k) &\\ + (y_N + y_{N-1} + \cdots + y_k)&\end{aligned} \quad (4)$$

Let us state this result, due to Ky Fan, in the following form.

Theorem 8. *Let us define the matrix function*

$$S_k(A) = \lambda_N + \lambda_{N-1} + \cdots + \lambda_k \quad (5)$$

for a symmetric matrix A. Then

$$S_k[\lambda A + (1 - \lambda)B] \geq \lambda S_k(A) + (1 - \lambda)S_k(B) \quad (6)$$

for $0 \leq \lambda \leq 1$, $k = 1, 2, \ldots, N$.

From this follows, upon replacing A by $-A$,

Theorem 9. *If*

$$T_k(A) = \lambda_1 + \lambda_2 + \cdots + \lambda_k \quad (7)$$

where A is symmetric, then

$$T_k[\lambda A + (1 - \lambda)B] \leq \lambda T_k(A) + (1 - \lambda)T_k(B) \quad (8)$$

for $0 \leq \lambda \leq 1$, $k = 1, 2, \ldots, N$.

10. An Alternate Route. Let us obtain these results in a different fashion, one that will allow us to exhibit some results of independent interest. We begin by demonstrating Theorem 10.

Theorem 10. *We have, if A is positive definite*

$$\begin{aligned}\lambda_1\lambda_2 \cdots \lambda_{N-k+1} &= \max_R |(z^i, Az^j)| \\ \lambda_N\lambda_{N-1} \cdots \lambda_k &= \min_R |(z^i, Az^j)|\end{aligned} \quad (1)$$

where R is the z region defined by

$$(z^i, z^j) = \delta_{ij} \qquad i, j = 1, 2, \ldots, N - k + 1 \quad (2)$$

In other words, the minimization is over all sets of $N - k + 1$ *orthonormal vectors.*

Proof. Consider the determinant

$$D_2(x,y) = \begin{vmatrix} (x,Ax) & (x,Ay) \\ (x,Ay) & (y,Ay) \end{vmatrix} = \begin{vmatrix} \sum_{k=1}^{N} \lambda_k u_k^2 & \sum_{k=1}^{N} \lambda_k u_k v_k \\ \sum_{k=1}^{N} \lambda_k u_k v_k & \sum_{k=1}^{N} \lambda_k v_k^2 \end{vmatrix} \tag{3}$$

upon setting $x = \sum_{k=1}^{N} u_k x^k$, $y = \sum_{k=1}^{N} v_k x^k$, where the x^k are the characteristic vectors of A.

The identity of Lagrange yields

$$D_2(x,y) = \sum_{i,j=1}^{N} \lambda_i \lambda_j (u_i v_j - u_j v_i)^2 \tag{4}$$

No terms with $i = j$ enter, since then $u_i v_j - u_j v_i = 0$.

It follows that

$$\lambda_1 \lambda_2 \sum_{i,j=1}^{N} (u_i v_j - u_j v_i)^2 \geq D_2(x,y)$$
$$\geq \lambda_N \lambda_{N-1} \sum_{i,j=1}^{N} (u_i v_j - u_j v_i)^2 \tag{5}$$

Hence

$$\lambda_1 \lambda_2 \begin{vmatrix} \sum_{i=1}^{N} u_i^2 & \sum_{i=1}^{N} u_i v_i \\ \sum_{i=1}^{N} u_i v_i & \sum_{i=1}^{N} v_i^2 \end{vmatrix} \geq D_2(x,y) \geq \lambda_N \lambda_{N-1} \begin{vmatrix} \sum_{i=1}^{N} u_i^2 & \sum_{i=1}^{N} u_i v_i \\ \sum_{i=1}^{N} u_i v_i & \sum_{i=1}^{N} v_i^2 \end{vmatrix} \tag{6}$$

or

$$\lambda_1 \lambda_2 \geq D_2(x,y) \geq \lambda_N \lambda_{N-1} \tag{7}$$

This proves the result for $k = 1$. To derive the general result, we use the identity for the Gramian given in Sec. 5 of Chap. 3, rather than the Lagrangian identity used above.

EXERCISE

1. Use Theorem 9, together with Theorem 4, to establish Theorem 7.

11. A Simpler Expression for $\lambda_N \lambda_{N-1} \cdots \lambda_k$. Since determinants are relatively complicated functions, let us obtain a simpler representation than that given in Theorem 10. We wish to prove Theorem 11.

Theorem 11. *If A is positive definite,*

$$\lambda_N\lambda_{N-1} \cdots \lambda_k = \min_R\, (z^1,Az^1)(z^2,Az^2) \cdots (z^{N-k+1},Az^{N-k+1}) \tag{1}$$

where R is as defined in (10.2).

Proof. We know that (z^i,Az^j), $i, j = 1, 2, \ldots, N - k + 1$, is positive definite if A is positive definite. Hence, Theorem 5 yields

$$|(z^i,Az^j)| \leq \prod_{i=1}^{N-k+1} (z^i,Az^i) \tag{2}$$

Thus

$$\min_R\, |(z^i,Az^j)| \leq \min_R \prod_{i=1}^{N-k+1} (z^i,Az^i) \tag{3}$$

Choosing $z^i = x^i$, $i = N, N-1, \ldots, N-k+1$, we see that

$$\min_R \prod_{i=1}^{N-k+1} (z^i,Az^i) \leq \lambda_N\lambda_{N-1} \cdots \lambda_k \tag{4}$$

This result combined with Theorem 9 yields Theorem 10.

EXERCISES

1. Prove Theorem 7, using the foregoing result.

2. Consider the case $k = N$ and specialize z^1 so as to show that

$$\lambda_N \leq \min_i a_{ii}$$

3. Considering the general case, show that

$$\lambda_N\lambda_{N-1} \cdots \lambda_k \leq a_{NN}a_{N-1,N-1} \cdots a_{kk}$$

4. Show that $\min_B \operatorname{tr}(AB)/N = |A|^{1/N}$, where the minimum is taken over all B satisfying the conditions $B \geq 0$, $|B| = 1$, and A is positive definite.

5. Hence, show that $|A + B|^{1/N} \geq |A|^{1/N} + |B|^{1/N}$ if A and B are positive definite (*A. Minkowski*).

12. Arithmetic-Geometric Mean Inequality. For our subsequent purposes, we require the fundamental arithmetic-geometric mean inequality.

Theorem 12. *If* $x_i \geq 0$, *then*

$$\sum_{i=1}^{N} x_i/N \geq (x_1x_2 \cdots x_N)^{1/N} \tag{1}$$

Proof. Although the result can easily be established via calculus, the following proof is more along the lines we have been pursuing.† Starting

† This proof is due to Cauchy.

with the relation $(a_1^{1/2} - a_2^{1/2})^2 \geq 0$, or

$$\frac{a_1 + a_2}{2} \geq (a_1 a_2)^{1/2} \tag{2}$$

the result for $N = 2$, we set

$$a_1 = (b_1 + b_2)/2 \qquad a_2 = (b_3 + b_4)/2 \tag{3}$$

The result is

$$\begin{aligned} \frac{b_1 + b_2 + b_3 + b_4}{4} &\geq \left(\frac{b_1 + b_2}{2}\right)^{1/2} \left(\frac{b_3 + b_4}{2}\right)^{1/2} \\ &\geq (b_1 b_2 b_3 b_4)^{1/4} \end{aligned} \tag{4}$$

Continuing in this fashion, we see that (1) is valid for N a power of 2. To complete the proof, we use a backward induction, namely, we show that the result is valid for $N - 1$ if it holds for N. To do this, set

$$y_1 = x_2, \ldots, y_{N-1} = x_{N-1} \qquad y_N = \frac{x_1 + x_2 + \cdots + x_{N-1}}{N - 1} \tag{5}$$

and substitute in (1). The result is the same inequality for $N - 1$.

Examining the steps of the proof, we see that there is strict inequality unless $x_1 = x_2 = \cdots = x_N$.

13. Multiplicative Inequalities from Additive. In a preceding section, we showed how to deduce additive inequalities from multiplicative results. Let us now indicate how the converse deduction can be made.

Theorem 13. *The inequality*

$$\sum_{i=k}^{N} \lambda_i \leq \sum_{i=k}^{N} (x^i, Ax^i) \tag{1}$$

$k = 1, 2, \ldots, N$, *valid for any set of orthonormal vectors* $\{x^i\}$, *yields the inequality*

$$\prod_{i=k}^{N} \lambda_i \leq \prod_{i=k}^{N} (x^i, Ax^i) \tag{2}$$

Proof. Consider the sum

$$\begin{aligned} c_k\lambda_k + c_{k+1}\lambda_{k+1} + \cdots + c_N\lambda_N &= c_k(\lambda_k + \lambda_{k+1} + \cdots + \lambda_N) \\ &\quad + (c_{k+1} - c_k)(\lambda_{k+1} + \cdots + \lambda_N) + \cdots + (c_N - c_{N-1})\lambda_N \end{aligned} \tag{3}$$

If we impose the restriction that

$$0 \leq c_k \leq c_{k+1} \leq \cdots \leq c_N \tag{4}$$

we have, upon using (1), the inequality

$$\begin{aligned} c_k\lambda_k + c_{k+1}\lambda_{k+1} + \cdots + c_N\lambda_N &\leq c_k(x^k, Ax^k) \\ &\quad + c_{k+1}(x^{k+1}, Ax^{k+1}) + \cdots + c_N(x^N, Ax^N) \end{aligned} \tag{5}$$

for any set of orthonormal vectors $\{x^i\}$ and scalar constants satisfying (2).

We deduce from (5) that

$$\min_R \left(\sum_{i=k}^{N} c_i\lambda_i \right) \leq \min_R \left[\sum_{i=k}^{N} c_i(x^i,Ax^i) \right] \tag{6}$$

where R is the region in c space defined by (4).

Now it follows from Theorem 12 that

$$\frac{1}{N-k+1} \sum_{i=k}^{N} c_i\lambda_i \geq (c_k c_{k+1} \cdots c_N)^{1/(N-1+1)} (\lambda_k \lambda_{k+1} \cdots \lambda_N)^{1/(N-k+1)} \tag{7}$$

with equality only if all $\lambda_i c_i$ are equal, whence

$$c_i = (\lambda_k \lambda_{k+1} \cdots \lambda_N)^{1/(N-k+1)}/\lambda_i \tag{8}$$

Since $\lambda_k \geq \lambda_{k+1} \geq \cdots \geq \lambda_N$, it follows that $0 < c_k \leq c_{k+1} \leq \cdots \leq c_N$.

This means that the minimum over all c_i is equal to the minimum over R for the left side of (6). Let us temporarily restrict ourselves to orthonormal vectors $\{x^i\}$ satisfying the restriction

$$(x^k,Ax^k) \leq (x^{k+1},Ax^{k+1}) \leq \cdots \leq (x^N,Ax^N) \tag{9}$$

Then (6) yields the inequality

$$\lambda_k\lambda_{k+1} \cdots \lambda_N \leq (x^k,Ax^k)(x^{k+1},Ax^{k+1}) \cdots (x^N,Ax^N) \tag{10}$$

for these $\{x^i\}$. However, since it is clear that any set of x^i can be reordered to satisfy (9), the inequality in (10) holds for all orthonormal $\{x^i\}$.

MISCELLANEOUS EXERCISES

1. Let $n(A) = \left(\sum_{i,j} |a_{ij}|^2 \right)^{1/2}$. Show that

$$\left(\sum_{j=1}^{N} |\lambda_j|^{2k} \right)^{1/2} \leq n(A^k) \leq c_0 N^{m-1} \left(\sum_{j=1}^{N} |\lambda_j|^{2k} \right)^{1/2}$$

where m is the maximum multiplicity of any characteristic root of A (*Gautschi*).

2. If λ_i, μ_i, ν_i are the characteristic values arranged in decreasing order of A^*A, B^*B, and $(A+B)^*(A+B)$, respectively, then

$$\sum_{i=1}^{k} \nu_i^{1/2} \leq \sum_{i=1}^{k} \lambda_i^{1/2} + \sum_{i=1}^{k} \mu_i^{1/2} \qquad k = 1, 2, \ldots, N \qquad (Ky\ Fan)$$

Under the same hypothesis, one has

$$(\nu_{i+j-1})^{1/2} \leq \lambda_i^{1/2} + \mu_i^{1/2} \qquad (Ky\ Fan)$$

3. If $H = A + iB$ is a Hermitian matrix, then it is positive definite if and only if the characteristic values of $iA^{-1}B$ are real and less than or equal to 1 (*Robertson-O. Taussky*).

4. If $H = A + iB$ is positive definite, where A and B are real, then $|A| \geq |H|$, with equality if and only if $B = 0$ (*O. Taussky*).

5. If H_1 is a positive definite Hermitian matrix and H_2 is a Hermitian matrix, then $H_1 + H_2$ is positive definite if and only if the characteristic values of $H_1^{-1}H_2$ are all greater than -1 (*Ky Fan-O. Taussky*).

6. Let K_1 be a positive definite matrix and K_2 such that K_1K_2 is Hermitian. Then K_1K_2 is positive definite if and only if all characteristic values of K_2 are real and positive (*Ky Fan-O. Taussky*).

7. If A and B are symmetric matrices, the characteristic roots of $AB - BA$ are pure complex. Hence, show that $\operatorname{tr}((AB)^2) \leq \operatorname{tr}(A^2B^2)$.

8. If A, B are two matrices, the square of the absolute value of any characteristic root of AB is greater than or equal to the product of the minimum characteristic root of AA' by the minimum characteristic root of BB'.

9. Let $H = A + iB$ be positive definite Hermitian; then $|A| > |B|$ (*Robertson*).

10. If $A = B + C$, where B is positive definite and C is skew-symmetric, then $|A| \geq |B|$ (*O. Taussky*).

11. If A is symmetric, A and $I - A$ are non-negative definite, and O is orthogonal, then $|I - AO| > |I - A|$ (*O. Taussky*).

12. Establish Schur's inequality[1]

$$\sum_{i=1}^{N} |\lambda_i|^2 \leq \sum_{i,j} |a_{ij}|^2$$

13. Let A be symmetric and k be the number of zero characteristic roots. Then

$$\sum_{i=1}^{N} \lambda_i = \sum_{i=1}^{N-k} \lambda_i \qquad \sum_{i=1}^{N} \lambda_i^2 = \sum_{i=1}^{N-k} \lambda_i^2$$

where the summation is now over nonzero roots. From the Cauchy inequality

$$\left(\sum_{i=1}^{N-k} \lambda_i\right)^2 \leq (N - k)\left(\sum_{i=1}^{N-k} \lambda_i^2\right)$$

deduce that

$$(\operatorname{tr} A)^2 \leq (N - k) \operatorname{tr}(A^2)$$

whence

$$k \leq \frac{N \operatorname{tr}(A^2) - (\operatorname{tr} A)^2}{\operatorname{tr}(A^2)}$$

14. From the foregoing, conclude that the rank of a symmetric matrix A is greater than or equal to $(\operatorname{tr} A)^2/\operatorname{tr}(A^2)$.

15. Obtain similar relations in terms of $\operatorname{tr}(A^k)$ for $k = 1, 2, 3, \ldots$. (The foregoing trick was introduced by Schnirelman in connection with the problem of representing every integer as a sum of at most a fixed number of primes.)

16. Let λ_N be the smallest characteristic value and λ_1 the largest characteristic value of the positive definite matrix A. Then

$$(x,x)^2 \leq (Ax,x)(A^{-1}x,x) \leq \frac{(\lambda_1 + \lambda_N)^2}{4\lambda_1\lambda_N} (x,x)^2$$

[1] I. Schur, *Math. Ann.*, vol. 66, pp. 488–510, 1909.

This is a special case of more general results given by W. Greub and W. Rheinboldt.[1]

17. If X is positive definite, then $X + X^{-1} \geq 2I$.

18. If X is positive definite, there is a unique Y which minimizes tr (XY^{-1}) subject to the conditions that Y be positive definite and have prescribed diagonal elements.

19. This matrix Y satisfies an equation of the form $X = Y\Lambda Y$, where Λ is a diagonal matrix of positive elements λ_i.

20. The minimum value of tr (XY^{-1}) is $\sum_{j=1}^{N} \lambda_j b_{jj}$ (*P. Whittle*).

21. If B and C are both positive definite, is $BC + CB \geq 2B^{1/2}CB^{1/2}$?

22. If A, B, and C are positive definite, is $(A + B)^{1/2}C(A + B)^{1/2} \leq A^{1/2}CA^{1/2} + B^{1/2}CB^{1/2}$?

23. Let A and B be $N \times N$ complex matrices with the property that $I - A^*A$ and $I - B^*B$ are both positive semidefinite. Then $|I - A^*B|^2 \geq |I - A^*A|\,|I - B^*B|$. (*Hua, Acta Math. Sinica, vol.* 5, *pp.* 463–470, 1955; *see Math. Rev., vol.* 17, *p.* 703, 1956.) For extensions, using the representation of Sec. 10 of Chap. 6, see R. Bellman, Representation Theorems and Inequalities for Hermitian Matrices, *Duke Math. J.*, 1959, and, for others using a different method, M. Marcus, On a Determinantal Inequality, *Amer. Math. Monthly*, vol. 65, pp. 266–268, 1958.

24. Let A, $B \geq 0$. Then tr $((AB)^{2^n+1}) \leq$ tr $((A^2B^2)^{2^n})$, $n = 0, 1, \ldots$ (*Golden*).

25. Let X^+ denote the Penrose-Moore generalized inverse. Define the *parallel sum* of two non-negative definite matrices A and B by the expression

$$A:B = A(A + B)^+B$$

Then $A:B = B:A$, $(A:B)C = A:(B:C)$.

26. If $Ax = \lambda x$, $Bx = \mu x$, then $(A:B)x = (\lambda:\mu)x$.

27. If $a_i, b_i > 0$, then $\left(\sum_i a_i\right):\left(\sum_i b_i\right) \geq \sum_i (a_i:b_i)$ and if A, $B \geq 0$, then tr $(A:B) \leq$ (tr A):(tr B), with equality if and only if $A = c_1B$, c_1 a scalar.

28. If A, $B \geq 0$, then $|A:B| \leq |A|:|B|$.

29. Can one derive the result of Exercise 27 from the result of Exercise 28 and conversely? For the results in Exercises 25–28, and many further results, see W. N. Anderson, Jr., and R. J. Duffin, Series and Parallel Addition of Matrices, *J. Math. Anal. Appl.*, to appear.

30. Let A_1, $A_2 > 0$ and $M(A_1,A_2)$ denote the convex hull of the set of matrices C such that $C \geq A_1$ and $C \geq A_2$. Similarly define $M(A_1,A_2,A_3)$ with $A_3 > 0$. Is it true that $M(A_1,A_2,A_3) = M(M(A_1,A_2),A_3)$, with the obvious definition of the right-hand side?

31. A matrix is said to be *doubly stochastic* if its elements are nonnegative and the row and column sums are equal to one. A conjecture of Van der Waerden then states that per$(A) \geq N!/N^N$ with equality only if $a_{ij} = 1/N$, N the dimension of A. Here, per(A) is the expression obtained from $|A|$ by changing all minus signs into plus signs. Establish the conjecture for $N = 2, 3$. See Marcus and Minc.[2] For an application of this result to some interesting problems in statistical mechanics, see Hammersley.[3]

[1] W. Greub and W. Rheinboldt, On a Generalization of an Inequality of L. V. Kantorovich, *Proc. Am. Math. Soc.*, 1959.

[2] M. Marcus and H. Minc, *Am. Math. Monthly*, vol. 72, pp. 577–591, 1965.

[3] J. M. Hammersley, An Improved Lower Bound for the Multidimensional Dimer Problem, *Proc. Cambridge Phil. Soc.*, vol. 64, pp. 455–463, 1968.

Bibliography and Discussion

§1. The modern theory of inequalities, and the great continuing interest in this field, stems from the classic volume

G. H. Hardy, J. E. Littlewood, and G. Polya, *Inequalities*, Cambridge University Press, New York, 1934.

An account of more recent results, together with work in different directions, will be found in

E. F. Beckenbach and R. Bellman, *Inequalities*, Springer-Verlag, 1961.

A much more extensive account of inequalities pertaining to matrices and characteristic root will be found therein.

The principal results of this chapter have been motivated by papers by Ky Fan,

Ky Fan, Some Inequalities Concerning Positive-definite Matrices, *Proc. Cambridge Phil. Soc.*, vol. 51, pp. 414–421, 1955.

Ky Fan, On a Theorem of Weyl Concerning the Eigenvalues of Linear Transformations, I, *Proc. Natl. Acad. Sci. U.S.*, vol. 35, pp. 652–655, 1949.

Ky Fan, On a Theorem of Weyl Concerning the Eigenvalues of Linear Transformations, II, *Proc. Natl. Acad. Sci. U.S.*, vol. 36, pp. 31–35, 1950.

Ky Fan, Maximum Properties and Inequalities for the Eigenvalues of Completely Continuous Operators, *Proc. Natl. Acad. Sci. U.S.*, vol. 37, pp. 760–766, 1951.

Ky Fan, Problems 4429 and 4430, *Am. Math. Monthly*, vol. 58, p. 194, 1951; Solutions in vol. 60, pp. 48–50, 1953. (Theorems 7 and 11 are to be found in these pages.)

The methods employed are, in the main, quite different. Our aim has been to show the utility of representation theorems in the derivation of inequalities.

§2 to §4. An intensive discussion of these classical inequalities will be found in the book by Hardy, Littlewood, and Polya cited above.

§5. The result appears to be due to Ky Fan; cf. the third and fifth references cited above; cf. also the papers by Oppenheim.

A. Oppenheim, Inequalities Connected with Definite Hermitian Forms, *J. London Math. Soc.*, vol. 5, pp. 114–119, 1930.

A. Oppenheim, Inequalities Connected with Definite Hermitian Forms, II, *Am. Math. Monthly*, vol. 61, pp. 463–466, 1954.

See also further references given there.

Applications of a similar nature, using the representation of Sec. 10 of Chap. 6, may be found in

R. Bellman, Hermitian Matrices and Representation Theorems, *Duke Math. J.*, 1959.

where some results due to Hua are generalized; see

L. K. Hua, Inequalities Involving Determinants, *Acta Math. Sinica*, vol. 5, pp. 463–470, 1955; *Math. Revs.*, vol. 17, p. 703, 1956.

Hua uses a different type of integral representation, derived from the general theory of group representations.

§6. See

E. F. Beckenbach, An Inequality for Definite Hermitian Determinants, *Bull. Am. Math. Soc.*, pp. 325–329, 1929.

§7. Hadamard's inequality is one of the most proved results in analysis with well over one hundred proofs in the literature. The proof here follows:

R. Bellman, Notes on Matrix Theory—II, *Am. Math. Monthly*, vol. LX, pp. 174–175, 1953.

Hadamard has a quite interesting remark concerning his inequality in his fascinating monograph,

J. Hadamard, *The Psychology of Invention in the Mathematical Field*, Princeton University Press, Princeton, N.J., 1949.

For the original result, see

J. Hadamard, Résolution d'une question relative aux déterminants, *Bull. sci. math.*, vol. 2, pp. 240–248, 1893.

For one extension, see

E. Fischer, Über den Hadamardschen Determinantensatz, *Arch. Math. u. Phys.*, (3), vol. 13, pp. 32–40, 1908.

and for a far-reaching extension, see

I. Schur, Über endliche Gruppen und Hermitesche Formen, *Math. Z.*, vol. 1, pp. 184–207, 1918.

See also

J. Williamson, Note on Hadamard's Determinant Theorem, *Bull. Am. Math. Soc.*, vol. 53, pp. 608–613, 1947.

H. S. Wilf, Hadamard Determinants, Mobius Functions and the Chromatic Number of a Graph, *Bull. Am. Math. Soc.*, vol. 74, pp. 960–964, 1968.

§8. For this result and the related Theorem 11, see the fifth reference of Ky Fan given above. The proof here follows

R. Bellman, I. Glicksberg, and O. Gross, Notes on Matrix Theory—VI, *Am. Math. Monthly*, vol. 62, pp. 571–572, 1955.

The result of Exercise 2 is due to Bergstrøm,

H. Bergstrøm, A Triangle Inequality for Matrices, *Den 11te Skandinaviske Matematiker-kongressen, Trondheim*, 1949, *Oslo*, pp. 264–267, 1952.

while the method of proof follows

R. Bellman, Notes on Matrix Theory—IV: An Inequality Due to Bergstrøm, *Am. Math. Monthly*, vol. 62, pp. 172–173, 1955.

The result of Exercise 3 is given in

R. Bellman, Notes on Matrix Theory—IX, *Am. Math. Monthly*, vol. 64, pp. 189–191, 1957.

where two methods are used, one depending upon the generalization of the Ingham-Siegel representation referred to above and the other upon Bergstrøm's inequality.

§9. This section illustrates an important technique for obtaining two results from one. Occasionally, the multiplicative results are easier to obtain ab initio. The result follows

R. Bellman, Notes on Matrix Theory—VII, *Am. Math. Monthly*, vol. 62, pp. 647–648, 1955.

§12. The book by Hardy, Littlewood, and Polya contains a history of the arithmetic-geometric mean inequality together with a large number of proofs. The one given here is our favorite because of its dependence upon *backward* induction.

§13. We have omitted the proof of inequality (1) which is given in Ky Fan's second paper. The present derivation of (2) from (1) is given in

R. Bellman, Notes on Matrix Theory—XV: Multiplicative Properties from Additive Properties, *Am. Math. Monthly*, vol. 65, pp. 693–694, 1958.

The topic of inequalities for various functions of characteristic values has been thoroughly explored by a number of authors; cf. the following articles where a large number of additional references will be found.

A. Ostrowski, Sur quelques applications des fonctions convexes et concaves au sens de I. Schur, *J. math. pures et appl.*, vol. 31, no. 9, pp. 253–292, 1952.

M. D. Marcus and L. Lopes, Inequalities for Symmetric Functions and Hermitian Matrices, *Can. J. Math.*, vol. 8, pp. 524–531, 1956.

M. D. Marcus and J. L. McGregor, Extremal Properties of Hermitian Matrices, *Can. J. Math.*, vol. 8, 1956.

M. Marcus and B. N. Moyls, *Extreme Value Properties of Hermitian Matrices*, Department of Mathematics, University of British Columbia, Vancouver, Canada, 1956.

M. Marcus, B. N. Moyls, and R. Westwick, Some Extreme Value Results for Indefinite Hermitian Metrics II, *Illinois J. Math.*, vol. 2, pp. 408–414, 1958.

We have omitted all discussion of the interesting and important question of inequalities for the characteristic roots in terms of the elements of A. Thorough discussions of this field may be found in the papers by

A. Brauer, Limits for the Characteristic Roots of a Matrix, IV, *Duke Math. J.*, vol. 19, pp. 75–91, 1952.

E. T. Browne, Limits to the Characteristic Roots of Matrices, *Am. Math. Monthly*, vol. 46, pp. 252–265, 1939.

W. V. Parker, The Characteristic Roots of Matrices, *Duke Math. J.*, vol. 12, pp. 519–526, 1945.

Next, let us mention the paper

N. G. DeBruijn and D. Van Dantzig, Inequalities Concerning Minors and Eigenvalues, *Nieuw. Arch. Wisk.*, vol. 4, no. 3, pp. 18–35, 1956.

where a systematic derivation of a large number of determinantal inequalities will be found.

Next, a number of interesting matrix inequalities are given in the papers by Masani and Wiener and by Helson and Lowdenslager referred to at the end of Chap. 11, and in

Ky Fan and A. J. Hoffman, Some Metric Inequalities in the Space of Matrices, *Proc. Amer. Math. Soc.*, vol. 6, pp. 111–116, 1955.

Finally, see

O. Taussky, Bibliography on Bounds for Characteristic Roots of Finite Matrices, *National Bureau of Standards Rept.* 1162, September, 1951.

O. Taussky, Positive-Definite Matrices and Their Role in the Study of the Characteristic Roots of General Matrices, *Advances in Mathematics*, vol. 2, pp. 175–186, Academic Press Inc., New York, 1968.

A recent paper containing inequalities quite different from those described above is

L. Mirsky, Inequalities for Normal and Hermitian Matrices, *Duke Math. J.*, vol. 24, pp. 591–599, 1957.

See also

E. V. Haynsworth, Note on Bounds for Certain Determinants, *Duke Math. J.*, vol. 24, pp. 313–320, 1957.

M. Marcus and H. Minc, Extensions of Classical Matrix Inequalities, *Lin. Algebra Appl.*, vol. 1, pp. 421–444, 1968.

H. J. Ryser, Inequalities of Compound and Induced Matrices with Applications to Combinatorial Analysis, *Illinois J. Math.*, vol. 2, pp. 240–253, 1958.

H. J. Ryser, Compound and Induced Matrices in Combinatorial Analysis, *Proc. Symp. Appl. Math., Combin. Anal.*, pp. 149–167, 1960.

R. C. Thompson, Principal Submatrices of Normal and Hermitian Matrices, *Illinois J. Math.*, vol. 10, pp. 296–308, 1966.

R. C. Thompson and P. McEnteggert, Principal Submatrices, II: The Upper and Lower Quadratic Inequalities, *Lin. Algebra Appl.*, vol. 1, pp. 211–243, 1968.

9

Dynamic Programming

1. Introduction. In Chap. 1, we encountered quadratic forms in connection with maximization and minimization problems, and observed that systems of linear equations arose in connection with the determination of the extrema of quadratic functions.

Although simple conceptually, the solution of linear equations by means of determinants is not feasible computationally for reasons we have discussed in previous pages. Consequently, we must devise other types of algorithms to compute the solution. This suggests that it might be well to develop algorithms connected directly with the original maximization problem without considering the intermediate linear equations at all.

In this chapter, we shall study a number of problems in which this can be done. Our basic tool will be the functional equation technique of dynamic programming.

2. A Problem of Minimum Deviation. Given a sequence of real numbers $\{a_k\}$, $k = 1, 2, \ldots, N$, we wish to determine another sequence $\{x_k\}$, $k = 1, 2, \ldots, N$, "close" to the a_k and "close" to each other. Specifically, we wish to minimize the quadratic expression

$$Q(x) = \sum_{k=1}^{N} c_k(x_k - x_{k-1})^2 + \sum_{k=1}^{N} d_k(x_k - a_k)^2 \tag{1}$$

where c_k and d_k are prescribed positive quantities, and x_0 is a given constant.

Proceeding in the usual fashion, taking partial derivatives, we obtain the system of linear equations

$$\begin{aligned}
&c_1(x_1 - x_0) - c_2(x_2 - x_1) + d_1(x_1 - a_1) = 0 \\
&\cdot \\
&\cdot \\
&\cdot \\
&c_k(x_k - x_{k-1}) - c_{k+1}(x_{k+1} - x_k) + d_k(x_k - a_k) = 0 \qquad (2) \\
&\cdot \\
&\cdot \\
&\cdot \\
&c_N(x_N - x_{N-1}) + d_N(x_N - a_N) = 0
\end{aligned}$$

If N is large, the solution of this system requires some care.

In place of using any of a number of existing techniques to resolve this system of equations, we shall pursue an entirely different course.

3. Functional Equations. Let us consider the sequence of minimization problems:

Minimize over all x_k

$$Q_r(x) = \sum_{k=r}^{N} c_k(x_k - x_{k-1})^2 + \sum_{k=r}^{N} d_k(x_k - a_k)^2 \tag{1}$$

with $x_{r-1} = u$, a given quantity.

It is clear that the minimum value depends upon u, and, of course, upon r. Let us then introduce the sequence of functions $\{f_r(u)\}$, where $-\infty < u < \infty$ and $r = 1, 2, \ldots, N$, defined by the relation

$$f_r(u) = \min_x Q_r(x) \tag{2}$$

We see that

$$f_N(u) = \min_{x_N} [c_N(x_N - u)^2 + d_N(x_N - a_N)^2] \tag{3}$$

a function which can be readily determined.

To derive a relation connecting $f_r(u)$ with $f_{r+1}(u)$, for $r = 1, 2, \ldots, N - 1$, we proceed as follows:

$$\begin{aligned} f_r(u) &= \min_{x_r} \min_{x_{r+1}} \cdots \min_{x_N} \left[\sum_{k=r}^{N} c_k(x_k - x_{k-1})^2 + \sum_{k=r}^{N} d_k(x_k - a_k)^2 \right] \\ &= \min_{x_r} \Bigg\{ c_r(x_r - u)^2 + d_r(x_r - a_r)^2 \\ &\quad + \min_{x_{r+1}} \cdots \min_{x_N} \left[\sum_{k=r+1}^{N} c_k(x_k - x_{k-1})^2 + \sum_{k=r+1}^{N} d_k(x_k - a_k)^2 \right] \Bigg\} \\ &= \min_{x_r} [c_r(x_r - u)^2 + d_r(x_r - a_r)^2 + f_{r+1}(x_r)] \end{aligned} \tag{4}$$

Since $f_N(u)$ is determined by (3), this recurrence relation in principle determines $f_{N-1}(u)$, $f_{N-2}(u)$, and so on, back to $f_1(u)$, the function we originally wanted.

Using a digital computer, this series of equations can be used to compute the members of the sequence $\{f_r(u)\}$.

In place of solving a particular problem for a single value of N, our aim has been to imbed this variational problem within a family of problems of the same general type. Even though an individual problem may be complex, it may be quite simply related to others members of the family. This turns out to be the case in a large class of variational questions of which the one treated above is a quite special example.

4. Recurrence Relations. Making use of the analytic structure of each member of the sequence $f_r(u)$, we can do much better. Let us begin by proving inductively that each member of the sequence is a quadratic function of u, having the form

$$f_r(u) = u_r + v_r u + w_r u^2 \tag{1}$$

where u_r, v_r, and w_r are constants, dependent on r, but not on u.

The result is easily seen to be true for $r = N$, since the minimum value in (3.1) is furnished by the value

$$x_N = \frac{c_N u + d_N a_N}{c_N + d_N} \tag{2}$$

Using this value of x_N in the right-hand side of (3.3), we see that $f_N(u)$ is indeed quadratic in u.

The same analysis shows that $f_{N-1}(u)$, and so inductively that each function, is quadratic. This means that the functions $f_r(u)$ are determined once the sequence of coefficients $\{u_r,v_r,w_r\}$ has been found. Since we know u_N, v_N, and w_N, it suffices to obtain recurrence relations connecting u_r, v_r, and w_r with u_{r+1}, v_{r+1}, w_{r+1}. To accomplish this, we turn to (3.4) and write

$$u_r + v_r u + w_r u^2 = \min_{x_r} [c_r(x_r - u)^2 + d_r(x_r - a_r)^2 + u_{r+1} + v_{r+1}x_r + w_{r+1}x_r^2] \tag{3}$$

The minimizing value of x_r is seen to be given by

$$(c_r + d_r + w_{r+1})x_r = c_r u + d_r a_r - \frac{v_{r+1}}{2} \tag{4}$$

Substituting this value in (3) and equating coefficients of powers of u, we obtain the desired recurrence relations, which are nonlinear. We leave the further details as an exercise for the reader.

5. A More Complicated Example. In Sec. 2, we posed the problem of approximating to a varying sequence $\{a_k\}$ by a sequence $\{x_k\}$ with smaller variation. Let us now pursue this a step further, and consider the problem of minimizing the quadratic function

$$Q(x) = \sum_{k=1}^{N} c_k(x_k - x_{k-1})^2 + \sum_{k=1}^{N} d_k(x_k - a_k)^2 + \sum_{k=1}^{N} e_k(x_k - 2x_{k-1} + x_{k-2})^2 \tag{1}$$

where $x_0 = u$, $x_{-1} = v$.

As above, we introduce the function of two values $f_r(u,v)$ defined by the relation

$$f_r(u,v) = \min_x \Big[\sum_{k=r}^{N} c_k(x_k - x_{k-1})^2 + \sum_{k=r}^{N} d_k(x_k - a_k)^2 + \sum_{k=r}^{N} e_k(x_k - 2x_{k-1} + x_{k-2})^2 \Big] \quad (2)$$

$r = 1, 2, \ldots, N, -\infty < u, v < \infty$, with $x_{r-1} = u, x_{r-2} = v$.

It is easy to see, as above, that

$$f_r(u,v) = \min_{x_r} [c_r(x_r - u)^2 + d_r(x_r - a_r)^2 + e_r(x_r - 2u + v)^2 + f_{r+1}(x_r,u)] \quad (3)$$

$r = 1, 2, \ldots, N$, and that

$$f_r(u,v) = u_r + v_r u + w_r u^2 + v'_r v + w'_r v^2 + z_r uv$$

where the coefficients depend only upon r.

Combining these relations, we readily derive recurrence relations connecting successive members of sequence $\{u_r,v_r,w_r,u'_r,v'_r,z_r\}$. Since we easily obtain the values for $r = N$, we have a simple method for determining the sequence.

6. Sturm-Liouville Problems. A problem of great importance in theoretical physics and applied mathematics is that of determining the values of λ which permit the homogeneous equation

$$\begin{aligned} u'' + \lambda\phi(t)u &= 0 \\ u(0) = u(1) &= 0 \end{aligned} \quad (1)$$

to possess a nontrivial solution. The function $\phi(t)$ is assumed to be real, continuous, and uniformly positive over [0,1].

In Sec. 12 of Chap. 7, we sketched one way of obtaining approximate solutions to this problem. Let us now present another, once again leaving aside the rigorous aspects. Observe, however, that the questions of degree of approximation as $N \to \infty$ are of great importance, since the efficacy of the method rests upon the answers to these questions.

In place of seeking to determine a function $u(t)$ for $0 \leq t < 1$, we attempt to determine a sequence $\{u_k = u(k\Delta)\}$, $k = 0, 1, \ldots, N - 1$, where $N\Delta = 1$. The second derivative u'' is replaced by the second difference

$$\frac{u_{k+1} - 2u_k + u_{k-1}}{\Delta^2} \quad (2)$$

with the result that the differential equation in (1) is replaced by a difference equation which is equivalent to the system of linear homogeneous equations

$$\begin{aligned} u_2 - 2u_1 + \lambda\Delta^2\phi_1 u_1 &= 0 \\ u_3 - 2u_2 + u_1 + \lambda\Delta^2\phi_2 u_2 &= 0 \\ &\cdot \\ &\cdot \\ &\cdot \\ -2u_{N-1} + u_{N-2} + \lambda\Delta^2\phi_{N-1}u_{N-1} &= 0 \end{aligned} \tag{3}$$

where $\phi_k = \phi(k\Delta)$.

We see then that the original problem has been approximated to by a problem of quite familiar type, the problem of determining the characteristic values of the symmetric matrix

$$A = \begin{bmatrix} -2 + \Delta^2\phi_1 & 1 & & & \\ 1 & -2 + \Delta^2\phi_2 & 1 & & \\ & & \cdot & & \\ & & & \cdot & \\ & & & & \cdot \\ & 1 & & -2 + \Delta^2\phi_{N-2} & 1 \\ & & & 1 & -2 + \Delta^2\phi_{N-1} \end{bmatrix} \tag{4}$$

We shall discuss this type of matrix, a particular example of what is called a Jacobi matrix, again below.

It is essential to note that this is merely one technique for reducing the differential equation problem to a matrix problem. There are many others.

EXERCISES

1. Show that the equation in (6.1), together with the boundary condition, is equivalent to the integral equation

$$u = \lambda \int_0^1 K(t,s)\phi(s)u(s)\,ds$$

where

$$\begin{aligned} K(t,s) &= t(1-s) \qquad 1 \geq t \geq s \geq 0 \\ &= s(1-t) \qquad 0 \leq t \leq s \leq 1 \end{aligned}$$

2. Use this representation to obtain another approximating matrix problem of the form

$$u_i = \lambda \sum_{j=1}^{N} k_{ij}\phi_j u_j \qquad i = 1, 2, \ldots, N$$

7. Functional Equations. Since A, as given by (6.4), is real and symmetric, we know that the characteristic values can be determined from the problem:

Minimize

$$\sum_{k=1}^{N} (u_k - u_{k-1})^2 \tag{1}$$

subject to the conditions

$$u_0 = u_N = 0 \tag{2a}$$

$$\sum_{k=1}^{N-1} \phi_k u_k^2 = 1 \tag{2b}$$

In place of this problem, consider the problem of determining the minimum of

$$\sum_{k=r}^{N} (u_k - u_{k-1})^2 \tag{3}$$

subject to the conditions

$$u_{r-1} = v \tag{4a}$$

$$u_N = 0 \tag{4b}$$

$$\sum_{k=r}^{N-1} \phi_k u_k^2 = 1 \tag{4c}$$

for $r = 1, 2, \ldots, N-1$, where $-\infty < v < \infty$.

The minimum is then for each r a function of v, $f_r(v)$. We see that

$$f_{N-1}(v) = (1/\sqrt{\phi_{N-1}} - v)^2 + \left(\frac{1}{\phi_{N-1}}\right) \tag{5}$$

and that, arguing as before, we have the relation

$$f_r(v) = \min_{u_r} [(u_r - v)^2 + (1 - \phi_r u_r^2) f_{r+1}(u_r/\sqrt{1 - \phi_r u_r^2})] \tag{6}$$

for $r = 1, 2, \ldots, N-1$.

The sequence of functions $\{f_r(v)\}$ apparently possesses no simple analytic structure. However, the recurrence relation in (6) can be used to compute the sequence numerically.

EXERCISE

1. Treat in the same fashion the problem of determining the maximum and minimum of

   (a) $(ax_1)^2 + (x_1 + ax_2)^2 + \cdots (x_1 + x_2 + \cdots + x_{N-1} + ax_N)^2$ subject to $x_1^2 + x_2^2 + \cdots + x_N^2 = 1$

   (b) $x_1^2 + (x_1 + ax_2)^2 + \cdots (x_1 + ax_2 + a^2x_3 + \cdots + a^{N-1}x_N)^2$ subject to $x_1^2 + x_2^2 + \cdots + x_N^2 = 1$

   (c) $x_1^2 + (x_1 + ax_2)^2 + [x_1 + ax_2 + (a + b)x_3]^2 + \cdots + \{x_1 + ax_2 + (a + b)x_3 + \cdots + [a + (N-2)b]x_N\}^2$ subject to $x_1^2 + x_2^2 + \cdots + x_N^2 = 1$

For a discussion of these questions by other means, see A. M. Ostrowski, On the Bounds for a One-parameter Family of Matrices, *J. für Math.*, vol. 200, pp. 190–200, 1958.

8. Jacobi Matrices. By a *Jacobi matrix*, we shall mean a matrix with the property that the only nonzero elements appear along the main diagonal and the two contiguous diagonals. Thus

$$a_{ij} = 0 \qquad |i - j| \geq 2 \tag{1}$$

Let us now consider the question of solving the system of equations

$$Ax = c \tag{2}$$

where A is a positive definite Jacobi matrix. As we know, this problem is equivalent to that of minimizing the inhomogeneous form

$$Q(x) = (x,Ax) - 2(c,x) \tag{3}$$

Writing this out, we see that we wish to minimize

$$\begin{aligned} Q(x) = {} & a_{11}x_1^2 + 2a_{12}x_1x_2 + a_{22}x_2^2 + 2a_{23}x_2x_3 + \cdots \\ & + 2a_{N-1,N}x_{N-1}x_N + a_{NN}x_N^2 - 2c_1x_1 - 2c_2x_2 - \cdots - 2c_Nx_N \end{aligned} \tag{4}$$

Consider then the problem of minimizing

$$\begin{aligned} Q_k(x,z) = {} & a_{11}x_1^2 + 2a_{12}x_1x_2 + a_{22}x_2^2 + 2a_{23}x_2x_3 + \cdots \\ & + 2a_{k-1\,k}x_{k-1}x_k + a_{kk}x_k^2 - 2c_1x_1 - 2c_2x_2 - \cdots - 2zx_k \end{aligned} \tag{5}$$

for $k = 1, 2, \ldots$, with

$$Q_1(x,z) = a_{11}x_1^2 - 2zx_1 \tag{6}$$

Define

$$f_k(z) = \min_x Q_k(x,z) \tag{7}$$

Then it is easy to see that

$$f_k(z) = \min_{x_k} [a_{kk}x_k^2 - 2zx_k + f_{k-1}(c_{k-1} - a_{k-1,k}x_k)] \tag{8}$$

for $k = 2, 3, \ldots$.

Once again, we observe that each function $f_k(z)$ is a quadratic in z,

$$f_k(z) = u_k + v_kz + w_kz^2 \qquad k = 1, 2, \ldots \tag{9}$$

with
$$f_1(z) = \frac{z^2}{a_{11}}$$

Substituting in (8), we obtain recurrence relations connecting u_k, v_k, and w_k with u_{k-1}, v_{k-1}, and w_{k-1}. We leave the derivation of these as exercises for the reader.

EXERCISES

1. Generalize the procedure above in the case where A is a matrix with the property that $a_{ij} = 0$, $|j - i| \geq 3$.

2. Obtain relations corresponding to those given in Sec. 7 for the largest and smallest values of a symmetric matrix A with $a_{ij} = 0$, $|j - i| \geq 2$; and $a_{ij} = 0$, $|j - i| \geq 3$.

9. Analytic Continuation. We were able to apply variational techniques to the problem of solving the equation $Ax = c$ at the expense of requiring that A be symmetric and positive definite. It is tempting to assert that the explicit solutions we have found are valid first of all for symmetric matrices which are not necessarily positive definite, provided only that none of the denominators which occur is zero, and then suitably interpreted for not necessarily symmetric matrices.

The argument would proceed along the following lines. The expressions for the x_i are linear functions of the c_i and rational functions of the a_{ij}. An equality of the form

$$\sum_{j=1}^{N} a_{1j}x_j = c_1 \tag{1}$$

valid in the domain of a_{ij} space where A is positive definite should certainly be valid for all a_{ij} which satisfy the symmetry condition.

It is clear, however, that the rigorous presentation of an argument of this type would require a formidable background of the theory of functions of *several* complex variables. Consequently, we shall abandon this path, promising as it is, and pursue another which requires only the rudimentary facts concerning analytic continuation for functions of *one* complex variable.

Consider in place of the symmetric matrix A the matrix $zI + A$, where z is a scalar. If z is chosen to be a sufficiently large positive number, this matrix will be positive definite, and will be a Jacobi matrix if A is a Jacobi matrix.

We can now apply the principle of analytic continuation. The x_i, as given by the formulas derived from the variational technique, are rational functions of z, analytic for the real part of z sufficiently large. Hence, identities valid in this domain must hold for $z = 0$, provided none of the functions appearing has singularities at $z = 0$. These singularities can arise only from zeros of the denominators. In other words, the formulas are valid whenever they make sense.

10. Nonsymmetric Matrices. In order to apply variational techniques to nonsymmetric matrices, we must use a different device.

Consider the expression

$$f(x,y) = (x,Bx) + 2(x,Ay) + (y,By) - 2(a,x) - 2(b,y) \tag{1}$$

where we shall assume that B is symmetric and positive definite and that A is merely real.

Minimizing over x and y, we obtain the variational equations

$$\begin{aligned} Bx + Ay &= a \\ A'x + By &= b \end{aligned} \tag{2}$$

In order to ensure that $f(x,y)$ is positive definite in x and y, we must impose some further conditions upon B. Since we also wish to use analytic continuation, perhaps the simplest way to attain our ends is to choose $B = zI$, where z is a sufficiently large positive quantity.

The functional equation technique can now be invoked as before. We leave it to the reader to work through the details and to carry out the analytic continuation.

11. Complex A. Let us now see what can be done in case A is symmetric, but complex, $A = B + iC$, B and C real. The equation $Ax = c$ takes the form

$$(B + iC)(x + iy) = a + ib \tag{1}$$

or, equating real and complex parts,

$$\begin{aligned} Bx - Cy &= a \\ Cx + By &= b \end{aligned} \tag{2}$$

Observe that these relations establish a natural correspondence between the N-dimensional complex matrix $B + iC$ and the $2N$-dimensional matrix

$$\begin{bmatrix} B & -C \\ C & B \end{bmatrix} \tag{3}$$

a relationship previously noted in the exercises.

In order to relate this matrix, and the system of equations in (2) to a variational problem, we consider the quadratic form

$$f(x,y) = (x,Cx) + 2(x,By) - (y,Cy) - 2(b,x) - 2(a,y) \tag{4}$$

If we impose the condition that C is positive definite, it follows that $f(x,y)$ is convex in x and concave in y and hence, as a consequence of general theorems, which we shall mention again in Chap. 16, that

$$\min_x \max_y f(x,y) = \max_y \min_x f(x,y) \tag{5}$$

Since $f(x,y)$ is a quadratic form, there is no need to invoke these general results since both sides can be calculated explicitly and seen to be equal. We leave this as an exercise for the reader.

Before applying the functional equation technique, we need the further result that the minimum over the x_i and the maximum over the y_i can be taken in any order. In particular,

$$\min_{(x_1,x_2,\ldots,x_N)} \max_{(y_1,y_2,\ldots,y_N)} = \min_{x_1} \max_{y_1} \min_{(x_2,\ldots,x_N)} \max_{(y_2,\ldots,y_N)} \tag{6}$$

This we also leave to the enterprising reader. Problems of this nature are fundamental in the theory of games, which we shall discuss briefly in Chap. 16.

Combining the foregoing results, we can treat the case of complex A. If necessary, we consider $A + izI$, where z is a sufficiently large positive quantity, once again introduced for the purposes of analytic continuation.

12. Slightly Intertwined Systems. Let us now consider the problem of resolving a set of linear equations of the form

$$\begin{aligned}
a_{11}x_1 + a_{12}x_2 + a_{13}x_3 &= c_1 \\
a_{21}x_1 + a_{22}x_2 + a_{23}x_3 &= c_2 \\
a_{31}x_1 + a_{32}x_2 + a_{33}x_3 + b_1x_4 &= c_3 \\
b_1x_3 + a_{44}x_4 + a_{45}x_5 + a_{46}x_6 &= c_4 \\
a_{54}x_4 + a_{55}x_5 + a_{56}x_6 &= c_5 \\
a_{64}x_4 + a_{65}x_5 + a_{66}x_6 + b_2x_7 &= c_6 \\
&\;\;\vdots \\
b_{N-1}x_{3N-3} + a_{3N-2,3N-2}x_{3N-2} + a_{3N-2,3N-1}x_{3N-1} + a_{3N-2,3N}x_{3N} &= c_{3N-2} \\
a_{3N-1,3N-2}x_{3N-2} + a_{3N-1,3N-1}x_{3N-1} + a_{3N-1,3N}x_{3N} &= c_{3N-1} \\
a_{3N,3N-2}x_{3N-2} + a_{3N,3N-1}x_{3N-1} + a_{3N,3N}x_{3N} &= c_{3N}
\end{aligned} \tag{1}$$

If the coefficients b_i were all zero, this problem would separate into N simple three-dimensional problems. It is reasonable to assume that there is some way of utilizing the near-diagonal structure. We shall call matrices of the type formed by the coefficients above *slightly intertwined.* They arise in a variety of physical, engineering, and economic analyses of multicomponent systems in which there is "weak coupling" between different parts of the system.

To simplify the notation, let us introduce the matrices

$$A_k = (a_{i+3k-3,\,j+3k-3}) \qquad i, j = 1, 2, 3 \tag{2}$$

and the vectors

$$x^k = \begin{bmatrix} x_{3k-2} \\ x_{3k-1} \\ x_{3k} \end{bmatrix} \qquad c^k = \begin{bmatrix} c_{3k-2} \\ c_{3k-1} \\ c_{3k} \end{bmatrix} \tag{3}$$

$k = 1, 2, \ldots$.

We shall assume initially that the matrix of coefficients in (1) is a positive definite. Hence, the solution of the linear system is equivalent to determining the minimum of the inhomogeneous quadratic form

$$\begin{aligned}(x^1,Ax^1) + (x^2,Ax^2) + \cdots + (x^N,Ax^N) \\ - 2(c^1,x^1) - 2(c^2,x^2) - \cdots - 2(c^N,x^N) \\ + 2b_1x_3x_4 + 2b_2x_6x_7 + \cdots + 2b_{N-1}x_{3N-3}x_{3N-2}\end{aligned} \tag{4}$$

This we shall attack by means of functional equation techniques. Introduce the sequence of functions $\{f_N(z)\}$, $-\infty < z < \infty$, $N = 1, 2, \ldots$, defined by the relation

$$f_N(z) = \min_{x_i} \left[\sum_{i=1}^{N} (x^i,A_ix^i) - 2 \sum_{i=1}^{N} (c^i,x^i) + 2 \sum_{i=1}^{N-1} b_ix_{1+3i}x_{3i} + 2zx_{3N} \right] \tag{5}$$

Proceeding in the usual fashion, we obtain the recurrence relation

$$f_N(z) = \min_{R_N} [(x^N,A_Nx^N) + 2zx_{3N} - 2(c^N,x^N) + f_{N-1}(b_{N-1}x_{3N-2})] \tag{6}$$

where R_N is three-dimensional region $-\infty < x_{3N}, x_{3N-1}, x_{3N-2} < \infty$.

13. Simplifications—I. We can write (12.6) in the form

$$f_N(z) = \min_{x_{3N-2}} [\min_{x_{3N},x_{3N-1}} [(x^N,A_Nx^N) + 2zx_{3N} - 2(c^N,x^N)] + f_{N-1}(b_{N-1}x_{3N-2})] \tag{1}$$

Introduce the sequence of functions

$$g_N(z,y) = \min_{x_{3N},x_{3N-1}} [(x^N,A^Nx^N) + 2zx_{3N} - 2(c^N,x^N)] \tag{2}$$

where $x_{3N-2} = y$. Then

$$f_N(z) = \min_{y} [g_N(z,y) + f_{N-1}(b_{N-1}y)] \tag{3}$$

a simple one-dimensional recurrence relation.

14. Simplifications—II. We can, as in the section on Jacobi matrices, go even further, if we observe that $f_N(z)$ is for each N a quadratic in z,

$$f_N(z) = u_N + 2v_Nz + w_Nz^2 \tag{1}$$

where u_N, v_N, w_N are independent of z. Using this relation in (3), we readily obtain recurrence relations for $\{u_N,v_N,w_N\}$.

EXERCISES

1. Obtain recurrence relations which enable us to compute the largest and smallest characteristic roots of slightly intertwined matrices.

2. Extend the foregoing technique to the case where the A_k are not all of the same dimension.

15. The Equation $Ax = y$. Let us now pursue a simple theme. If A is a positive definite matrix, we know that the solution of

$$Ax = y \tag{1}$$

given by $x = A^{-1}y$ can also be obtained as the solution of the problem of minimizing the quadratic form

$$Q(x) = (x,Ax) - 2(x,y) \tag{2}$$

The minimum has the value $-(y,A^{-1}y)$. Comparing the two approaches to the problem of solving (1), we can obtain some interesting identities.

Introduce the function of N variables

$$f_N(y_1,y_2, \ldots ,y_N) = \min_{x_i} \left[\sum_{i,j=1}^{N} a_{ij}x_ix_j - 2 \sum_{i=1}^{N} x_iy_i \right] \tag{3}$$

Write

$$\sum_{i,j=1}^{N} a_{ij}x_ix_j - 2 \sum_{i=1}^{N} x_iy_i = a_{NN}x_N^2 + \sum_{i,j=1}^{N-1} a_{ij}x_ix_j - 2 \sum_{i=1}^{N-1} x_i(y_i - a_{iN}x_N) - 2x_Ny_N \tag{4}$$

From this expression, we obtain the functional equation

$$f_N(y_1,y_2, \ldots ,y_N) = \min_{x_N} [a_{NN}x_N^2 - 2x_Ny_N + f_{N-1}(y_1 - a_{1N}x_N, y_2 - a_{2N}x_N, \ldots ,y_{N-1} - a_{N-1\,N}x_N)] \tag{5}$$

In order to use this relation in some constructive form, we recall that $f_N(y_1,y_2, \ldots ,y_N) = -(y,A^{-1}y)$. Hence, write

$$f_N(y_1,y_2, \ldots ,y_N) = \sum_{i,j=1}^{N} c_{ij}(N)y_iy_j \tag{6}$$

Returning to (5), we obtain the relation

$$\sum_{i,j=1}^{N} c_{ij}(N)y_iy_j = \min_{x_N} \left[a_{NN}x_N^2 - 2x_Ny_N + \sum_{i,j=1}^{N-1} c_{ij}(N-1)(y_i - a_{iN}x_N)(y_j - a_{jN}x_N) \right] \tag{7}$$

or, collecting terms,

$$\sum_{i,j=1}^{N} c_{ij}(N)y_iy_j = \min_{x_N} \left[x_N^2 \left\{ a_{NN} + \sum_{i,j=1}^{N-1} a_{iN}a_{jN}c_{ij}(N-1) \right\} + x_N \left\{ -2y_N - 2 \sum_{i,j=1}^{N-1} [y_ja_{iN}]c_{ij}(N-1) \right\} + \sum_{i,j=1}^{N-1} c_{ij}(N-1)y_iy_j \right] \tag{8}$$

The minimization can now be readily performed,

$$x_N = \frac{y_N + \sum_{i,j=1}^{N-1} a_{iN} y_j c_{ij}(N-1)}{a_{NN} + \sum_{i,j=1}^{N-1} a_{iN} a_{jN} c_{ij}(N-1)} \tag{9}$$

while the minimum value itself is given by

$$\frac{\left(\sum_{i,j=1}^{N-1} c_{ij}(N-1) y_i y_j\right)\left(a_{NN} + \sum_{i,j=1}^{N-1} a_{iN} a_{jN} c_{ij}(N-1)\right) - \left(y_N + \sum_{i,j=1}^{N-1} y_j a_{iN} c_{ij}(N-1)\right)^2}{a_{NN} + \sum_{i,j=1}^{N-1} a_{iN} a_{jN} c_{ij}(N-1)}$$

Equating coefficients of $y_i y_j$ in (10) and $\sum_{i,j=1}^{N} c_{ij}(N) y_i y_j$, we obtain recurrence relations connecting the sequences $\{c_{ij}(N)\}$ and $\{c_{ij}(N-1)\}$.

16. Quadratic Deviation. The problem of minimizing over all x_k the quadratic form

$$Q_N(x) = \int_0^T \left(f(t) - \sum_{k=1}^{N} x_k g_k(t)\right)^2 dt \tag{1}$$

is, in principle, quite easily resolvable. Let $\{h_k(t)\}$ denote the orthonormal sequence formed from $\{g_k(t)\}$ by means of the Gram-Schmidt orthogonalization procedure, where, without loss of generality, the $g_k(t)$ are taken to be linearly independent. Then

$$\begin{aligned} \min_x Q_N(x) &= \min_y \int_0^T \left(f(t) - \sum_{k=1}^{N} y_k h_k(t)\right)^2 dt \\ &= \int_0^T f^2\,dt - \sum_{k=1}^{N} \left(\int_0^T f h_k(t)\,dt\right)^2 \end{aligned} \tag{2}$$

This result may be written

$$\int_0^T f^2\,dt - \int_0^T \int_0^T f(s) f(t) k_N(s,t)\,ds\,dt \tag{3}$$

where

$$k_N(s,t) = \sum_{k=1}^{N} h_k(s) h_k(t) \tag{4}$$

Let us now obtain a recurrence relation connecting the members of the sequence $\{k_N(s,t)\}$. Introduce the quadratic functional, a function of the function $f(t)$,

$$\phi_N(f) = \min_x \int_0^T \left(f - \sum_{k=1}^N x_k g_k\right)^2 dt \tag{5}$$

Employing the functional equation technique, we have for $N = 2, 3, \ldots,$

$$\phi_N(f) = \min_{x_N} \phi_{N-1}(f - x_N g_N) \tag{6}$$

and

$$\phi_1(f) = \min_{x_1} \int_0^T (f - x_1 g_1)^2 \, dt \tag{7}$$

In order to use (6), we employ the technique used above repeatedly—we take advantage of the quadratic character of $\phi_N(f)$. Hence,

$$\begin{aligned}\phi_N(f) &= \min_{x_N} \left[\int_0^T (f - x_N g_N)^2 \, dt \right.\\ &\qquad \left. - \int_0^T \int_0^T (f(s) - x_N g_N(s))(f(t) - x_N g_N(t)) k_N(s,t) \, ds \, dt \right]\\ &= \min_{x_N} \left\{ \int_0^T f^2 \, dt - \int_0^T \int_0^T k_{N-1}(s,t) f(s) f(t) \, ds \, dt \right.\\ &\qquad - 2x_N \left[\int_0^T f g_N \, dt - \int_0^T \int_0^T g_N(s) f(t) k_{N-1}(s,t) \, ds \, dt \right]\\ &\qquad \left. + x_N^2 \left[\int_0^T g_N^2 \, dt - \int_0^T \int_0^T k_{N-1}(s,t) g_N(s) g_N(t) \, ds \, dt \right] \right\}\end{aligned} \tag{8}$$

Obtaining the minimum value of x_N, and determining the explicit value of the minimum, we obtain the recurrence relation

$$\begin{aligned}k_N(s,t) &= k_{N-1}(s,t) + \frac{g_N(s) g_N(t)}{d_N} - \frac{2 g_N(s) \int_0^T g_N(s_1) k_{N-1}(t,s_1) \, ds_1}{d_N}\\ &\qquad + \frac{1}{d_N} \int_0^T \int_0^T g_N(s_1) g_N(t_1) k_{N-1}(s_1,s) k_{N-1}(t_1,t) \, ds_1 \, dt_1\end{aligned} \tag{9}$$

where

$$d_N = \int_0^T g_N^2(s) \, ds - \int_0^T \int_0^T k_{N-1}(s,t) g_N(s) g_N(t) \, ds \, dt \tag{10}$$

17. A Result of Stieltjes. Since we have been emphasizing in the preceding sections the connection between the solution of $Ax = b$ and the minimization of $(x,Ax) - 2(b,x)$ when A is positive definite, let us use the same idea to establish an interesting result of Stieltjes. We shall obtain a generalization in Chap. 16.

Theorem 1. *If A is a positive definite matrix with the property that $a_{ij} < 0$ for $i \neq j$, then A^{-1} has all positive elements.*

Proof. Consider the problem of minimizing $Q(x) = (x,Ax) - 2(b,x)$, where the components of b are all positive. Assume that $x_1, x_2, \ldots, x_k < 0$, $x_{k+1}, x_{k+2}, \ldots, x_N \geq 0$, at the minimum point. Writing out $(x,Ax) - 2(b,x)$ in the form

$$a_{11}x_1^2 + a_{22}x_2^2 + a_{NN}x_N^2 + \sum_{i,j=1}^{k} a_{ij}x_ix_j + \sum_{i=1}^{k}\sum_{j=k+1}^{N} a_{ij}x_ix_j$$
$$+ \sum_{i=k+1}^{N}\sum_{j=1}^{N} a_{ij}x_ix_j + \sum_{i,j=k+1}^{N} a_{ij}x_ix_j - 2\sum_{i=1}^{k} b_ix_i - 2\sum_{i=k+1}^{N} b_ix_i \quad (1)$$

we see, in view of the negativity of a_{ij} for $i \neq j$, that we can obtain a smaller value of $Q(x)$ by replacing x_i by $-x_i$ for $i = 1, 2, \ldots, k$, and leaving the other values unchanged, provided at least one of the x_i, $i = k + 1, \ldots, N$, is positive. In any case, we see that all the x_i can be taken to be non-negative.

To show that they are actually all positive, if the b_i are positive, we observe that one at least must be positive. For if

$$x_1 = x_2 = \cdots = x_{N-1} = 0$$

at the minimum point, then x_N determined as the value which minimizes $x_N^2 - 2b_Nx_N$ is equal to b_N and thus positive. Since $Ax = b$ at the minimum point, we have

$$a_{ii}x_i = b_i - \sum_{j \neq i} a_{ij}x_j \qquad i = 1, 2, \ldots, N - 1 \quad (2)$$

which shows that $x_i > 0$.

We see then that $A^{-1}b$ is a vector with positive components whenever b is likewise. This establishes the non-negativity of the elements of A^{-1}. To show that A^{-1} actually has all positive elements, we must show that $A^{-1}b$ has positive components whenever b has non-negative components with at least one positive.

Turning to (2), we see that the condition $a_{ij} < 0$, $i \neq j$, establishes this.

MISCELLANEOUS EXERCISES

1. Given two sequences $\{a_k\}$ and $\{b_k\}$, $k = 0, 1, 2, \ldots, N$, we often wish to determine a finite sequence $\{x_k\}$, $k = 0, 1, \ldots, M$ which expresses b_k most closely in the form $b_k = \sum_{l=0}^{M} x_l a_{k-l}$.

To estimate the closeness of fit, we use the sum

$$Q_{N,M}(x) = \sum_{k=0}^{N} \left(b_k - \sum_{l=0}^{M} x_l a_{k-l}\right)^2 \qquad N > M \geq 1$$

Consider the quadratic form in $y_0, y_1, \ldots, y_N$ defined by

$$f_{N,M}(y_0,y_1, \ldots ,y_N) = \min_x [(y_0 - x_0a_0)^2 + (y_1 - x_0a_1 - x_1a_0)^2 + \cdots + (y_N - x_0a_N - x_1a_{N-1} - \cdots - x_Ma_{N-M})^2]$$

Show that

$$f_{N,0}(y_0,y_1, \ldots ,y_N) = \left(\sum_{k=0}^{N} y_k^2\right)\left(\sum_{k=0}^{N} a_k^2\right) - \left(\sum_{k=0}^{N} a_ky_k\right)^2$$

and that, generally,

$$f_{N,M}(y_0,y_1, \ldots ,y_N) = \min_{x_0} [(y_0 - x_0a_0)^2 + f_{N-1,M-1}(y_1 - x_0a_1, y_2 - x_0a_2, \ldots , y_N - x_0a_N)]$$

2. Write

$$f_{N,M}(y_0,y_1, \ldots ,y_N) = \sum_{i,j=0}^{N} c_{ij}(N,M)y_iy_j$$

and use the preceding recurrence relation to obtain the $c_{ij}(N,M)$ in terms of the $c_{ij}(N - 1, M - 1)$.

3. Write $f_{N,M}(y_0,y_1, \ldots ,y_N) = (y,A_{MN}y)$ and obtain the recurrence relation in terms of matrices.

For an alternate discussion of this problem which is paramount in the Kolmogorov-Wiener theory of prediction, see N. Wiener, *The Extrapolation, Interpolation and Smoothing of Stationary Time Series and Engineering Applications*, John Wiley & Sons, Inc., 1942, and particularly the Appendix by N. Levinson.

4. Let A and B be two positive definite matrices of order N with $AB \neq BA$ and c a given N-dimensional vector. Consider the vector $x_N = Z_NZ_{N-1} \cdots Z_2Z_1c$, where each matrix Z_i is either an A or a B. Suppose that the Z_i, $i = 1, 2, \ldots, N$, are to be chosen so as to maximize the inner product (x_N,b) where b is a fixed N-dimensional vector.

Define the function

$$f_N(c) = \max_{\{z_i\}} (x_N,b)$$

for $N = 1, 2, \ldots$, and all c. Then

$$f_1(c) = \max ((Ac,b),(Bc,b)),$$
$$f_N(c) = \max (f_{N-1}(Ac),f_{N-1}(Bc)) \qquad N = 2, 3, \ldots .$$

5. Does there exist a scalar λ such that $f_N(c) \sim \lambda^N g(c)$ as $N \to \infty$?

6. What is the answer to this question if $AB = BA$?

7. Suppose that C is a given positive definite matrix, and the Z_i are to be chosen so that $Z_NZ_{N-1} \cdots Z_2Z_1C$ has the maximum maximum characteristic root. Let $g_N(C)$ denote this maximum maximorum. Then

$$g_1(C) = \max [\phi(AC),\phi(BC)],$$
$$g_N(C) = \max [g_{N-1}(AC),g_{N-1}(BC)] \qquad N = 2, 3, \ldots ,$$

where $\phi(X)$ is used to denote the maximum characteristic root of X.

8. Does there exist a scalar λ such that $g_N(C) \sim \lambda^N h(c)$ as $N \to \infty$? See C. Bohm, Sulla minimizzazione di una funzione del prodotto di enti non commutati, *Lincei-Rend. Sci. fis. Mat. e Nat.*, vol. 23, pp. 386–388, 1957.

Bibliography

§1. The theory of dynamic programming in actuality is the study of multistage decision processes. It turns out that many variational problems can be interpreted in these terms. This interpretation yields a new approach which is useful for both analytic and computational purposes. In this chapter, pursuant to our general program, we are interested only in analytic aspects. For a detailed discussion of the general theory, the reader is referred to

R. Bellman, *Dynamic Programming*, Princeton University Press, Princeton, N.J., 1957.

§2. The problem treated here is a simple example of what is called a "smoothing" problem; cf.

R. Bellman, Dynamic Programming and Its Application to Variational Problems in Mathematical Economics, *Calculus of Variations; Proceedings of Symposia in Applied Mathematics*, vol. VIII, pp. 115–139, McGraw-Hill Book Company, Inc., New York, 1958.

For a discussion of smoothing problems of a different type, see the following:

I. J. Schoenberg, On Smoothing Operations and Their Generating Functions, *Bull. Am. Math. Soc.*, vol. 59, pp. 199–230, 1953.

N. Wiener, *Cybernetics*, John Wiley & Sons, Inc., New York, 1951.

§3. The derivation of this recurrence relation is a particular application of the principle of optimality; see the book referred to above in §1. The results discussed in §3 to §5 were first presented in

R. Bellman, On a Class of Variational Problems, *Quart. Appl. Math.*, vol. XIV, pp. 353–359, 1957.

§7. These results were presented in

R. Bellman, Eigenvalues and Functional Equations, *Proc. Am. Math. Soc.*, vol. 8, pp. 68–72, 1957.

§8. This discussion follows

R. Bellman, On a Class of Variational Problems, *Quart. Appl. Math.* vol. XIV, pp. 353–359, 1957.

For a number of results dependent upon the fact that Jacobi matrices arise in the treatment of random walk processes, see

S. Karlin and J. McGregor, Coincident Properties of Birth and Death Processes, *Technical Rept.* 9, Stanford University, 1958.

P. Dean, The Spectral Distribution of a Jacobian Matrix, *Proc. Cambridge Phil. Soc.*, vol. 52, pp. 752–755, 1956.

§9. This discussion, given for the case of finite matrices, can be readily extended to cover more general operators. An application of this technique is contained in

R. Bellman and S. Lehman, Functional Equations in the Theory of Dynamic Programming—X: Resolvents, Characteristic Functions and Values, *Proc. Natl. Acad. Sci. U.S.*, vol. 44, pp. 905–907, 1958.

§10–§11. The contents of these sections are taken from a more general setting in

R. Bellman and S. Lehman, Functional Equations in the Theory of Dynamic Programming—IX: Variational Analysis, Analytic Continuation and Imbedding of Operators, *Duke Math. J.*, 1960.

For an extensive discussion of matters of this type, see

C. L. Dolph, A Saddle Point Characterization of the Schwinger Stationary Points in Exterior Scattering Problems, *J. Soc. Ind. Appl. Math.*, vol. 5, pp. 89–104, 1957.

C. L. Dolph, J. E. McLaughlin, and I. Marx, Symmetric Linear Transformations and Complex Quadratic Forms, *Comm. Pure and Appl. Math.*, vol. 7, pp. 621–632, 1954.

§12 to **§14.** The results are taken from

R. Bellman, On Some Applications of Dynamic Programming to Matrix Theory, *Illinois J. Math.*, vol. 1, pp. 297–301, 1957.

§15 to **§16.** The results follow

R. Bellman, *Dynamic Programming and Mean Square Deviation*, The RAND Corporation, Paper P 1147, September 13, 1957.

§17. This result was given by

T. J. Stieltjes, Sur les racines de $X_n = 0$, *Acta Math.*, vol. 9, 1886–1887.

A generalization of this result will be given in Chap. 16.

For another application of this device, see

R. Bellman, On the Nonnegativity of Green's Functions, *Boll. Unione Matematica*, vol. 12, pp. 411–413, 1957.

For some further applications of functional equation techniques to quadratic forms, see

R. E. Kalman and R. W. Koepcke, Optimal Synthesis of Linear Sampling Control Systems Using Generalized Performance Indexes, *Am. Soc. Mech. Eng., Instrument and Regulators Conferences,* Newark, April 2–4, 1958.

R. Bellman and J. M. Richardson, On the Application of Dynamic Programming to a Class of Implicit Variational Problems, *Quart. Appl. Math.,* 1959.

M. Freimer, Dynamic Programming and Adaptive Control Processes, *Lincoln Laboratory Report,* 1959.

H. A. Simon, Dynamic Programming under Uncertainty with a Quadratic Criterion Function, *Econometrica,* vol. 24, pp. 74–81, 1956.

See Chap. 17 for many further results and references.

The study of routing problems introduces a new type of matrix composition. Write $C = M(A,B)$ if $c_{ij} = \min_k (a_{ik} + b_{kj})$. See

R. Bellman, K. L. Cooke, and Jo Ann Lockett, *Algorithms, Graphs, and Computers,* Academic Press Inc., New York, 1970.

C. Berge, *The Theory of Graphs and Its Applications,* John Wiley & Sons, New York, 1962.

J. F. Shapiro, Shortest Route Methods for Finite State Deterministic Dynamic Programming Problems, *SIAM J. Appl. Math.,* vol. 16, pp. 1232–1250, 1968.

10

Matrices and Differential Equations

1. Motivation. In this chapter which begins the second main portion of the book, we shall discuss the application of matrix theory to the solution of linear systems of differential equations of the form

$$\frac{dx_i}{dt} = \sum_{j=1}^{N} a_{ij}x_j \qquad x_i(0) = c_i \qquad i = 1, 2, \ldots, N \tag{1}$$

where the a_{ij} are constants.

In order to understand the pivotal positions that equations of this apparently special type occupy, let us explain a bit of the scientific background. Consider a physical system S whose state at any time t is assumed to be completely described by means of the N functions $x_1(t)$, $x_2(t), \ldots, x_N(t)$. Now make the further assumption that the rate of change of all these functions at any time t depends only upon the values of these functions at this time.

This is always an approximation to the actual state of affairs, but a very convenient and useful one.

The analytic transliteration of this statement is a set of differential equations of the form

$$\frac{dx_i}{dt} = f_i(x_1,x_2, \ldots ,x_N) \qquad i = 1, 2, \ldots, N \tag{2}$$

with an associated set of initial conditions

$$x_i(0) = c_i \qquad i = 1, 2, \ldots, N \tag{3}$$

The vector $c = (c_1,c_2, \ldots ,c_N)$ represents the initial state of the system.

Sets of constants, $\{c_i\}$, for which

$$f_i(c_1,c_2, \ldots ,c_N) = 0 \qquad i = 1, 2, \ldots, N \tag{4}$$

play a particularly important role. They are obviously *equilibrium states* since S cannot depart from them without the intervention of external forces.

Whenever such states exist, it is of great interest to study the behavior of the system in the neighborhood of these states. In other words, we are examining the *stability* of the system under small disturbances. If the perturbed system eventually returns to the equilibrium state, we say that it is *stable;* otherwise, we say that it is *unstable.* These considerations are of great practical significance.

In order to carry out this study, we set

$$x_i = c_i + y_i \tag{5}$$

where the y_i are taken to be small quantities. Substituting in (2), we obtain the equations

$$\begin{aligned}\frac{dy_i}{dt} = \frac{dx_i}{dt} &= f_i(c_1 + y_1, c_2 + y_2, \ldots, c_N + y_N) \qquad i = 1, 2, \ldots, N \\ &= f_i(c_1,c_2, \ldots, c_N) + \sum_{j=1}^{N} a_{ij}y_j + \cdots \end{aligned} \tag{6}$$

where

$$a_{ij} = \frac{\partial f_i}{\partial x_j} \qquad \text{at } x_1 = c_1, x_2 = c_2, \ldots, x_N = c_N \tag{7}$$

and the three dots signify terms involving higher powers of the y_i.

The behavior of S in the neighborhood of the equilibrium state, $\{c_i\}$, is thus determined, to an approximation whose accuracy must be carefully examined, by the linear system with constant coefficients given in (1).

We have then a powerful motivation for the study of linear systems of this type. Our aim is to determine analytic representations of the solution which will permit us to ascertain its limiting behavior as $t \to \infty$.

These questions of stability will be taken up again in Chap. 13.

2. Vector-matrix Notation. To study (1.1), we introduce the vectors y and c, possessing the components y_i and c_i, respectively, and the matrix $A = (a_{ij})$. It is clear from the way that the difference of two vectors is defined that the appropriate way to define the derivative of a vector is the following:

$$\frac{dy}{dt} = \begin{bmatrix} \frac{dy_1}{dt} \\ \frac{dy_2}{dt} \\ \cdot \\ \cdot \\ \cdot \\ \frac{dy_N}{dt} \end{bmatrix} \tag{1}$$

Similarly, the integral of $y(t)$ is defined to be

$$\int^t y(s)\,ds = \begin{bmatrix} \int^t y_1(s)\,ds \\ \int^t y_2(s)\,ds \\ \cdot \\ \cdot \\ \cdot \\ \int^t y_N(s)\,ds \end{bmatrix} \tag{2}$$

The derivatives and integrals of matrices are defined analogously. It follows that (1.1) can be written

$$\frac{dy}{dt} = Ay \qquad y(0) = c \tag{3}$$

The matrix A will, in general, not be symmetric. Consequently, the techniques and results of the first part of this volume can be expected to play a small role. We shall have to develop some new methods for treating general square matrices.

A vector whose components are functions of t will be called a vector function, or briefly, a function of t. It will be called *continuous* if its components are continuous functions of t in the interval of interest. We shall use similar terms in describing matrix functions.

EXERCISES

1. Show that

(a) $\frac{d}{dt}(x,y) = \left(\frac{dx}{dt},y\right) + \left(x,\frac{dy}{dt}\right)$

(b) $\frac{d}{dt}(Ax) = \left(\frac{dA}{dt}\right)x + A\frac{dx}{dt}$

(c) $\frac{d}{dt}(AB) = \left(\frac{dA}{dt}\right)B + A\left(\frac{dB}{dt}\right)$

(d) $\frac{d}{dt}(X^{-1}) = -X^{-1}\left(\frac{dX}{dt}\right)X^{-1}$

(e) $\frac{d}{dt}(X^n) = \left(\frac{dX}{dt}\right)X^{n-1} + X\left(\frac{dX}{dt}\right)X^{n-2} + \cdots + X^{n-1}\frac{dX}{dt}$

2. Obtain an equation for the derivative of $X^{1/2}$.

3. Norms of Vectors and Matrices. We could, if we so desired, use the scalar function $(x,\bar{x})$ as a measure of the magnitude of x. However, it is more convenient to use not these Euclidean norms, but the simpler function

$$\|x\| = \sum_{i=1}^{N} |x_i| \tag{1}$$

and, for matrices,

$$\|A\| = \sum_{i,j=1}^{N} |a_{ij}| \tag{2}$$

It is readily verified that

$$\begin{aligned} \|x + y\| &\leq \|x\| + \|y\| & \|A + B\| &\leq \|A\| + \|B\| \\ \|Ax\| &\leq \|A\|\,\|x\| & \|AB\| &\leq \|A\|\,\|B\| \\ \|c_1 x\| &= |c_1|\,\|x\| & \|c_1 A\| &= |c_1|\,\|A\| \end{aligned} \tag{3}$$

The reason why we have chosen the foregoing norms for vectors and matrices is that the verification of the results in (3) is particularly simple. As we shall see in the exercises below, there are a large number of choices of norms which are equally useful when dealing with finite dimensional vectors and matrices. It is only when we turn to infinite dimensional vectors and matrices that the choice of a norm becomes critical.

EXERCISES

1. Show that we could define as norms satisfying (3) the functions

$$\|x\| = \left(\sum_{i=1}^{N} |x_i|^2\right)^{1/2} = (x,\bar{x})^{1/2}$$

$$\|A\| = \left(\sum_{i,j=1}^{N} |a_{ij}|^2\right)^{1/2} = \text{tr }(AA^*)^{1/2}$$

2. If we set $\|x\| = \max_i |x_i|$, what definition should we take for $\|A\|$ in order that all of the inequalities of (3) be valid?

3. Let $\|x\|$ be a vector norm satisfying the vector relations in (3) and the condition that $\|x\| \geq 0$ and $\|x\| = 0$ if and only if $x = 0$. Show that $\|A\| = \max_{\|x\|=1} \|Ax\|$ is a matrix norm which satisfies the remaining relations in (3). (This is the standard way of *inducing* a norm for transformations, given a norm for vectors.)

4. If we use the norm for x appearing in Exercise 2, which norm do we obtain for A using the technique of Exercise 3?

5. Show that convergence of a sequence of vectors $\{x^n\}$ to a vector x implies and is implied by convergence of the kth components of the members of the sequence $\{x^n\}$ to the kth component of x.

6. Show that convergence of a sequence of vectors $\{x^n\}$ in one norm satisfying the conditions of Exercise 3 implies convergence in any other norm satisfying these conditions.

7. Show that

$$\|\textstyle\int x(t)\,dt\| \leq \int \|x(t)\|\,dt$$
$$\|\textstyle\int A(t)\,dt\| \leq \int \|A(t)\|\,dt$$

8. Show that $\|A^n\| \leq \|A\|^n$ for any norm satisfying the condition in (3).

9. Is there any norm satisfying the conditions in (3) for which $\|AB\| = \|A\|\,\|B\|$?

4. Infinite Series of Vectors and Matrices. In the course of establishing the existence of solutions of the linear vector equation appearing above, we shall have need of infinite series of vectors and matrices. By the vector $\sum_{n=0}^{\infty} x^n$, we shall mean the vector whose ith component is the sum of the series $\sum_{n=0}^{\infty} x_i^n$. Thus, the convergence of the vector series is equivalent to the simultaneous convergence of the N series, $\sum_{n=0}^{\infty} x_i^n$. It follows that a sufficient condition for convergence of the vector series $\sum_{n=0}^{\infty} x^n$ is that the scalar series $\sum_{n=0}^{\infty} \|x^n\|$ converge.

Similarly, a matrix series of the form $\sum_{n=0}^{\infty} A_n$ represents N^2 infinite series, and a sufficient condition for convergence is that $\sum_{n=0}^{\infty} \|A_n\|$ converge.

5. Existence and Uniqueness of Solutions of Linear Systems. With these preliminaries, we are ready to demonstrate the following basic result.

Theorem 1. *If $A(t)$ is continuous for $t \geq 0$, there is a unique solution to the vector differential equation*

$$\frac{dx}{dt} = A(t)x \qquad x(0) = c \tag{1}$$

This solution exists for $t \geq 0$, and may be written in the form

$$x = X(t)c \tag{2}$$

where $X(t)$ is the unique matrix satisfying the matrix differential equation

$$\frac{dX}{dt} = A(t)X \qquad X(0) = I \tag{3}$$

Proof. We shall employ the method of successive approximations to establish the existence of a solution of (3). In place of (3), we consider the integral equation

$$X = I + \int_0^t A(s)X\, ds \tag{4}$$

Define the sequence of matrices $\{X_n\}$ as follows:

$$X_0 = I \tag{5}$$
$$X_{n+1} = I + \int_0^t A(s)X_n\,ds \qquad n = 0, 1, \ldots$$

Then we have

$$X_{n+1} - X_n = \int_0^t A(s)(X_n - X_{n-1})\,ds \qquad n = 1, 2, \ldots \tag{6}$$

Let

$$m = \max_{0 \le t \le t_1} \|A(s)\| \tag{7}$$

Here and in what follows, we are employing the norms defined in (3.1) and (3.2). Using (6), we obtain

$$\begin{aligned}\|X_{n+1} - X_n\| &= \left\|\int_0^t A(s)(X_n - X_{n-1})\,ds\right\| \\ &\le \int_0^t \|A(s)\|\,\|X_n - X_{n-1}\|\,ds \\ &\le m\int_0^t \|X_n - X_{n-1}\|\,ds\end{aligned} \tag{8}$$

for $0 \le t \le t_1$. Since, in this same interval,

$$\|X_1 - X_0\| \le \int_0^t \|A(s)\|\,ds \le mt \tag{9}$$

we have inductively from (8),

$$\|X_{n+1} - X_n\| \le \frac{m^{n+1}t^{n+1}}{(n+1)!} \qquad \text{for } 0 \le t \le t_1 \tag{10}$$

Hence, the series $\sum_{n=0}^{\infty} (X_{n+1} - X_n)$ converges uniformly for $0 \le t \le t_1$. Consequently, X_n converges uniformly to a matrix $X(t)$ which satisfies (4), and thus (3).

Since, by assumption, $A(t)$ is continuous for $t \ge 0$, we may take t_1 arbitrarily large. We thus obtain a solution valid for $t \ge 0$.

It is easily verified that $x = X(t)c$ is a solution of (1), satisfying the required initial condition.

Let us now establish uniqueness of this solution. Let Y be another solution of (3). Then Y satisfies (4), and thus we have the relation

$$X - Y = \int_0^t A(s)(X(s) - Y(s))\,ds \tag{11}$$

Hence

$$\|X - Y\| \le \int_0^t \|A(s)\|\,\|X(s) - Y(s)\|\,ds \tag{12}$$

Since Y is differentiable, hence continuous, define

$$m_1 = \max_{0 \le t \le t_1} \|X - Y\| \tag{13}$$

From (12), we obtain

$$\|X - Y\| \leq m_1 \int_0^t \|A(s)\| \, ds \qquad 0 \leq t \leq t_1 \tag{14}$$

Using this bound in (12), we obtain

$$\begin{aligned} \|X - Y\| &\leq m_1 \int_0^t \|A(s)\| \left(\int_0^s \|A(s_1)\| \, ds_1 \right) ds \\ &\leq \frac{m_1 \left(\int_0^t \|A(s)\| \, ds \right)^2}{2} \end{aligned} \tag{15}$$

Iterating, we obtain

$$\|X - Y\| \leq \frac{m_1 \left(\int_0^t \|A(s)\| \, ds \right)^{n+1}}{(n+1)!} \tag{16}$$

Letting $n \to \infty$, we see that $\|X - Y\| \leq 0$. Hence $X \equiv Y$.

Having obtained the matrix X, it is easy to see that $X(t)c$ is a solution of (1). Since the uniqueness of solutions of (1) is readily established by means of the same argument as above, it is easy to see $X(t)c$ is *the* solution.

EXERCISE

1. Establish the existence of a solution of (1) under the condition that $A(t)$ is a Riemann-integrable function over any finite interval. In this case, need the differential equation be satisfied everywhere? Examine, in particular, the case where $A(t) = A$, $0 \leq t \leq t_0$, $A(t) = B$, $t > t_0$.

6. The Matrix Exponential. Consider now the particular case where $A(t)$ is a constant matrix. In the scalar case, the equation

$$\frac{du}{dt} = au \qquad u(0) = c \tag{1}$$

has the solution $u = e^{at}c$. It would be very convenient to find an analogous solution of the matrix equation

$$\frac{dX}{dt} = AX \qquad X(0) = C \tag{2}$$

having the form $X = e^{At}C$.

By analogy with the scalar case, or from the method of successive approximations used in Sec. 5, we are led to define the matrix exponential function by means of the infinite series

$$e^{At} = I + At + \cdots + \frac{A^n t^n}{n!} + \cdots \tag{3}$$

Let us now demonstrate

Theorem 2. *The matrix series defined above exists for all A for any fixed value of t, and for all t for any fixed A. It converges uniformly in any finite region of the complex t plane.*

Proof. We have

$$\frac{\|A^n t^n\|}{n!} \leq \frac{\|A\|^n |t|^n}{n!} \tag{4}$$

Since $\|A\|^n|t|^n/n!$ is a term in the series expansion of $e^{\|A\| \, |t|}$, we see that the series in (3) is dominated by a uniformly convergent series, and hence is itself uniformly convergent in any finite region of the t plane.

EXERCISE

1. Using the infinite series representation, show that $d/dt(e^{At}) = Ae^{At} = e^{At}A$.

7. Functional Equations—I. The scalar exponential function satisfies the fundamental functional equation

$$e^{a(s+t)} = e^{as}e^{at} \tag{1}$$

Unless there is an analogue of this for the matrix exponential, we have no right to use the notation of (3).

Let us now demonstrate that

$$e^{A(s+t)} = e^{As}e^{At} \tag{2}$$

Using the series expansions for the three exponentials and the fact that absolutely convergent series may be rearranged in arbitrary fashion, we have

$$\begin{aligned} e^{As}e^{At} &= \left(\sum_{k=0}^{\infty} \frac{A^k s^k}{k!}\right)\left(\sum_{l=0}^{\infty} \frac{A^l t^l}{l!}\right) \\ &= \sum_{n=0}^{\infty} A^n \left(\sum_{k+l=n} \frac{s^k t^l}{k!l!}\right) \\ &= \sum_{n=0}^{\infty} A^n \frac{(s+t)^n}{n!} = e^{A(s+t)} \end{aligned} \tag{3}$$

From (2) we obtain, upon setting $s = -t$, the important result that

$$e^{A(-t+t)} = I = e^{-At}e^{At} \tag{4}$$

Hence, e^{At} is never singular and its inverse is e^{-At}. This is a matrix analogue of the fact that the scalar exponential never vanishes.

8. Functional Equations—II. The proof of the functional equation in Sec. 7 was a verification rather than a derivation. In order to understand the result, let us turn to the differential equation

$$\frac{dX}{dt} = AX \tag{1}$$

Observe that e^{At} is a solution with the boundary condition $X(0) = I$, and that $e^{A(s+t)}$ is a solution with the boundary condition $X(0) = e^{As}$. Hence, from the uniqueness theorem, we may conclude that

$$e^{A(s+t)} = e^{At}e^{As} \tag{2}$$

9. Functional Equations—III. Having derived the functional equation discussed above, the question naturally arises as to the relation between $e^{(A+B)t}$ and $e^{At}e^{Bt}$. Since

$$\begin{aligned} e^{(A+B)t} &= I + (A+B)t + \frac{(A+B)^2}{2}t^2 + \cdots \\ e^{At}e^{Bt} &= \left(I + At + \frac{A^2t^2}{2} + \cdots\right)\left(I + Bt + \frac{B^2t^2}{2} + \cdots\right) \\ &= I + (A+B)t + \frac{A^2t^2}{2} + ABt^2 + \frac{B^2t^2}{2} + \cdots \end{aligned} \tag{1}$$

we see that

$$e^{(A+B)t} - e^{At}e^{Bt} = (BA - AB)\frac{t^2}{2} + \cdots \tag{2}$$

Consequently, $e^{(A+B)t} = e^{At}e^{Bt}$ for *all* t only if $AB = BA$, which is to say if A and B commute. It is easy to see that this is a sufficient condition.

10. Nonsingularity of Solution. We observed in Sec. 7 that e^{At} is never singular. Let us now point out that this is a special case of the general result that the solution of the equation

$$\frac{dX}{dt} = A(t)X \qquad X(0) = I \tag{1}$$

is nonsingular in any interval $0 \leq t \leq t_1$ in which $\int_0^{t_1} \|A(t)\|\,dt$ exists.

There are several ways of establishing this result. The first is the most interesting, while two other methods will be given in the exercises below. The first method is based upon the following identity of Jacobi:

$$|X(t)| = e^{\int_0^t \operatorname{tr}(A(s))ds} \tag{2}$$

To derive this result, let us consider the derivative of the scalar function $|X(t)|$. To simplify the notational problem, consider the two-dimensional case. We have

$$|X(t)| = \begin{vmatrix} x_1 & y_1 \\ x_2 & y_2 \end{vmatrix} \tag{3}$$

where

$$\begin{aligned} \frac{dx_1}{dt} &= a_{11}x_1 + a_{12}x_2 & \frac{dy_1}{dt} &= a_{11}y_1 + a_{12}y_2 \\ \frac{dx_2}{dt} &= a_{21}x_1 + a_{22}x_2 & \frac{dy_2}{dt} &= a_{21}y_1 + a_{22}y_2 \end{aligned} \tag{4}$$

and

$$\begin{aligned} x_1(0) &= 1 & y_1(0) &= 0 \\ x_2(0) &= 0 & y_2(0) &= 1 \end{aligned} \tag{5}$$

Then

$$\begin{aligned} \frac{d}{dt}|X(t)| &= \begin{vmatrix} \frac{dx_1}{dt} & \frac{dy_1}{dt} \\ x_2 & y_2 \end{vmatrix} + \begin{vmatrix} x_1 & y_1 \\ \frac{dx_2}{dt} & \frac{dy_2}{dt} \end{vmatrix} \\ &= \begin{vmatrix} a_{11}x_1 + a_{12}x_2 & a_{11}y_1 + a_{12}y_2 \\ x_2 & y_2 \end{vmatrix} \\ &+ \begin{vmatrix} x_1 & y_1 \\ a_{21}x_1 + a_{22}x_2 & a_{21}y_1 + a_{22}y_2 \end{vmatrix} \\ &= a_{11}\begin{vmatrix} x_1 & y_1 \\ x_2 & y_2 \end{vmatrix} + a_{22}\begin{vmatrix} x_1 & y_1 \\ x_2 & y_2 \end{vmatrix} \\ &= (\text{tr } A(t))|X(t)| \end{aligned} \tag{6}$$

Thus

$$|X(t)| = e^{\int_0^t \text{tr } A(s)ds} \tag{7}$$

since $|X(0)| = 1$.

EXERCISES

1. Consider the equations $\frac{dX}{dt} = A(t)X$, $X(0) = I$, and $dY/dt = -YA(t)$, $Y(0) = I$. Show that $Y = X^{-1}$, and conclude that $X(t)$ is nonsingular in any interval where $\|A(t)\|$ is integrable.

2. Consider the second-order differential equation $u'' + p(t)u' + q(t)u = 0$. Write $u' = v$ and obtain a first-order system corresponding to this second-order equation, namely,

$$\begin{aligned} \frac{du}{dt} &= v \\ \frac{dv}{dt} &= -p(t)v - q(t)u \end{aligned}$$

3. Consider the integral $J = \int_0^t w(u'' + p(t)u' + q(t)u)\, dt$. Integrating by parts we obtain

$$J = [\cdot\ \cdot\ \cdot] + \int_0^t u(w'' + p_1(t)w' + q_1(t)w)\, dt$$

What is the connection between the vector-matrix system for $w'' + p_1(t)w' + q_1(t)w = 0$ and the vector-matrix system for $u'' + p(t)u' + q(t)u = 0$?

4. Consider the following proof of the nonsingular nature of $X(t)$. If $|X(t)| = 0$ at a point t_2, $0 \leq t_2 \leq t_1$, there exists a nontrivial relation between the vectors constituting the columns of $X(t)$, $c_1x^1 + c_2x^2 + \cdots + c_Nx^N = 0$. Since $c_1x^1 + c_2x^2 + \cdots + c_Nx^N$ is a solution of the equation $dx/dt = A(t)x$, if it is zero at one point, it is zero for all t in $[0,t_1]$. The relation $|X(t)| = 0$, however, clearly does not hold at $t = 0$; a contradiction.

11. Solution of Inhomogeneous Equation—Constant Coefficients. Let us now consider the problem of solving the inhomogeneous system

$$\frac{dx}{dt} = Ax + f(t) \qquad x(0) = c \tag{1}$$

The utility of the matrix exponential notation shows to advantage here. We have

$$e^{-At}\left(\frac{dx}{dt} - Ax\right) = \frac{d}{dt}(e^{-At}x) = e^{-At}f \tag{2}$$

Hence,

$$e^{-At}x = c + \int_0^t e^{-As}f(s)\,ds \tag{3}$$

or

$$x = e^{At}c + \int_0^t e^{A(t-s)}f(s)\,ds \tag{4}$$

Observe how the use of matrix exponential function permits us to obtain the solution of (1) in exactly the same fashion as if we were dealing with a scalar equation.

12. Inhomogeneous Equation—Variable Coefficients. Consider the case where A is time-dependent. We wish to solve

$$\frac{dx}{dt} = A(t)x + f(t) \qquad x(0) = c \tag{1}$$

Let us use the Lagrange variation of parameters technique and attempt to find a solution of the form $x = Xy$ where $X = X(t)$ is the solution of the equation $dX/dt = A(t)X$, $X(0) = I$. Substituting in (1), we obtain the equation

$$X'y + Xy' = A(t)Xy + Xy' = A(t)Xy + f(t) \tag{2}$$

Hence

$$Xy' = f(t) \tag{3}$$

whence

$$y' = X^{-1}(t)f(t) \tag{4}$$

or

$$y = c + \int_0^t X^{-1}(s)f(s)\,ds \tag{5}$$

Consequently,

$$x = X(t)c + \int_0^t X(t)X^{-1}(s)f(s)\,ds \tag{6}$$

a generalization of the result of (11.4).

13. Inhomogeneous Equation—Adjoint Equation. Let us now pursue a different approach, one of great importance in the general theory of linear functional equations. Take $Y(t)$ to be a variable matrix, as yet unspecified, and integrate between 0 and t the equation

$$Y(t)\frac{dx}{dt} = Y(t)A(t)x + Y(t)f(t) \tag{1}$$

The result is, upon integrating by parts,

$$Y(t)x(t) - Y(0)c - \int_0^t \frac{dY}{ds}x(s)\,ds = \int_0^t Y(s)A(s)x(s)\,ds + \int_0^t Y(s)f(s)\,ds \tag{2}$$

Without loss of generality, we can take $c = 0$, since we can obtain the solution of (12.1) by adding to the solution of this special case the vector $X(t)c$. Since our aim is to solve for $x(t)$, suppose that we make the most convenient assumptions that we can, namely, that

$$\frac{dY}{ds} = -Y(s)A(s) \qquad 0 \le s \le t \tag{3a}$$

$$Y(t) = I \tag{3b}$$

If we can satisfy both of these equations, we can write x in the simple form

$$x = \int_0^t Y(s)f(s)\,ds \tag{4}$$

The equation in (3) is called the *adjoint equation*. We know from the general existence and uniqueness theorem established previously that a unique solution to (3) exists. The matrix Y will now be a function of s and t.

EXERCISE

1. Show that $Y(s) = X(t)^{-1}X(s)$.

14. Perturbation Theory. An interesting application of the formula for the solution of the inhomogeneous equation is in the direction of perturbation theory. Given the matrix exponential $e^{A+\epsilon B}$, we wish to evaluate it as a power series in ϵ,

$$e^{A+\epsilon B} = e^A + \sum_{n=1}^{\infty} \epsilon^n Q_n(A,B) \tag{1}$$

The problem is readily resolved if A and B commute, since then

$$e^{A+\epsilon B} = e^A e^{\epsilon B}$$

Let us then consider the interesting case where $AB \neq BA$.

If we write

$$e^{A+\epsilon B} = I + \sum_{n=1}^{\infty} \frac{(A + \epsilon B)^n}{n!} \tag{2}$$

and attempt to collect the terms in ϵ, we soon see that it is quite difficult to do this in a systematic fashion. In place of this direct procedure, we pursue the following route. The matrix $e^{A+\epsilon B}$ is the solution of the differential equation

$$\frac{dX}{dt} = (A + \epsilon B)X \qquad X(0) = I \tag{3}$$

evaluated at the point $t = 1$.

Let us write this equation in the form

$$\frac{dX}{dt} = AX + \epsilon BX \qquad X(0) = I \tag{4}$$

It follows from (11.4) that X satisfies the linear integral equation

$$X = e^{At} + \epsilon \int_0^t e^{A(t-s)}BX(s)\, ds \tag{5}$$

Solving this Volterra integral equation by iteration, we obtain an infinite series of the form

$$X = e^{At} + \epsilon \int_0^t e^{A(t-s)}Be^{As}\, ds + \cdots \tag{6}$$

Hence, $e^{A+\epsilon B}$ has as the first two terms of its series expansion

$$e^{A+\epsilon B} = e^A + \epsilon \int_0^1 e^{A(1-s)}Be^{As}\, ds + \cdots \tag{7}$$

EXERCISES

1. Set $X(t) = e^{At} + \sum_{n=1}^{\infty} \epsilon^n P_n(t)$ and use (5) to determine recurrence relations connecting $P_n(t)$ and $P_{n-1}(t)$.

2. Assuming for the moment that $e^{At}e^{Bt}$ can be written in the form e^C (a result we shall establish below), where $C = C_1t + C_2t^2 + C_3t^3 + \cdots$, determine the coefficient matrices C_1, C_2, C_3.

3. Assuming that $e^{A+\epsilon B}$ can be written in the form

$$e^{A+\epsilon B} = e^A e^{\epsilon C_1} e^{\epsilon^2 C_2} e^{\epsilon^3 C_3} \cdots$$

determine the matrices C_1, C_2, C_3.†

(The perturbation expansion in (7) possesses the great defect that it takes a matrix $e^{A+\epsilon B}$, which will be unitary if A and B are skew-Hermitian, and replaces it by an approximation which is nonunitary. The perturbation expansion given above does not suffer from this.)

† See also F. Fer, *Acad. Roy. Belg. Cl. Sci.*, vol. 44, no. 5, pp. 818–829, 1958.

15. Non-negativity of Solution. The following question arises in mathematical economics. Consider the equation

$$\frac{dx}{dt} = Ax + f(t) \qquad x(0) = c \tag{1}$$

where A is a constant matrix. What are necessary and sufficient conditions upon A in order that all the components of x be non-negative for $t \geq 0$ whenever the components of c are non-negative and the components of $f(t)$ are non-negative for $t \geq 0$?

Referring to (11.4), we see that a sufficient condition is that all elements of e^{At} be non-negative for $t \geq 0$, and it is easily seen that this condition is necessary as well.

It is rather surprising that there is a very simple criterion for this condition.

Theorem 3. *A necessary and sufficient condition that all elements of e^{At} be non-negative for $t \geq 0$ is that*

$$a_{ij} \geq 0 \qquad i \neq j \tag{2}$$

Proof. Since

$$e^{At} = I + At + \cdots \tag{3}$$

it is clear that the condition in (2) is necessary for the result to be true for small t. To establish the sufficiency, let us show that $a_{ij} > 0$, $i \neq j$, implies that the elements of e^{At} are positive for all t. It is clear that $a_{ij} > 0$ implies that e^{At} has positive elements for small t. Since

$$e^{At} = (e^{At/n})^n \tag{4}$$

for any integer n, the fact that the product of two positive matrices is positive yields the requisite positivity.

Since the elements of e^{At} are continuous functions of the a_{ij}, we see that the positivity of the elements of e^{At} for $a_{ij} > 0$, $i \neq j$, implies the non-negativity of the elements of e^{At} for $a_{ij} \geq 0$, $i \neq j$.

Another more direct proof proceeds as follows. Let c_1 be a scalar so that all the elements of $A + c_1I$ are non-negative. Then, clearly, all the elements of $e^{(A+c_1I)t}$ are non-negative. Also, the elements of e^{-c_1It} are non-negative since exponentials are always non-negative. Since

$$\begin{aligned} e^{At} &= e^{(A+c_1I)t-c_1It} \\ &= e^{(A+c_1I)t}e^{-c_1It} \end{aligned} \tag{5}$$

observing that $A + c_1I$ and $-c_1I$ commute, we have the desired non-negativity.

EXERCISES

1. Prove the foregoing result by using the system of differential equations

$$\frac{dx_i}{dt} = \sum_{j=1}^{N} a_{ij}x_j \qquad x_i(0) = c_i \qquad i = 1, 2, \ldots, N$$

2. Show that the result $a_{ij}(t) \geq 0$, $i \neq j$, is sufficient to ensure that the solution of the corresponding equation with variable coefficients is non-negative if the initial conditions are non-negative.

16. Polya's Functional Equation. We have seen that $Y(t) = e^{At}$ satisfies the functional equation

$$Y(s + t) = Y(s)Y(t) \qquad -\infty < s,t < \infty \qquad Y(0) = I \tag{1}$$

An interesting and important question is whether or not there are any other types of solutions of this fundamental matrix equation.

If $Y(t)$ has a derivative for all finite t, the answer is simple. Differentiating first with respect to s and then with respect to t, we obtain the two equations

$$\begin{aligned} Y'(s + t) &= Y(s)Y'(t) \\ Y'(s + t) &= Y'(s)Y(t) \end{aligned} \tag{2}$$

Hence

$$Y(s)Y'(t) = Y'(s)Y(t) \tag{3}$$

From (1) we see that $Y(0) = Y(-t)Y(t)$, which shows that $Y(t)$ cannot be singular for any value of t. Thus (3) yields

$$Y^{-1}(s)Y'(s) = Y'(t)Y^{-1}(t) \tag{4}$$

for all s and t. Thus we must have $Y'(t)Y^{-1}(t) = A$, a constant matrix. The equation

$$Y'(t) = AY(t) \tag{5}$$

then yields $Y(t) = e^{At}Y(0) = e^{At}$.

Let us now prove a stronger result.

Theorem 4. *Let $Y(t)$ be a continuous matrix function of t satisfying the functional equation in (1) for $0 \leq s,t$, $s + t \leq t_0$. Then in $[0,t_0]$, $Y(t)$ is of the form e^{At} for some constant matrix A.*

Proof. Consider the Riemann integral

$$\int_0^t Y(s)\,ds = \lim_{\delta \to 0} \sum_{k=0}^{N} Y(k\delta)\delta \tag{6}$$

where N is determined by the condition that $(N + 1)\delta = t$. Since, from (1) for $\delta > 0$,

$$Y(k\delta) = Y((k - 1)\delta)Y(\delta) = Y(\delta)^k \tag{7}$$

we see that

$$\int_0^t Y(s)\,ds = \lim_{\delta\to 0} \sum_{k=0}^{N} Y(\delta)^k \delta \tag{8}$$

We also have

$$\frac{[Y(\delta) - I]}{\delta} \sum_{k=0}^{N} Y(k\delta)\delta = Y((N+1)\delta) - I \tag{9}$$

which leads to the result

$$\lim_{\delta\to 0} \left\{ \left(\frac{Y(\delta) - I}{\delta}\right) \left(\sum_{k=0}^{N} Y(k\delta)\delta\right) \right\} = Y(t) - I \tag{10}$$

We have assumed that $Y(0) = I$. Hence, for small t,

$$\int_0^t Y(s)\,ds = tI + o(t) \tag{11}$$

and thus $\int_0^t Y(s)\,ds$ is nonsingular for small t. It follows that for fixed small t and sufficiently small δ, the matrix $\sum_{k=0}^{N} Y(k\delta)\delta$ is nonsingular and

$$\lim_{\delta\to 0} \left(\sum_{k=0}^{N} Y(k\delta)\delta\right)^{-1} = \left(\int_0^t Y(s)\,ds\right)^{-1} \tag{12}$$

Hence (10) yields the fact that $[Y(\delta) - I]/\delta$ has a limit as $\delta \to 0$. Call this limit A. Then

$$A = \lim_{\delta\to 0} \left(\frac{Y(\delta) - I}{\delta}\right) = [Y(t) - I] \left[\int_0^t Y(s)\,ds\right]^{-1} \tag{13}$$

for small t. From this we obtain

$$A \int_0^t Y(s)\,ds = Y(t) - I \tag{14}$$

and, finally,

$$Y'(t) = AY \qquad Y(0) = I \tag{15}$$

for small t. Since $Y(t) = e^{At}$ for small t, the functional equation yields $Y(t) = e^{At}$ for all t in $[0,t_0]$.

EXERCISES

1. What is wrong with the following continuation of (10): From (10) we have

$$\lim_{\delta\to 0} \left[\frac{Y(\delta) - I}{\delta}\right] \lim_{\delta\to 0} \left[\sum_{k=0}^{N} Y(k\delta)\delta\right] = Y(t) - I$$

$$\lim_{\delta\to 0} \left[\frac{Y(\delta) - I}{\delta}\right] \int_0^t Y(s)\,ds = Y(t) - I$$

etc.?

2. Let Y and Z denote the solutions of

$$\frac{dY}{dt} = A(t)Y \qquad Y(0) = I \qquad \frac{dZ}{dt} = ZB(t) \qquad Z(0) = I$$

Then the solution of

$$\frac{dX}{dt} = A(t)X + XB(t) \qquad X(0) = C$$

is given by $X = YCZ$.

3. Write the linear equation $u'' + p(t)u' + q(t)u = 0$ in the form

$$\begin{bmatrix} u' \\ v' \end{bmatrix} = \begin{bmatrix} 0 & 1 \\ -q(t) & -p(t) \end{bmatrix} \begin{bmatrix} u \\ v \end{bmatrix}$$

Use the preceding result to show that the linear equation whose general solution is $u = a_1u_1^2 + a_2u_1u_2 + a_3u_2^2$, where u_1 and u_2 are a set of linearly independent solutions of the second-order equation, is

$$u''' + 3p(t)u'' + [2p^2(t) + p'(t) + 4q(t)]u' + [4p(t)q(t) + 2q'(t)]u = 0$$

17. The Equation $dX/dt = AX + XB$. Let us now demonstrate Theorem 5.

Theorem 5. *The solution of*

$$\frac{dX}{dt} = AX + XB \qquad X(0) = C \tag{1}$$

is given by

$$X = e^{At}Ce^{Bt} \tag{2}$$

The proof follows immediately by direct verification. The result, although simple (and a special case of Exercise 2 of Sec. 16), plays an important role in various parts of mathematical physics.

EXERCISES

1. Obtain the solution given above by looking for a solution of the form $X = e^{At}Y$.

2. Let $Y(t)$ be a square root of $X(t)$. Show that

$$\left(\frac{dY}{dt}\right) Y + Y \left(\frac{dY}{dt}\right) = \frac{dX}{dt}$$

18. The Equation $AX + XB = C$. Using the foregoing result, we can establish Theorem 6.

Theorem 6. *If the expression*

$$X = -\int_0^\infty e^{At}Ce^{Bt}\,dt \tag{1}$$

exists for all C, it represents the unique solution of

$$AX + XB = C \tag{2}$$

Proof. Consider the equation

$$\frac{dZ}{dt} = AZ + ZB \qquad Z(0) = C \tag{3}$$

Let us integrate both sides between 0 and ∞, under the assumption that $\lim_{t \to \infty} Z(t) = 0$. The result is

$$-C = A\left(\int_0^\infty Z\,ds\right) + \left(\int_0^\infty Z\,ds\right)B \tag{4}$$

We see then that

$$-\int_0^\infty Z\,ds = -\int_0^\infty e^{At}Ce^{Bt}\,dt$$

satisfies (2).

In Chap. 12, we shall discuss in some detail the existence of the integral in (1) and the solutions of (2).

The uniqueness of solution follows from the linearity of the equation in (2). Considering this matrix equation as N^2 linear equations for the elements x_{ij} of X, we see that the existence of a solution for *all* C implies that the determinant of the coefficients of the x_{ij} is nonzero. Hence, there is a *unique* solution.

MISCELLANEOUS EXERCISES

1. Consider the sequence of matrices $\{X_n\}$ defined by

$$X_{n+1} = X_n(2I - AX_n) \qquad X_0 = B$$

Under what conditions on A and B does this sequence converge, and, if so, to what does it converge?

2. If $AB - BA = I$, and c_1 and c_2 are scalars, then

$$e^{c_1A+c_2B} = e^{c_1A}e^{c_2B}e^{c_1c_2/2}$$

See R. A. Sack,[1] where the general problem of determining the expansion of $f(A + B) = f(A) + \cdots$, for the case where $AB \neq BA$, is discussed. See also W. O. Kermack and W. H. McCrea[2] and R. Kubo,[3] where related expansions in terms of commutators appear.

3. What is the analogue of the result in Exercise 2 if $AB - BA = c_1A$?

4. If A is positive definite, $B(t)$ is positive definite for $t \geq 0$, and $|A| \leq |B(t)|$ for all $t \geq 0$, then $|A| \leq \left|\int_0^\infty B(t)g(t)\,dt\right|$ for any non-negative function $g(t)$ such that $\int_0^\infty g(t)\,dt = 1$.

[1] R. A. Sack, Taylor's Theorem for Shift Operators, *Phil. Mag.*, vol. 3, pp. 497–503, 1958.

[2] W. O. Kermack and W. H. McCrea, *Proc. Edinburgh Math. Soc.*, vol. 2 (220).

[3] R. Kubo, *J. Chem. Phys.*, vol. 20, p. 770, 1952.

5. If A is symmetric positive definite and $A \leq I$, then the sequence of matrices defined by the recurrence relation

$$X_{n+1} = X_n + \tfrac{1}{2}(A - X_n^2) \qquad X_0 = 0$$

converges to the positive definite square root of A.†

6. The problem of determining the values of λ satisfying the determinantal equation $|I - \lambda F_1 - \lambda^2 F_2 - \cdots - \lambda^k F_k| = 0$, where the F_i are square matrices of order N, is equivalent to determining the characteristic values of

$$A = \begin{bmatrix} F_1 & F_2 & \cdots & & F_k \\ I & 0 & \cdots & & 0 \\ 0 & I & \cdots & & 0 \\ \cdots & & 0 & I & 0 \end{bmatrix}$$

See K. G. Guderley, On Nonlinear Eigenvalue Problems for Matrices, *J. Ind. and Appl. Math.*, vol. 6, pp. 335–353, 1958.

7. Let H be a function of a parameter t, $H = H(t)$. Then

$$\frac{\partial}{\partial t} e^{-sH} = -\int_0^s e^{-(s-s_1)H} \frac{\partial H}{\partial t} e^{-s_1 H}\, ds_1$$

8. Show that

$$[A,e^{-sH}] = -\int_0^s e^{-(s-s_1)H}[A,H]e^{-s_1 H}\, ds_1$$

9. If $[A,H(t)] = \partial H/\partial t$, then $[A,f(H)] = \partial f(H)/\partial t$.

10. Show that $e^A B e^{-A} = B + [A,B] + [A,[A,B]]/2! + \cdots$.

11. Write $e^Z = e^{tA}e^{tB}$, $Z = \sum_{n=1}^{\infty} Z_n t^n$, $Z' = \sum_{n=1}^{\infty} nZ_n t^{n-1}$. Show that

$$\int_0^1 e^{rZ}Z'(t)e^{-rZ}\, dr = A + e^{tA}Be^{-tA}$$

12. Obtain recurrence relations for the Z_n in this fashion, for example $Z_1 = A + B$, $Z_2 = [A,B]/2$, $Z_3 = [[A,[A,B]] + [[A,B],B]]/12$, . . . (*Baker-Campbell-Hausdorff formula*). For the results of Exercises 7–12 and many further results, see Wilcox[1] and Eichler.[2]

13. Show that $e^{t(A+B)} = \lim_{n\to\infty} (e^{At/n}e^{Bt/n})^n$. The result is given in Trotter,[3] and has interesting applications to the study of the Schrödinger equation and Feynman integrals. See Faris.[4]

† C. Visser, Notes on Linear Operators, *Proc. Acad. Sci. Amsterdam*, vol. 40, pp. 270–272, 1937.

[1] R. M. Wilcox, Exponential Operators and Parameter Differentiation in Quantum Physics, *J. Math. Phys.*, vol. 8, pp. 962–982, 1967.

[2] M. Eichler, A New Proof of the Baker-Campbell-Hausdorff Formula, *J. Math. Soc. Japan*, vol. 20, pp. 23–25, 1968.

[3] H. F. Trotter, On the Products of Semigroups of Operators, *Proc. Am. Math. Soc.*, pp. 545–551, 1959.

[4] W. G. Faris, The Trotter Product Formula for Perturbations of Semibounded Operators, *Bull. Am. Math. Soc.*, vol. 73, pp. 211–215, 1967.

14. Let $x = y + Ax$ where A depends on a parameter t. Write $x = y + Ry$, defining the resolvent matrix. Show that $R' = (I + R)A'(I + R)$.

15. Assume that $A(t)$ has distinct characteristic values for $0 \le t \le t_1$, $\lambda_1, \lambda_2, \ldots, \lambda_N$ and let $x_1, x_2, \ldots, x_N$, the N associated characteristic vectors, be orthonormal. Show that

$$\lambda_n' = (A'x_n,x_n) \qquad n = 1, 2, \ldots, N$$
$$x_n' = \sum_{i \neq n} \frac{(A'x_n,x_i)x_i}{\lambda_n - \lambda_i}$$

Discuss the use of these results for computational purposes (*Kalaba-Schmaedeke-Vereeke*).

16. Consider the matrix operation

$$\frac{df(T)}{dT} = \left(\eta_{ij} \frac{\partial f(T)}{\partial t_{ij}}\right)$$

where $f(T)$ is a complex-valued function of the N^2 elements of T and $\eta_{ij} = 1$, $i = j$ $= \frac{1}{2}$, $i \neq j$. Show that

$$\frac{d}{dT}(f(T)g(T)) = \frac{df(T)}{dT} g(T) + f(T)\frac{dg(T)}{dT}$$
$$\frac{d}{dT}\varphi(f(T)) = \varphi'(f(T))\frac{df(T)}{dT}$$
$$\frac{d}{dT}|T| = |T|T^{-1}$$

The operator d/dT was introduced by Garding in the study of hyperbolic partial differential equations and has been used by Koecher and Maass in the study of Siegel modular functions. See Bellman and Lehman[1] for another application and references to the foregoing; also see Vetter.[2]

17. Introduce the norms

$$\|A\|_1 = \max_{1 \le j \le N} \sum_{i=1}^{N} |a_{ij}|$$
$$\|A\|_2 = \max_{(x,\bar{x})=1} (\overline{Ax},Ax)^{1/2}$$
$$\|A\|_\infty = \max_{1 \le i \le N} \sum_{j=1}^{N} |a_{ij}|$$
$$\alpha(A) = \sum_{i,j} |a_{ij}|$$
$$N(A) = \left(\sum_{i,j} |a_{ij}|^2\right)^{1/2}$$
$$M(A) = N \max_{i,j} |a_{ij}|$$

[1] R. Bellman and R. S. Lehman, The Reciprocity Formula for Multidimensional Theta Functions, *Proc. Am. Math. Soc.*, vol. 12, pp. 954–961, 1961.

[2] W. J. Vetter, *An Operational Calculus for Matrices*, to appear.

For what A do the ratios of these norms achieve their upper and lower bounds. See B. J. Stone, *Best Possible Ratios of Certain Matrix Norms,* Technical Report 19, Stanford University, 1962.

18. Let $u(t)$ be a function of t n times differentiable over $[0,T]$ and consider the quadratic form

$$Q_n(a) = \int_0^T (u^{(n)} + a_1u^{(n-1)} + \cdots + a_nu)^2\,dt$$

Does $\min_a Q_n(a)$ necessarily approach zero as $n \to \infty$.

19. Consider the case where the interval is $[-\infty, \infty]$. Use Fourier integrals and the Parseval-Plancherel theorem to obtain an expression for $\min_a Q_n(a)$ in terms of orthogonal polynomials. (The foregoing is connected with *differential approximation* see Bellman and Kalaba[1] and Lew.[2])

20. Consider the matrix equation $X - UXV = W$. Show that a formal solution is $X = \sum_{k=0}^{\infty} U^kWV^k$. When does the series converge and actually represent a solution? See R. A. Smith, Matrix Equation $XA + BX = C$, *SIAM J. Appl. Math.*, vol. 16, pp. 198–201, 1968.

21. Consider the equation $X = C + \epsilon(AX + XB)$, ϵ a scalar parameter. Write $X = C + \sum_{n=1}^{\infty} \epsilon^n\varphi_n(A,B)$. Show that $\varphi_n = A\varphi_{n-1} + \varphi_{n-1}B$, $n \geq 1$, with $\varphi_0 = C$, and that

$$\varphi_n = A^nC + \binom{n}{1} A^{n-1}CB + \cdots + CB^n$$

22. Introduce a position operator P with the property that when it operates on a monomial consisting of powers of A and B in any order with C somewhere, it shifts all powers of A in front of C and all powers of B after C. Thus

$$P(A^{a_1}B^{b_1} \cdots A^{a_k}B^{b_k}CA^{a_{k+1}}B^{b_{k+1}} \cdots A^{a_N}B^{b_N}) = A^{\Sigma a_i}CB^{\Sigma b_i}$$

Further, define P to be additive,

$$P(m_1(A,B) + m_2(A,B)) = P(m_1(A,B)) + P(m_2(A,B))$$

where m_1 and m_2 are monomials of the foregoing types. Show that $\varphi_n(A,B) = P((A + B)^nC)$.

23. Hence, show that X, as defined by Exercise 21, may be written $X = P([I - \epsilon(A + B)]^{-1}C)$.

24. Similarly show that if $X = E + \epsilon(AX + XB + CXD)$, then $X = P([I - \epsilon(A + B + CF)]^{-1}E)$, with P suitably defined.

[1] R. Bellman and R. Kalaba, *Quasilinearization and Nonlinear Boundary-value Problems,* American Elsevier Publishing Company, Inc., New York, 1965.

[2] A. Lew, *Some Results in Differential Approximation,* University of Southern California Press, Los Angeles, USCEE-314, 1968.

25. Show that if all the characteristic roots of A are less than one in absolute value, the solution of $A^*XA - X = -Q$ is given by

$$X = (2\pi i)^{-1} \int (A^* - z^{-1}I)^{-1}Q(A - zI)^{-1}z^{-1}\,dz$$

where $\int$ denotes integration round the circle $|z| = 1$. Alternately,

$$X = (2\pi)^{-1} \int_{-\pi}^{\pi} (G^{-1})^*QG^{-1}\,d\theta$$

where $G = A - Ie^{i\theta}$. See R. A. Smith.[1]

Bibliography and Discussion

§1. The equilibrium or stability theory of differential systems was started independently, and almost simultaneously, by Poincaré and Lyapunov. Further references and discussion will be found in Chap. 13 and in the book

R. Bellman, *Stability Theory of Differential Equations*, Dover Publications, New York, 1969.

Some interesting transformations of matrix differential equations may be found in the following papers:

J. H. Barrett, Matrix Systems of Second Order Differential Systems, *Portugalicae Math.*, vol. 14, pp. 79–89, 1955.

J. H. Barrett, A Prufer Transformation for Matrix Differential Equations, *Proc. Am. Math. Soc.*, vol. 8, pp. 510–518, 1957.[2]

§2. The matrix exponential lies at the very heart of all advanced work in the field of linear functional equations. Suitably generalized to arbitrary operators, it is the foundation stone of the theory of semigroups, see

E. Hille, Functional Analysis and Semi-groups, *Am. Math. Soc. Pub.*, 1942.

and its revised form

[1] R. A. Smith, Matrix Calculations for Lyapunov Quadratic Forms, *J. Diff. Eq.*, vol. 2, pp. 208–217, 1966.

R. A. Smith, Bounds for Quadratic Lyapunov Functions, *J. Math. Anal. Appl.*, vol. 12, pp. 425–435, 1966.

[2] J. J. Levin, On the Matrix Riccati Equation, *Proc. Am. Math. Soc.*, vol. 10, pp. 519–524, 1959.

E. Hille and R. Phillips, Functional Analysis and Semi-groups, *Am. Math. Soc. Colloq. Pub.*, vol. 31, 1958.

In a second direction, the problem of expressing $e^{At}e^{Bt}$ in the form e^{Ct}, where A and B do not commute, has important ramifications not only in the theory of Lie groups and algebras, but also in modern quantum field theory. The interested reader may consult

W. Magnus, Algebraic Aspects of the Theory of Systems of Linear Differential Equations, *Research Rept.* BR-3, New York University, Institute of Mathematical Sciences, June, 1953; also *Comm. Pure Appl. Math.*, vol. 7, no. 4, 1954.

H. F. Baker, On the Integration of Linear Differential Equations, *Proc. London Math. Soc.*, vol. 34, pp. 347–360, 1902; vol. 35, pp. 333–374, 1903; second series, vol. 2, pp. 293–296, 1904.

H. F. Baker, Alternants and Continuous Groups, *Proc. London Math. Soc.*, second series, vol. 3, pp. 24–47, 1904.

H. B. Keller and J. B. Keller, On Systems of Linear Ordinary Differential Equations, *Research Rept.* EM-33, New York University, Washington Square College, Mathematics Research Group, 1957.

F. Hausdorff, Die symbolische exponential Formel in der Gruppentheorie, *Saechsischen Akademie der Wissenschaften zu Leipzig, Math.-phys. Klasse*, vol. 58, pp. 19–48, 1906.

K. Goldberg, The Formal Power Series for log (e^xe^y), *Duke Math. J.*, vol. 23, pp. 13–21, 1956.†

In another, and related direction, we meet the problem of expressing the solution of a linear system of the form $dX/dt = A(t)X$, $X(0) = I$, in the form of an "exponential." This question leads to the study of "product integrals"; see

L. Schlesinger, Neue Grundlagen für einen Infinitesimalkalkul der Matrizen, *Math. Z.*, vol. 33, pp. 33–61, 1931.

L. Schlesinger, Weitere Beitrage zum Infinitesimalkalkul der Matrizen, *Math. Z.*, vol. 35, pp. 485–501, 1932.

G. Rasch, Zur Theorie und Anwendung des Produktintegrals, *J. reine angew. Math.*, vol. 171, pp. 65–119, 1934.

B. W. Helton, Integral Equations and Product Integrals, *Pacific J. Math.*, vol. 16, pp. 297–322, 1968.

† See also Kuo-Tsai Chen, Integration of Paths, Geometric Invariants and a Generalized Baker-Hausdorff Formula, *Ann. Math.*, vol. 65, pp. 163–178, 1957.

For more recent developments, see

W. Magnus, On the Exponential Solution of Differential Equations for a Linear Operator, *Comm. Pure Appl. Math.*, vol. 7, pp. 649–673, 1954.

For an understanding of the physical interest in this problem, see

R. P. Feynman, An Operator Calculus Having Applications in Quantum Electrodynamics, *Phys. Rev.*, vol. 84, pp. 108–128, 1951.

Finally, let us mention that a great deal of effort has been devoted to the generalization of the foregoing results and methods to infinite systems of linear differential equations with constant coefficients. Systems of this type arise in a natural fashion in probability theory, particularly in its applications to physics and biology in the study of "birth-and-death" processes. These matters will be discussed again in Chap. 16. See

N. Arley and A. Borchsenius, On the Theory of Infinite Systems of Differential Equations and Their Applications to the Theory of Stochastic Processes and the Perturbation Theory of Quantum Mechanics, *Acta Math.*, vol. 76, pp. 261–322, 1945.

R. Bellman, The Boundedness of Solutions of Infinite Systems of Linear Differential Equations, *Duke Math J.*, vol. 14, pp. 695–706, 1947.

§3. Further results concerning norms of matrices, and further references, may be found in

A. S. Householder, The Approximate Solution of Matrix Problems, *J. Assoc. Comp. Mach.*, vol. 5, pp. 205–243, 1958.

J. von Neumann and H. Goldstine, Numerical Inverting of Matrices of High Order, *Bull. Am. Math. Soc.*, vol. 53, pp. 1021–1099, 1947.

A. Ostrowski, Über Normen von Matrizen, *Math. Z.*, Bd. 63, pp. 2–18, 1955.

K. Fan and A. J. Hoffman, Some Metric Inequalities in the Space of Matrices, *Proc. Am. Math. Soc.*, vol. 6, pp. 111–116, 1955.

T. E. Easterfield, Matrix Norms and Vector Measures, *Duke Math. J.*, vol. 24, pp. 663–671, 1957.

§5. The reader who is familiar with the theory of Lebesgue integration will see that the result of Theorem 1 can be obtained under much weaker conditions on $A(t)$. However, since we have no need for the stronger result, we have not stated it.

§14. Variation of the characteristic values and characteristic roots of A as the *dimension* changes can be discussed using the techniques presented in

R. Bellman and S. Lehman, Functional Equations in the Theory of Dynamic Programming-X: Resolvents, Characteristic Values and Functions, *Duke Math. J.*, 1960.

§15. Non-negativity results of this type play an important role in probability theory and mathematical economics, where the physical model makes the result intuitively clear. This point will be discussed again in Chap. 16. They also play a role in the study of various classes of nonlinear equations; see

R. Bellman, Functional Equations in the Theory of Dynamic Programming—V: Positivity and Quasi-linearity, *Proc. Natl. Acad. Sci. U.S.*, vol. 41, pp. 743–746, 1955.

R. Kalaba, On Nonlinear Differential Equations, the Maximum Operation, and Monotone Convergence, Ph.D. Thesis, New York University, February, 1958.

For an extensive discussion of the positivity of operators, see Chap. 5 of

E. F. Beckenbach and R. Bellman, *Inequalities*, Springer, 1961.

The first proof is due to S. Karlin, and the second proof to O. Taussky. The result was first presented in

R. Bellman, I. Glicksberg, and O. Gross, On Some Variational Problems Occurring in the Theory of Dynamic Programming, *Rend. Circ. Palermo*, Serie II, pp. 1–35, 1954.

§16. The result and proof follow

G. Polya, Über die Funktionalgleichung der Exponentialfunktion in Matrizkalkul, *Sitzber. Akad. Berlin*, pp. 96–99, 1928.

The nature of the solutions of $Y(s + t) = Y(s)Y(t)$ without the normalizing condition $Y(0) = I$ is also of interest. See Shaffer.[1]

§17. We are not aware of the origin of this result which may be found in the papers of many authors. For an extensive discussion of this and related equations, see

J. A. Lappo-Danilevsky, *Mémoires sur la théorie des systèmes des équations différentielles linéaires*, Chelsea Publishing Co., New York, 1953.

[1] C. V. Shaffer, On Singular Solutions to Polya's Functional Equation, *IEEE*, vol. AC-13, pp. 135–136, 1968.

The fact that the solution of the equation $dX/dt = AX + XB$ has the indicated form becomes quite natural when one examines the way in which it arises in quantum mechanics; cf.

D. ter Haar, *Elements of Statistical Mechanics*, Rinehart & Company, Inc., New York, p. 149, 1954.

Y. Nambu, *Progr. Theoret. Phys.* (Kyoto), vol. 4, p. 331, 1949.

H. Primas and H. Gunthard, Eine Methode zur direkten Berechnung . . . , *Helv. Phys. Acta*, vol. 31, pp. 413–434, 1958.

§18. For a discussion of the operator T defined by $TX = AX - XB$, see

M. Rosenbloom, On the Operator Equation $BX - XA = Q$, *Duke Math. J.*, vol. 23, pp. 263–269, 1956.

For a generalization to the operator S defined by $SX = \sum_j A_jXB_j$, see

G. Lumer and M. Rosenbloom, Linear Operator Equations, *Proc. Am. Math. Soc.*, vol. 10, pp. 32–41, 1959.

Exercises 2 and 3 are taken from

R. Bellman, On the Linear Differential Equation Whose Solutions Are the Products of Solutions of Two Given Linear Differential Equations, *Bol. unione mat. Italiana*, ser. III, anno XII, pp. 12–15, 1957.

The intimate connection between matrix theory and the theory of the behavior of the solutions of linear differential equations and thus with the study of linear oscillatory systems leads to the beautiful theory of Gantmacher and Krein,

V. Gantmacher and M. Krein, Sur les matrices complètement non négatives et oscillatoires, *Comp. math.*, vol. 4, pp. 445–476, 1937.

As pointed out at the end of the remarks on Chap. 16, these results have important applications in probability theory.

For a treatment of the equation

$$A\frac{d^2x}{dt^2} + B\frac{dx}{dt} + Cx = f$$

where A, B, and C are symmetric by variational techniques, under suitable hypotheses concerning A, B, and C, see

R. J. Duffin, A Minimax Theory for Overdamped Networks, *J. Rat. Mech. and Analysis,* vol. 4, pp. 221–233, 1955.

See also

A. M. Ostrowski, On Lancaster's Decomposition of a Matrix Differential Operator, *Arch. Rat. Mech. Anal.,* vol. 8, pp. 238–241, 1961.

H. Langer, Über Lancaster's Zerlegung von Matrizen-Scharen, *Arch. Rat. Mech. Anal.,* vol. 29, pp. 75–80, 1968.

The study of linear circuits leads in another way to an interesting domain of matrix theory, namely, the study of when a given matrix of rational functions can be considered to be the open-circuit impedance matrix of an n-port network. The concept of a *positive real* matrix enters here in a fundamental fashion. For work on this subject, and references, see

L. Weinberg and P. Slepian, *Positive Real Matrices,* Hughes Research Laboratories, Culver City, Calif., 1958.

11

Explicit Solutions and Canonical Forms

1. Introduction. Although the results obtained in Chap. 10 are of great elegance, it must be confessed that they do not satisfactorily resolve the problem of obtaining explicit scalar solutions.

In this chapter, we shall pursue this question and show that it leads us to the question of determining the characteristic roots and vectors of matrices which are not necessarily symmetric. As in the case of symmetric matrices, this leads us to the study of various canonical forms for matrices.

2. Euler's Method. In Chap. 10, we demonstrated the fact that the vector-matrix equation

$$\frac{dx}{dt} = Ax \qquad x(0) = c \tag{1}$$

possessed a unique solution which could be represented in the form

$$x = e^{At}c \tag{2}$$

Here, we traverse another road. Following the method of Euler, let us begin by looking for special solutions of the form $x = e^{\lambda t}c^1$, where λ is a scalar and c^1 a vector, ignoring for the moment the initial condition $x(0) = c$. Substituting, we see that λ and c^1 are bound by the relation

$$\lambda c^1 = Ac^1 \tag{3}$$

Since c^1 is to be nontrivial, λ must be a root of the characteristic equation

$$|A - \lambda I| = 0 \tag{4}$$

and c^1 must be an associated characteristic vector.

From this quite different direction then, we are led to the basic problem treated in the first part of the book—the determination of characteristic roots and vectors. Here, however, the analysis is both more complicated and less complete due to the fact that the matrices we encounter are not necessarily symmetric.

EXERCISE

1. Find the characteristic roots and vectors of the matrix

$$A = \begin{bmatrix} \lambda & 1 & & & 0 \\ & \lambda & 1 & & \\ & & \cdot & & \\ & & & \cdot & \\ & & & \cdot & 1 \\ 0 & & & & \lambda \end{bmatrix}$$

3. Construction of a Solution. Let us now see if we can construct a solution of the initial value problem along the preceding lines. As before, denote the characteristic roots by $\lambda_1, \lambda_2, \ldots, \lambda_N$. Since these will, in general, be complex, there is now no attached ordering.

To simplify our initial discussion, let us assume initially that the characteristic roots are *distinct*. Let $c^1, c^2, \ldots, c^N$, be a set of associated characteristic vectors, also, in general, complex.

To put together the required solution, we use the *principle of superposition.* Since $e^{\lambda_k t}c^k$ is a solution for $k = 1, 2, \ldots, N$, the linear combination

$$x = \sum_{k=1}^{N} a_k e^{\lambda_k t} c^k \tag{1}$$

is a solution of $dx/dt = Ax$ for any set of scalars a_k.

The question is now whether or not these scalars can be chosen so that $x(0) = c$. This condition leads to the vector equation

$$c = \sum_{k=1}^{N} a_k c^k \tag{2}$$

This system has a unique solution provided that C is nonsingular, where

$$C = (c^1 c^2 \cdots c^N) \tag{3}$$

and, as before, this represents the matrix whose columns are the vectors c^i.

4. Nonsingularity of C. There are several ways of establishing the fact that C is nonsingular. The first method proceeds as follows. Suppose that $|C| = 0$. Then there exist a nontrivial set of scalars $b_1, b_2, \ldots, b_N$ such that

$$0 = \sum_{k=1}^{N} b_k c^k \tag{1}$$

It follows that

$$z = \sum_{k=1}^{N} b_k c^k e^{\lambda_k t} \tag{2}$$

is a solution of $dx/dt = Ax$ satisfying the initial condition $x(0) = 0$. The uniqueness theorem asserts that actually $z = 0$ for all t.

We must then show that the relation

$$0 = \sum_{k=1}^{N} b_k c^k e^{\lambda_k t} \tag{3}$$

cannot hold for all t if the $\{b_k\}$ are a nontrivial set of scalars, none of the c^k is trivial, and the λ_i are all distinct.

Without loss of generality, assume that $b_N \neq 0$. Divide through by $e^{\lambda_1 t}$ in (3) and differentiate the resultant expression with respect to t. The new expression has the form

$$\sum_{k=2}^{N} b_k c^k (\lambda_k - \lambda_1) e^{(\lambda_k - \lambda_1)t} = 0 \tag{4}$$

Since the λ_k have been assumed distinct, the quantities $\lambda_k - \lambda_1$, $k = 2, 3, \ldots, N$, are all distinct. We may now appeal to an inductive argument, or merely repeat the above procedure.

It is clear that in this way we eventually reach a relation of the form

$$b_N c^N (\lambda_N - \lambda_1)(\lambda_N - \lambda_2) \cdots (\lambda_N - \lambda_{N-1}) e^{(\lambda_N - \lambda_{N-1})t} = 0 \tag{5}$$

Since c^N is nontrivial, we must have $b_N = 0$, a contradiction.

5. Second Method. Beginning with the relation in (4.1), multiply both sides by A. The result is

$$0 = \sum_{k=1}^{N} b_k \lambda_k c^k \tag{1}$$

Repeating this operation $(N - 1)$ times, the result is the system of linear equations

$$0 = \sum_{k=1}^{N} b_k \lambda_k^r c^k \qquad r = 0, 1, 2, \ldots, N - 1 \tag{2}$$

Considering only the ith components of the c^k, the quantities c_i^k, we obtain the scalar equations

$$0 = \sum_{k=1}^{N} b_k \lambda_k^r c_i^k \qquad r = 0, 1, 2, \ldots, N - 1 \tag{3}$$

Since not all of the b_k are zero, and since all of the c_i^k cannot be zero as i and k take the values 1, 2, . . . , N, we see that (3) possesses a nontrivial solution for some i.

It follows then that the determinant of the coefficients

$$|\lambda_k^r| \qquad k = 1, 2, \ldots, N, r = 0, 1, 2, \ldots, N-1 \tag{4}$$

must be zero. This, however, is not true, as we shall see in a moment.

6. The Vandermonde Determinant. The determinant

$$|\lambda_k^r| = \begin{vmatrix} 1 & 1 & \cdots & 1 \\ \lambda_1 & \lambda_2 & & \lambda_N \\ \lambda_1^2 & \lambda_2^2 & & \\ \cdot & \cdot & & \cdot \\ \cdot & \cdot & & \cdot \\ \cdot & \cdot & & \cdot \\ \lambda_1^{N-1} & \lambda_2^{N-1} & \cdots & \lambda_N^{N-1} \end{vmatrix} \tag{1}$$

is a very famous one, bearing the name, *Vandermonde determinant.*

We wish to show that $|\lambda_k^r| \neq 0$ if $\lambda_i \neq \lambda_j$ for $i \neq j$. The simplest way to do this is to evaluate the determinant. Regard $|\lambda_k^r|$ as a polynomial of degree $N-1$ in λ_1. As a polynomial of degree $N-1$, $|\lambda_k^r|$ has the roots $\lambda_1 = \lambda_2, \lambda_1 = \lambda_3, \ldots, \lambda_1 = \lambda_N$, since the determinant is zero whenever two columns are equal.

It follows that

$$|\lambda_k^r| = (\lambda_1 - \lambda_2)(\lambda_1 - \lambda_3) \cdots (\lambda_1 - \lambda_N)q(\lambda_2,\lambda_3, \ldots ,\lambda_N) \tag{2}$$

where q is a function depending only upon $\lambda_2, \lambda_3, \ldots, \lambda_N$. Similarly, viewed as a polynomial of degree $N-1$ in λ_2, $|\lambda_k^r|$ must possess the factor $(\lambda_2 - \lambda_1)(\lambda_2 - \lambda_3) \cdots (\lambda_2 - \lambda_N)$. Continuing in this way, we see that

$$|\lambda_k^r| = \prod_{1 \leq i < j \leq N} (\lambda_j - \lambda_i)\phi(\lambda_1,\lambda_2, \ldots ,\lambda_N) \tag{3}$$

where ϕ is a polynomial in the λ_i. Comparing, however, the degrees in the λ_i of the two sides, we see that ϕ must be a constant. Examining the coefficients of $\lambda_1\lambda_2^2 \cdots \lambda_N{}_1^{N-1}$ on both sides, we see that $\phi = 1$.

From this explicit representation, we see that $|\lambda_k^r| \neq 0$ if $\lambda_i \neq \lambda_j$ for $i \neq j$.

EXERCISE

1. Using similar reasoning, evaluate the Cauchy determinant

$$\left|\frac{1}{\lambda_i + \mu_j}\right| \qquad i, j = 1, 2, \ldots, N$$

7. Explicit Solution of Linear Differential Equations—Diagonal Matrices. Let us now consider another approach, quite different from that given in the preceding sections.

Given the equation

$$\frac{dx}{dt} = Ax \qquad x(0) = c \tag{1}$$

make a change of variable, $x = Ty$, where T is a constant nonsingular matrix to be chosen in a moment. The equation for y is then

$$\frac{dy}{dt} = T^{-1}ATy \qquad y(0) = T^{-1}c \tag{2}$$

What choice of T will simplify this equation to the point where the solution can be immediately obtained? Suppose that we can find a matrix T such that

$$T^{-1}AT = \begin{bmatrix} \mu_1 & & & & 0 \\ & \mu_2 & & & \\ & & \cdot & & \\ & & & \cdot & \\ & & & & \cdot \\ 0 & & & & \mu_N \end{bmatrix} \tag{3}$$

a diagonal matrix.

If this can be done, the equation in (2) decomposes into N independent equations of the form

$$\frac{dy_i}{dt} = \mu_i y_i \qquad y_i(0) = c_i' \qquad i = 1, 2, \ldots, N \tag{4}$$

having the elementary solutions $y_i = e^{\mu_i t}c_i'$. Once y has been determined, x is readily determined.

8. Diagonalization of a Matrix. Let us now discuss the possibility of the diagonalization of A, a problem of great interest and difficulty. As we know from the first part of the book, if A is symmetric, real or complex, a matrix T possessing the required property can be found, with $T^{-1} = T'$. As we shall see, there are a number of other important classes of matrices which can be diagonalized, and, what is more important, there are other types of canonical representations which are equally useful.

Consider, however, the general case. To begin with, we know that the $\{\mu_i\}$ must be the same set as the $\{\lambda_i\}$ since the characteristic roots of $T^{-1}AT$ are the same as those of A.

It follows then, as in previous chapters, that the columns of T must be characteristic vectors of A.

Conversely, if the λ_i are distinct and T is the matrix formed using the associated characteristic vectors as columns, it follows that

$$AT = T \begin{bmatrix} \lambda_1 & & & & 0 \\ & \lambda_2 & & & \\ & & \cdot & & \\ & & & \cdot & \\ & & & & \cdot \\ & 0 & & & \\ & & & & \lambda_N \end{bmatrix} \tag{1}$$

We have thus established the following important result.

Theorem 1. *If the characteristic roots of A are distinct, there exists a matrix T such that*

$$T^{-1}AT = \begin{bmatrix} \lambda_1 & & & & 0 \\ & \lambda_2 & & & \\ & & \cdot & & \\ & & & \cdot & \\ & & & & \cdot \\ & 0 & & & \\ & & & & \lambda_N \end{bmatrix} \tag{2}$$

As we shall see, the assumption that the roots are distinct is now quite an important one. In the general case, no such simple representation holds, as we shall see in Sec. 10.

EXERCISES

1. Show that the Cayley-Hamilton theorem is valid for matrices with distinct characteristic roots.

2. Show that the assumption concerning N distinct characteristic roots may be replaced by one requiring N linearly independent characteristic vectors.

9. Connection between Approaches. As we have seen, one method of solution of the linear equation

$$\frac{dx}{dt} = Ax \qquad x(0) = c \tag{1}$$

leads to the expression $x = e^{At}c$, while on the other hand, a method based upon characteristic roots and vectors produces scalar exponentials.

To obtain the connection between these approaches (still under the assumption that $\lambda_i \neq \lambda_j$), which must exist because of uniqueness of solution, in the equation

$$\frac{dX}{dt} = AX \qquad X(0) = I \tag{2}$$

make the change of variable $X = TY$, where T is as in Sec. 8. Then Y satisfies the equation

$$\frac{dY}{dt} = T^{-1}ATY \qquad Y(0) = T^{-1} \tag{3}$$

or

$$\frac{dY}{dt} = \begin{bmatrix} \lambda_1 & & & 0 \\ & \lambda_2 & & \\ & & \ddots & \\ 0 & & & \lambda_N \end{bmatrix} Y \qquad Y(0) = T^{-1} \tag{4}$$

It follows that

$$Y = \begin{bmatrix} e^{\lambda_1 t} & & & 0 \\ & e^{\lambda_2 t} & & \\ & & \ddots & \\ 0 & & & e^{\lambda_N t} \end{bmatrix} T^{-1} \tag{5}$$

whence

$$X = e^{At} = T \begin{bmatrix} e^{\lambda_1 t} & & & 0 \\ & e^{\lambda_2 t} & & \\ & & \ddots & \\ 0 & & & e^{\lambda_N t} \end{bmatrix} T^{-1} \tag{6}$$

Let us again note that this representation has been established under the assumption that the characteristic roots of A are distinct.

EXERCISES

1. Establish the representation in (6) directly from the exponential series and the representation for A.

2. Use (6) to show that $|e^A| = e^{\operatorname{tr}(A)}$.

3. Use the method of continuity to show that this result is valid for any square matrix A.

10. Multiple Characteristic Roots. Let us now turn to a discussion of the case where A has multiple roots. In order to appreciate some of the difficulties that we may expect to encounter, let us begin by showing that it may not always be possible to obtain a canonical representation such as that given in (8.2).

Theorem 2. *There exist matrices which cannot be reduced to a diagonal form by means of a nonsingular matrix, as in* (8.2). *Equivalently, there exist matrices of dimension N which do not possess N linearly independent characteristic vectors.*

Proof. Consider the particular matrix

$$A = \begin{bmatrix} 1 & 1 \\ 0 & 1 \end{bmatrix} \tag{1}$$

If there exists a T such that

$$T^{-1}AT = \begin{bmatrix} \lambda_1 & 0 \\ 0 & \lambda_2 \end{bmatrix} \tag{2}$$

we know that the columns of T must be characteristic vectors of A. Let us then determine the characteristic roots and vectors.

The characteristic equation is

$$\begin{vmatrix} 1-\lambda & 1 \\ 0 & 1-\lambda \end{vmatrix} = (1-\lambda)^2 = 0 \tag{3}$$

Hence 1 is a double root. The characteristic vectors are determined by the equations

$$\begin{aligned} x_1 + x_2 &= x_1 \\ x_2 &= x_2 \end{aligned} \tag{4}$$

It follows that $x_2 = 0$, with x_1 arbitrary. Consequently, all characteristic vectors are scalar multiples of

$$c^1 = \begin{bmatrix} 1 \\ 0 \end{bmatrix} \tag{5}$$

This means that T must be a singular matrix.

Observe then the surprising fact that an arbitrary matrix need not possess a full complement of characteristic vectors. This makes us more appreciative of what a strong restriction on a matrix symmetry is.

The fact that diagonalization may not be always available greatly complicates the study of general matrices—and, in return, adds equally to their interest. As we shall see below, there are several methods we can use, based upon the use of alternate canonical forms and approximation theorems, to bypass some of these hurdles.

EXERCISES

1. Why can't we use a continuity argument to deduce a diagonal representation for general matrices from the result for matrices with distinct characteristic roots?

2. Derive the solution of the differential equation

$$\frac{dx_1}{dt} = x_1 + x_2 \qquad x_1(0) = c_1$$
$$\frac{dx_2}{dt} = x_2 \qquad x_2(0) = c_2$$

and thus show that the matrix

$$A = \begin{bmatrix} 1 & 1 \\ 0 & 1 \end{bmatrix}$$

cannot be reduced to diagonal form.

11. Jordan Canonical Form. Let us now discuss a canonical form for an arbitrary square matrix which is useful in the treatment of a number of problems. Since its proof is quite detailed and is readily available in a number of texts, and furthermore since its use can always be avoided by one means or another (at least in our encounters), we shall merely state the result without a proof.

Theorem 3. *Let us denote by $L_k(\lambda)$ a $k \times k$ matrix of the form*

$$L_k(\lambda) = \begin{bmatrix} \lambda & 1 & 0 & \cdots & 0 \\ 0 & \lambda & 1 & \cdots & 0 \\ & & & & \cdot \\ & & & & \cdot \\ & & & & \cdot \\ \cdot & \cdot & & \lambda & 1 \\ \cdot & \cdot & & & \\ \cdot & \cdot & & & \\ 0 & 0 & & \cdots & \lambda \end{bmatrix} \tag{1}$$

where $L_1(\lambda) = \lambda$. *There exists a matrix T such that*

$$T^{-1}AT = \begin{bmatrix} L_{k_1}(\lambda_1) & & & & 0 \\ & L_{k_2}(\lambda_2) & & & \\ & & \cdot & & \\ & & & \cdot & \\ 0 & & & & L_{k_r}(\lambda_r) \end{bmatrix} \tag{2}$$

with $k_1 + k_2 + \cdots + k_r = N$. *The* λ_i *are the characteristic roots of* A, *not necessarily distinct.*

This representation is called the *Jordan canonical form.*

For example, if A is of the third degree and has λ_1 as a root of multiplicity three, it may be reduced to one of the three forms

$$A_1 = \begin{bmatrix} \lambda_1 & 0 & 0 \\ 0 & \lambda_1 & 0 \\ 0 & 0 & \lambda_1 \end{bmatrix} \qquad A_2 = \begin{bmatrix} \lambda_1 & 0 & 0 \\ 0 & \lambda_1 & 1 \\ 0 & 0 & \lambda_1 \end{bmatrix}$$
$$A_3 = \begin{bmatrix} \lambda_1 & 1 & 0 \\ 0 & \lambda_1 & 1 \\ 0 & 0 & \lambda_1 \end{bmatrix} \tag{3}$$

EXERCISES

1. Using the Jordan canonical form, determine the form of e^{At} for general A.

2. Use this result to obtain the necessary and sufficient condition that $e^{At} \to 0$ as $t \to \infty$.

3. Prove that the A_i, $i = 1, 2, 3$, are distinct in the sense that there exists no T for which $T^{-1}A_iT = A_j$, for $i \neq j$. Give both an algebraic proof and one depending upon the solutions of the associated linear differential equations.

4. Show that $L_k(\lambda) - \lambda I$ when raised to the kth power yields the null matrix.

12. Multiple Characteristic Roots. Using the Jordan canonical form, we possess a systematic method for obtaining the explicit analytic structure of the solutions of linear differential equations with constant coefficients. Thus, if the equation is reduced to

$$\begin{aligned} \frac{dy_1}{dt} &= \lambda y_1 + y_2 \qquad y_1(0) = c_1 \\ \frac{dy_2}{dt} &= \lambda y_2 \qquad y_2(0) = c_2 \end{aligned} \tag{1}$$

we have $y_2 = e^{\lambda t}c_2$, and y_1 determined as the solution of

$$\frac{dy_1}{dt} = \lambda y_1 + e^{\lambda t}c_2 \qquad y_1(0) = c_1 \tag{2}$$

Using the integrating factor $e^{-\lambda t}$, this yields

$$\frac{d}{dt}(y_1 e^{-\lambda t}) = c_2 \tag{3}$$

whence

$$y_1 = c_1 e^{\lambda t} + c_2 t e^{\lambda t} \tag{4}$$

This is one way of seeing why terms of the form $te^{\lambda t}$, and generally terms of the form $t^k e^{\lambda t}$, enter into the solution in the case of multiple roots. Let us now discuss an alternative approach.

If A has distinct characteristic roots, $\lambda_1, \lambda_2, \ldots, \lambda_N$, we may write N particular solutions of $dy/dt = Ay$ in the form $y = c(\lambda)e^{\lambda t}$, $\lambda = \lambda_1, \lambda_2, \ldots, \lambda_N$, where $c(\lambda)$ is a polynomial function of λ. This we can see in the following way.

As we know, to obtain a solution of the equation, we set $y = e^{\lambda t}c$, where c is a scalar, and determine λ and c from the equation

$$Ac = \lambda c \qquad \text{or} \qquad (A - \lambda I)c = 0 \tag{5}$$

Once λ has been determined from the characteristic equation

$$|A - \lambda I| = 0$$

we must determine a solution of the linear system in (5). Let

$$b_{ij} = |A - \lambda I|_{ij} \tag{6}$$

denote the cofactor of the term $a_{ij} - \lambda\delta_{ij}$ in the determinantal expansion of $|A - \lambda I|$.

Then

$$c_i = b_{1i} \qquad i = 1, 2, \ldots, N \tag{7}$$

is a solution of (5), with the property that the c_i are polynomials in λ. It may happen that this is a trivial solution in the sense that all the c_i are zero. In that case, we may try the alternative sets of solutions

$$c_i = b_{ki} \qquad i = 1, 2, \ldots, N \tag{8}$$

for $k = 2, 3, \ldots, N$.

Let us now show that one of these sets of solutions must be nontrivial if the characteristic roots of A are all distinct. In particular, let us show that for any particular characteristic root, $\lambda = \lambda_1$, not all the expressions b_{ii} can be zero if λ_1 is not a multiple root.

Consider the characteristic equation

$$f(\lambda) = |A - \lambda I| = \begin{vmatrix} a_{11} - \lambda & a_{12} & \cdots & a_{1N} \\ a_{21} & a_{22} - \lambda & \cdots & a_{2N} \\ \vdots & & & \\ a_{N1} & a_{N2} & \cdots & a_{NN} - \lambda \end{vmatrix} = 0 \tag{9}$$

Then, using the rule for differentiating a determinant, we have

$$
f'(\lambda) = \begin{vmatrix} -1 & 0 & \cdots & 0 \\ a_{21} & a_{22} - \lambda & \cdots & a_{2N} \\ \cdot & & & \\ \cdot & & & \\ \cdot & & & \\ a_{N1} & a_{N2} & \cdots & a_{NN} - \lambda \end{vmatrix} + \begin{vmatrix} a_{11} - \lambda & a_{12} & \cdots & a_{1N} \\ 0 & -1 & \cdots & 0 \\ \cdot & & & \\ \cdot & & & \\ \cdot & & & \\ a_{N1} & a_{N2} & \cdots & a_{NN} - \lambda \end{vmatrix} + \cdots + \begin{vmatrix} a_{11} - \lambda & a_{12} & \cdots & a_{1N} \\ a_{21} & a_{22} - \lambda & \cdots & a_{2N} \\ \cdot & & & \\ \cdot & & & \\ \cdot & & & \\ 0 & 0 & \cdots & -1 \end{vmatrix}
$$

$$
= -b_{11} - b_{22} - \cdots - b_{NN} \tag{10}
$$

If, for $\lambda = \lambda_1$, we have $b_{ii}(\lambda_1) = 0$, $i = 1, 2, \ldots, N$, then $f'(\lambda_1) = 0$, which means that λ_1 is a multiple root, a contradiction.

It follows that we have a solution of the desired form which we use in the following fashion.

If λ_1 and λ_2 are two distinct characteristic roots, then

$$
z = \frac{c(\lambda_1)e^{\lambda_1 t} - c(\lambda_2)e^{\lambda_2 t}}{\lambda_1 - \lambda_2} \tag{11}
$$

is also a solution of the differential equation. The case where λ_1 is a multiple root can be considered to be a limiting case of the situation where λ_2 approaches λ_1. Taking the limit as $\lambda_2 \to \lambda_1$, we find, as a candidate for a solution, the expression

$$
z = \left[\frac{dc(\lambda)}{d\lambda} e^{\lambda t} + te^{\lambda t}c(\lambda) \right]_{\lambda=\lambda_1} \tag{12}
$$

We leave as an exercise for the reader the task of putting this on a rigorous foundation. Let us merely point out that there are two ways of establishing this result, direct verification, or as a consequence of general theorems concerning the continuous dependence of solutions upon the matrix A.

EXERCISE

1. How do we obtain solutions if λ_1 is a double root, using a direct algebraic approach?

13. Semidiagonal or Triangular Form—Schur's Theorem. Let us now prove one of the most useful reduction theorems in matrix theory.

Theorem 4. *Given any matrix A, there exists a unitary matrix T such that*

$$T^*AT = \begin{bmatrix} \lambda_1 & b_{12} & \cdots & b_{1N} \\ & \lambda_2 & & \\ & & \ddots & \\ 0 & & & \lambda_N \end{bmatrix} \tag{1}$$

where, as the notation indicates, the elements below the main diagonal are zero.

Proof. Let us proceed by induction, beginning with the 2×2 case. Let λ_1 be a characteristic root of A and c^1 an associated characteristic vector normalized by the condition $(c^1,\overline{c^1}) = 1$. Let T be a matrix whose first column is c^1 and whose second column is chosen so that T is unitary. Then evaluating the expression $T^{-1}AT$ as the product $T^{-1}(AT)$, we see that

$$T^{-1}AT = \begin{bmatrix} \lambda_1 & b_{12} \\ 0 & b_{22} \end{bmatrix} \tag{2}$$

The quantity b_{22} must equal λ_2 since $T^{-1}AT$ has the same characteristic roots as A.

Let us now show that we can use the reduction for Nth-order matrices to demonstrate the result for $(N + 1)$-dimensional matrices. As before, let c^1 be a normalized characteristic vector associated with λ_1, and let N other vectors $a^1, a^2, \ldots, a^N$, be chosen so that the matrix T_1, whose columns are $c^1, a^1, a^2, \ldots, a^N$, is unitary. Then, as for $N = 2$, we have

$$T_1^{-1}AT_1 = \begin{bmatrix} \lambda_1 & b_{12}{}^1 & \cdots & b_{1N+1}{}^1 \\ 0 & & & \\ \vdots & & B_N & \\ 0 & & & \end{bmatrix} \tag{3}$$

where B_N is an $N \times N$ matrix.

Since the characteristic equation of the right-hand side is

$$(\lambda_1 - \lambda)|B_N - \lambda I| = 0 \tag{4}$$

it follows that the characteristic roots of B_N are $\lambda_2, \lambda_3, \ldots, \lambda_{N+1}$, the remaining N characteristic roots of A. The inductive hypothesis asserts that there exists a unitary matrix T_N such that

$$T_N^{-1}B_NT_N = \begin{bmatrix} \lambda_2 & c_{12} & \cdots & c_{1N} \\ & \lambda_3 & & c_{2N} \\ & & \ddots & \vdots \\ 0 & & & \lambda_{N+1} \end{bmatrix} \tag{5}$$

Let T_{N+1} be the unitary $(N + 1) \times (N + 1)$ matrix formed as follows:

$$T_{N+1} = \begin{bmatrix} 1 & 0 & \cdots & 0 \\ 0 & & & \\ \vdots & & T_N & \\ 0 & & & \end{bmatrix} \tag{6}$$

Consider the expression

$$\begin{aligned}(T_1T_{N+1})^{-1}A(T_1T_{N+1}) &= T_{N+1}^{-1}(T_1^{-1}AT_1)T_{N+1} \\ &= \begin{bmatrix} \lambda_1 & b_{12} & \cdots & b_{1,N+1} \\ & \lambda_2 & & \vdots \\ & & \ddots & \vdots \\ 0 & & & \lambda_{N+1} \end{bmatrix}\end{aligned} \tag{7}$$

The matrix T_1T_{N+1} is thus the required unitary matrix which reduces A to semidiagonal form.

EXERCISES

1. Show that if A has k simple roots, there exists a matrix T such that

$$T^{-1}AT = \begin{bmatrix} \lambda_1 & 0 & \cdots & 0 & b_{1,k+1} & \cdots & b_{1,N} \\ 0 & \lambda_2 & \cdots & 0 & b_{2,k+1} & \cdots & b_{2,N} \\ \vdots & & & & \vdots & & \vdots \\ 0 & \cdots & \cdots & \lambda_k & b_{k,k+1} & & b_{k,N} \\ & & & & \lambda_{k+1} & & \\ & & & & & \ddots & \\ & 0 & & & 0 & & \lambda_N \end{bmatrix}$$

2. Using the semidiagonal form derived above, obtain the general solution of $dx/dt = Ax$, $x(0) = c$.

3. Determine a necessary and sufficient condition that no terms of the form $te^{\lambda t}$ occur.

4. Determine the general solution of $x'' + Ax = 0$, where A is positive definite.

5. Use the triangular form to obtain necessary and sufficient conditions that $e^{At} \to 0$ as $t \to \infty$.

6. Consider the *difference equation* $x(n + 1) = Ax(n)$, $n = 0, 1, \ldots,$ where $x(0) = c$. Show that the solution is given by $x(n) = A^n c$.

7. Find particular solutions by setting $x(n) = \lambda^n c$, and determine the set of possible λ and c.

8. Show that the general solution has the form

$$x(n) = \sum_{i=1}^{N} p_i(n)\lambda_i^n$$

where the $p_i(n)$ are vectors whose components are polynomials in n of degree at most $N - 1$.

9. What is the necessary and sufficient condition that every solution of $x(n + 1) = Ax(n)$ approach zero as $n \to \infty$?

10. Consider the difference equation $x((n + 1)\Delta) = x(n\Delta) + A\,\Delta x(n\Delta)$, $n = 0, 1, \ldots,$ $x(0) = c$, where Δ is a positive scalar. As $\Delta \to 0$, show that $x(n\Delta) \to x(t)$, the solution of the differential equation $dx/dt = Ax$, $x(0) = c$, provided that $n\Delta \to t$. Give two proofs, one using the explicit form of the solution, and the other without this aid.

14. Normal Matrices. Real matrices which commute with their transposes, or, more generally, complex matrices which commute with their conjugate transposes, are called *normal*. Explicitly, we must have

$$AA^* = A^*A \tag{1}$$

The utility of this concept lies in the following theorem.

Theorem 5. *If A is normal, it can be reduced to diagonal form by a unitary matrix.*

Proof. As we know, we can find a unitary transformation, T, which reduces A to semidiagonal form,

$$A = T \begin{bmatrix} \lambda_1 & b_{12} & \cdots & b_{1N} \\ & \lambda_2 & \cdots & b_{2N} \\ & & \ddots & \vdots \\ 0 & & & \lambda_N \end{bmatrix} T^{-1} = TST^{-1} \tag{2}$$

Then, since T is unitary,

$$A^* = T \begin{bmatrix} \overline{\lambda_1} & & & 0 \\ \overline{b_{12}} & \overline{\lambda_2} & & \\ \vdots & & \ddots & \\ \overline{b_{1N}} & \cdots & & \overline{\lambda_N} \end{bmatrix} T^{-1} = TS^*T^{-1} \tag{3}$$

Hence,

$$\begin{aligned} AA^* &= TSS^*T^{-1} \\ &= A^*A = TS^*ST^{-1} \end{aligned} \tag{4}$$

Thus, we must have

$$SS^* = S^*S \tag{5}$$

Equating corresponding elements in the products, we see that

$$b_{ij} = 0 \qquad 1 \le i < j \le N \tag{6}$$

which means that the semidiagonal form in (2) is actually diagonal.

EXERCISES

1. If A is a normal matrix with all characteristic roots real, it is Hermitian, and it can be reduced to diagonal form by an orthogonal transformation.

2. Show that $(Ax,Ax) = (A'x,A^*x)$ if A is normal.

3. Show that $A - \lambda I$ is normal if A is normal.

4. Using the foregoing exercises, or otherwise, show that if A is normal, then x is a characteristic vector of A if and only if it is a characteristic vector of A^*.

5. If A is normal, characteristic vectors x, y belonging to distinct characteristic values are orthogonal in the sense that $(x,\bar{y}) = 0$.

6. Establish Theorem 5 using this fact.

7. Prove that the converse of Theorem 5 is also true.

8. If B is normal and there exists an angle θ such that $Ae^{i\theta} + A^*e^{-i\theta} \geq 0$ where $A^2 = B$, then A is normal (*Putnam, Proc. Am. Math. Soc., vol.* 8, *pp.* 768–769, 1957).

9. If A and B commute, then A^* and B commute if A is normal.

15. An Approximation Theorem. Let us now state a very useful approximation theorem which can often be used to treat questions involving general square matrices.

Theorem 6. *We can find a matrix T such that*

$$T^{-1}AT = \begin{bmatrix} \lambda_1 & b_{12} & \cdots & b_{1N} \\ & \lambda_2 & & \\ & & \ddots & \\ 0 & & & \lambda_N \end{bmatrix} \tag{1}$$

with $\sum_{i,j} |b_{ij}| \leq \epsilon$, where ϵ is any preassigned positive constant.

Proof. Let T_1 be a matrix which reduces A to semidiagonal form. Then the change of variable $y = T_1 z$ converts $dy/dt = Ay$ into

$$
\begin{aligned}
\frac{dz_1}{dt} &= \lambda_1 z_1 + b_{12} z_2 + \cdots + b_{1N} z_N \\
\frac{dz_2}{dt} &= \lambda_2 z_2 + \cdots + b_{2N} z_N \\
&\ \ \vdots \\
\frac{dz_N}{dt} &= \lambda_N z_N
\end{aligned}
\tag{2}
$$

It is now easy to see how to choose T so as to obtain the stated result. In (2) make the change of variable

$$z_k = r_1{}^k z_k{}^1 \tag{3}$$

Then the new variables $z_i{}^1$ satisfy the system of equations

$$
\begin{aligned}
\frac{dz_1{}^1}{dt} &= \lambda_1 z_1{}^1 + r_1 b_{12} z_2{}^1 + \cdots + r_1{}^{N-1} b_{1N} z_N{}^1 \\
\frac{dz_2{}^1}{dt} &= \lambda_2 z_2{}^1 + \cdots + r_1{}^{N-2} b_{2N} z_N{}^1 \\
&\ \ \vdots \\
\frac{dz_N{}^1}{dt} &= \lambda_N z_N{}^1
\end{aligned}
\tag{4}
$$

With a suitable choice of r_1, the sum of the absolute values of the off-diagonal terms can be made as small as possible. The last change of variable is equivalent to a transformation $z = Ez^1$, where E is nonsingular. Hence we can take T to be T_1E.

At first sight, the above result may seem to contradict the result that a general matrix possessing multiple characteristic roots may not be reducible to diagonal form. The point is that T depends upon ϵ. If we attempt to let $\epsilon \to 0$, we find that either T approaches a singular matrix, or else possesses no limit.

16. Another Approximation Theorem. A further result which can similarly be used to reduce the proof of results involving general matrices to a proof involving diagonal matrices is given in Theorem 7.

Theorem 7. *Given any matrix A, we can find a matrix B with distinct characteristic roots such that $\|A - B\| \leq \epsilon$, where ϵ is any preassigned positive quantity.*

Proof. Consider the matrix $A + E$, where $E = (e_{ij})$ with the e_{ij} independent complex variables. If $A + E$ has a multiple characteristic root, then $f(\lambda) = |A + E - \lambda I|$ and $f'(\lambda)$ have a root in common. If $f(\lambda)$

and $f'(\lambda)$ have a root in common, the resultant of the two polynomials, $R(E)$, a polynomial in the e_{ij} must vanish. We wish to show that we can find a set of values of the e_{ij} for which $\sum_{i,j} |e_{ij}|$ is arbitrarily small, and for which $R(E) \neq 0$. The negation of this is the statement that $R(E)$ vanishes identically in the neighborhood of the origin in e_{ij} space. If this is true, the polynomial $R(E)$ vanishes identically, which means that $f(\lambda)$ always has a multiple root.

Consider, however, the following choice of the e_{ij},

$$\begin{aligned} e_{ij} &= -a_{ij} \qquad i \neq j \\ e_{ii} &= i - a_{ii} \qquad i = 1, 2, \ldots, N \end{aligned} \tag{1}$$

The matrix $A + E$ clearly does not have multiple roots. Hence, $R(E)$ does not vanish identically, and we can find a matrix $B = A + E$ with the desired properties.

EXERCISE

1. Construct a proof of this result which depends only upon the fact that A can be reduced to triangular form. (The point of this is that the foregoing proof, although short and rigorous, makes use of the concept and properties of the resultant of two polynomials which are not as easy to establish rigorously as might be thought.)

17. The Cayley-Hamilton Theorem. Using the approximation theorem established in Sec. 16, we can finally establish in full generality the famous Cayley-Hamilton theorem.

Theorem 8. *Every matrix satisfies its characteristic equation.*

Proof. Let $A + E$, with $\|E\| \leq \epsilon$, be a matrix with distinct characteristic roots. Then, as we know, $A + E$ satisfies its characteristic equation. The characteristic polynomial of $A + E$ is $f(\lambda,E) = |A + E - \lambda I|$, a polynomial in λ whose coefficients are polynomials in the elements of E, and thus continuous in the elements of E. Hence,

$$\lim_{E \to 0} f(A + E,E) = f(A) \tag{1}$$

Since $f(A + E,E) = 0$, we see that $f(A) = 0$.

EXERCISES

1. Establish the Cayley-Hamilton theorem using the Jordan canonical form.

2. Establish the Cayley-Hamilton theorem under the assumption that A possesses a full set of characteristic vectors which are linearly independent.

3. Use the Cayley-Hamilton theorem to derive a representation for A^{-1} as a polynomial in A, provided that A is nonsingular.

18. Alternate Proof of Cayley-Hamilton Theorem. Once we know what we wish to prove, it is much easier to devise a variety of short and elegant proofs. Let us present one of purely algebraic nature, making no appeal to continuity.

Consider the matrix inverse to $A - \lambda I$, $(A - \lambda I)^{-1}$, for λ not a characteristic value. We see that

$$(A - \lambda I)^{-1} = B(\lambda)/f(\lambda) \tag{1}$$

where the elements of $B(\lambda)$ are polynomials in λ of degree $N - 1$ and $f(\lambda)$ is the characteristic polynomial of $A - \lambda I$, the determinant $|A - \lambda I|$. Hence,

$$(A - \lambda I)B(\lambda) = f(\lambda)I \tag{2}$$

Write

$$\begin{aligned} B(\lambda) &= \lambda^{N-1}B_{N-1} + \lambda^{N-2}B_{N-2} + \cdots + B_0 \\ f(\lambda) &= (-1)^N\lambda^N + c_1\lambda^{N-1} + \cdots + c_N \end{aligned} \tag{3}$$

where the B_i are matrices independent of λ. Equating coefficients, we obtain a sequence of relations

$$\begin{aligned} -B_{N-1} &= (-1)^N I \\ AB_{N-1} - B_{N-2} &= c_1 I \\ AB_{N-2} - B_{N-3} &= c_2 I \end{aligned} \tag{4}$$

and so on. We see that each B_i is a polynomial in A with scalar coefficients, and hence that $B_iA = AB_i$ for each i. It follows then that the identity in (2) is valid not only for all scalar quantities λ, but also for all matrices which commute with A.

In particular, it is valid for $\lambda = A$. Substituting in (2), we see that $f(A) = 0$, the desired result.

19. Linear Equations with Periodic Coefficients. Let us now examine the problem of solving a linear differential equation with periodic coefficients,

$$\frac{dx}{dt} = P(t)x \qquad x(0) = c \tag{1}$$

where $P(t)$ is a continuous function of t satisfying the condition

$$P(t + 1) = P(t) \tag{2}$$

for all t. Surprisingly, the problem is one of extreme difficulty. Even the relatively simple scalar equation

$$\frac{d^2u}{dt^2} + (a + b \cos t)u = 0 \tag{3}$$

the Mathieu equation, poses major difficulties, and has essentially a theory of its own.

In obtaining a canonical representation of the solution of (1), we are led to discuss a problem of independent interest, namely the representation of a nonsingular matrix as an exponential.

As usual, it is convenient to discuss the matrix equation first.

We shall prove Theorem 9.

Theorem 9. *The solution of*

$$\frac{dX}{dt} = P(t)X \qquad X(0) = I \tag{4}$$

where $P(t)$ *is periodic of period* 1, *may be written in the form*

$$X(t) = Q(t)e^{Ct} \tag{5}$$

where $Q(t)$ *is periodic of period* 1.

Proof. It is clear from the periodicity of $P(t)$, that $X(t + 1)$ is a solution of the differential equation in (4) whenever $X(t)$ is. Since $X(1)$ is not singular, we see that $X(t + 1)X(1)^{-1}$ is a solution of (4) with the initial value I. It follows from the uniqueness theorem that

$$X(t + 1)X(1)^{-1} = X(t) \tag{6}$$

Suppose that it were possible to write

$$X(1) = e^{C} \tag{7}$$

Consider, then, the matrix $Q(t) = X(t)e^{-Ct}$. We have

$$\begin{aligned} Q(t + 1) &= X(t + 1)e^{-C(t+1)} = X(t + 1)e^{-C}e^{-Ct} \\ &= X(t + 1)X(1)^{-1}e^{-Ct} = X(t)e^{-Ct} = Q(t) \end{aligned} \tag{8}$$

This establishes (5).

It remains to prove that (7) is valid.

20. A Nonsingular Matrix Is an Exponential. Let us now establish Theorem 10.

Theorem 10. *A nonsingular matrix is an exponential.*

Let A be nonsingular with distinct characteristic roots, so that A may be written

$$A = T \begin{bmatrix} \lambda_1 & & & & 0 \\ & \lambda_2 & & & \\ & & \cdot & & \\ & & & \cdot & \\ & & & & \cdot \\ 0 & & & & \lambda_N \end{bmatrix} T^{-1} \tag{1}$$

Then $A = e^C$ where

$$C = T \begin{bmatrix} \log \lambda_1 & & & 0 \\ & \log \lambda_2 & & \\ & & \ddots & \\ 0 & & & \log \lambda_N \end{bmatrix} T^{-1} \tag{2}$$

which proves the result for this case.

If A has multiple characteristic roots, we can use the Jordan canonical form with equal effect. Let

$$A = T \begin{bmatrix} L_{k_1}(\lambda_1) & & & 0 \\ & L_{k_2}(\lambda_2) & & \\ & & \ddots & \\ 0 & & & L_{k_r}(\lambda_r) \end{bmatrix} T^{-1} \tag{3}$$

It is sufficient to show that each of the matrices $L_{k_i}(\lambda_i)$ has a logarithm. For if B_i is a logarithm of $L_{k_i}(\lambda_i)$, then

$$B = T \begin{bmatrix} B_1 & & & 0 \\ & B_2 & & \\ & & \ddots & \\ 0 & & & B_r \end{bmatrix} T^{-1} \tag{4}$$

is a logarithm of A.

We have noted above in Exercise 4 of Sec. 11 that

$$[L_k(\lambda) - \lambda I]^k = 0 \tag{5}$$

It follows then that the formal logarithm

$$B = \log L_k(\lambda) = \log [\lambda I + L_k(\lambda) - \lambda I]$$
$$= I \log \lambda + \sum_{n=1}^{\infty} \frac{(-1)^{n+1}}{n\lambda^n} [L_k(\lambda) - \lambda I]^n$$
$$= I \log \lambda + \sum_{n=1}^{k-1} \frac{(-1)^{n+1}}{n\lambda^n} [L_k(\lambda) - \lambda I]^n$$

exists and is actually a logarithm in the sense that $e^B = A$.

EXERCISES

1. A nonsingular matrix has a kth root for $k = 2, 3, \ldots$.
2. The solution of $dX/dt = Q(t)X$, $X(0) = I$, can be put in the form $e^P e^{P_1} \cdots e^{P_n} \cdots$, where $P = \int_0^t Q(s)\,ds$, and $P_n = \int_0^t Q_n\,ds$, with

$$Q_n = e^{-P_{n-1}}Q_{n-1}e^{P_{n-1}} + \int_0^{-1} e^{sP_{n-1}}Q_{n-1}e^{-sP_{n-1}}\,ds$$

The infinite product converges if t is sufficiently small (*F. Fer, Bull. classe sci., Acad. roy. Belg., vol.* 44, *no.* 5, *pp.* 818–829, 1958).

21. An Alternate Proof. Since we have not proved the Jordan representation, let us present an independent proof of an inductive nature. Assume that the result holds for $N \times N$ matrices and that we have already converted the nonsingular $(N + 1)$-dimensional matrix under consideration to triangular form,

$$A_{N+1} = \begin{bmatrix} A_N & a_N \\ 0 & \lambda_{N+1} \end{bmatrix} \tag{1}$$

where A_N is an N-dimensional matrix in triangular form and a_N denotes an N-dimensional column vector.

Let B_N be a logarithm of A_N, which is nonsingular if A_{N+1} is, and write

$$B_{N+1} = \begin{bmatrix} B_N & x \\ 0 & l \end{bmatrix} \tag{2}$$

where $l = \log \lambda_{N+1}$ and x is an unknown N-dimensional vector.

It remains to show that x may be determined so that $e^{B_{N+1}} = A_{N+1}$. It is easy to see inductively that

$$B_{N+1}{}^k = \begin{bmatrix} B_N{}^k & (B_N{}^{k-1} + lB_N{}^{k-2} + \cdots + l^{k-1}I)x \\ 0 & l^k \end{bmatrix} \tag{3}$$

for $k = 1, 2, \ldots$.

Hence

$$e^{B_{N+1}} = \begin{bmatrix} e^{B_N} & \sum_{k=0}^{\infty} (B_N{}^{k-1} + B_N{}^{k-2}l + \cdots + l^{k-1}I)x/k! \\ 0 & \lambda_{N+1} \end{bmatrix} \tag{4}$$

where the first two terms are taken to be 0 and 1.

If l is not a characteristic root of B_N, we have

$$\begin{aligned} C(l) &= \sum_{k=0}^{\infty} (B_N{}^{k-1} + B_N{}^{k-2}l + \cdots + l^{k-1}I)/k! \\ &= \sum_{k=0}^{\infty} (B_N{}^k - l^kI)(B_N - lI)^{-1}/k! = (e^{B_N} - e^lI)(B_N - lI)^{-1} \end{aligned} \tag{5}$$

Hence

$$|C(l)| = \frac{|e^{B_N} - e^lI|}{|B_N - lI|} = \prod_{k=1}^{N} \frac{e^{r_k} - e^l}{r_k - l} \tag{6}$$

where $r_1, r_2, \ldots, r_N$ are the characteristic roots of B_N.

Since $|C(l)|$ is a continuous function of l for all l, as we see from the series, it follows that (6) derived under the assumption that $l \neq r_k$ holds for all l.

If $l \neq r_k$ or any of the other values of the logarithms of the λ_k, it is clear that $|C(l)| \neq 0$. If $l = r_k$, then the factor $(e^{r_k} - e^l)/(r_k - l)$ reduces to $e^{r_k} \neq 0$. Restricting ourselves to principal values of the logarithms, we see that $C(l)$ is never singular.

Hence, we see that x may be determined so that $C(l)x = a_N$.

This completes the demonstration.

22. Some Interesting Transformations. The reduction of differential equations with variable matrix,

$$\frac{dx}{dt} = A(t)x \qquad x(0) = c \tag{1}$$

to canonical form is a problem of some difficulty. Since it lies more within the province of the theory of differential equations than of matrix theory, we shall not pursue it further here.

Some interesting transformations, however, arise at the very beginning of the investigation. Set $x = Ty$, where T is a function of t. Then y satisfies the equation

$$\frac{dy}{dt} = T^{-1}(AT - dT/dt)y \tag{2}$$

Write

$$f(A,T) = T^{-1}(AT - dT/dt) \tag{3}$$

Then we see from the derivation that $f(A,T)$ satisfies the functional equation

$$f(A,ST) = f(T^{-1}(AT - dT/dt), S) \tag{4}$$

EXERCISE

1. Use the scalar equation

$$\frac{d^n u}{dt^n} + a_1(t)\frac{d^{n-1}u}{dt^{n-1}} + \cdots + a_n(t)u = 0$$

and the transformations $u = sv$, $t = \phi(s)$, to derive similar classes of functional equations.

23. Biorthogonality. In the first part of the book, dealing with symmetric matrices, we observed that every N-dimensional vector could be written as a linear combination of N orthonormal characteristic vectors associated with the N characteristic roots of an $N \times N$ matrix A.

Thus, if

$$x = \sum_{i=1}^{N} a_i x^i \tag{1}$$

the coefficients are determined very simply by means of the formula

$$a_i = (x,x^i) \tag{2}$$

If A is not symmetric, we face two difficulties in obtaining an analogue of this result. In the first place, A may not have a full quota of characteristic vectors, and in the second place, these need not be mutually orthogonal.

To overcome the first difficulty, we need merely make an assumption that we consider only matrices that do possess N linearly independent characteristic vectors. To overcome the second difficulty, we shall use the adjoint matrix and the concept of biorthogonality, rather than orthogonality.

Let $y^1, y^2, y^3, \ldots, y^N$ be a set of linear independent characteristic vectors associated with the characteristic roots $\lambda_1, \lambda_2, \ldots, \lambda_N$ of A. Then

$$T = (y^1,y^2, \ldots ,y^N) \tag{3}$$

is nonsingular and possesses the property of transforming A to diagonal form,

$$T^{-1}AT = \begin{bmatrix} \lambda_1 & & & & 0 \\ & \lambda_2 & & & \\ & & \cdot & & \\ & & & \cdot & \\ & & & & \cdot \\ 0 & & & & \lambda_N \end{bmatrix} \tag{4}$$

From this, it follows that

$$T^*A^*(T^*)^{-1} = \begin{bmatrix} \bar{\lambda}_1 & & & & 0 \\ & \bar{\lambda}_2 & & & \\ & & \cdot & & \\ & & & \cdot & \\ & & & & \cdot \\ 0 & & & & \bar{\lambda}_N \end{bmatrix} \tag{5}$$

Hence, A^* also possesses the property of possessing N linearly independent characteristic vectors, furnished by the columns of $(T^*)^{-1}$ and associated with the characteristic roots $\bar{\lambda}_1, \bar{\lambda}_2, \ldots, \bar{\lambda}_N$. Call these characteristic vectors $z^1, z^2, \ldots, z^N$, respectively. To determine the coefficients in the representation

$$x = \sum_{i=1}^{N} a_i y^i \tag{6}$$

we proceed as follows. From

$$\begin{aligned} Ay^i &= \lambda_i y^i \\ A^*z^j &= \bar{\lambda}_j z^j \end{aligned} \tag{7}$$

we obtain the further relations

$$(Ay^i,\overline{z^j}) = (\lambda_i y^i,\overline{z^j}) \tag{8}$$

and thus

$$(y^i,\overline{A^*z^j}) = (y^i,\lambda_j \overline{z_j}) = (\lambda_i y^i,\overline{z^j})$$

Hence,

$$(\lambda_i - \lambda_j)(y^i,\overline{z_j}) = 0 \tag{9}$$

Hence, if $\lambda_i \neq \lambda_j$, we have

$$(y^i,\overline{z^j}) = 0 \tag{10}$$

It is clear that $(y^i,\overline{z^i}) \neq 0$, since $\overline{z^i}$, being nontrivial, cannot be orthogonal to all of the y^i.

Thus, in the special case where the λ_i are all distinct, we can write, for the coefficients in (6),

$$a_i = \frac{(x,\overline{z^i})}{(y^i,\overline{z^i})} \tag{11}$$

Two sets of vectors $\{y^i\}$, $\{z^i\}$, satisfying the condition

$$(y^i,\overline{z^j}) = 0 \qquad i \neq j \tag{12}$$

are said to be *biorthogonal.* It is clear that we can normalize the vectors so as to obtain the additional condition

$$(y^i,\overline{z^i}) = 1 \tag{13}$$

This technique for expanding x as a linear combination of the y^i bears out a comment made earlier that often the properties of a matrix are most easily understood in terms of the properties of its transpose.

EXERCISES

1. How does one treat the case of multiple characteristic roots?
2. What simplifications ensue if A is normal?

24. The Laplace Transform. Once we know that the solution of a linear differential equation with constant coefficients has the form

$$x = \sum_{k=1}^{N} e^{\lambda_k t} p_k(t) \tag{1}$$

where $p_k(t)$ is a vector whose components are polynomials of degree at most $N - 1$, we can use the Laplace transform to determine the solution without paying any attention to a number of details of rigor which usually necessitate an involved preliminary discussion.

The value of the Laplace transform lies in the fact that it can be used to transform a transcendental function into an algebraic function. Specifically, it transforms the simplest exponential function into the simplest rational function.

The integral

$$g(s) = \int_0^\infty e^{-st} f(t)\, dt \tag{2}$$

is called the *Laplace transform* of $f(t)$, provided that it exists. Since we shall be interested only in functions $f(t)$ having the form given in (1), it is clear that $g(s)$ will always exist if Re (s) is sufficiently large.

The basic formula is

$$\int_0^\infty e^{-st}e^{at}\,dt = \frac{1}{s-a} \tag{3}$$

for Re $(s) >$ Re (a). It follows that if we know that $f(t)$ has the form

$$f(t) = c_1 e^{\lambda_1 t} \tag{4}$$

then the equality

$$\int_0^\infty e^{-st}f(t)\,dt = c_2/(s-a) \tag{5}$$

means that $c_1 = c_2$ and $\lambda_1 = a$.

Similarly, from the additivity of the integral, it follows that if $f(t)$ is of the form given in (1), and if $g(s)$ has the form

$$g(s) = c_1/(s-a_1) + c_2/(s-a_2) \tag{6}$$

then we must have

$$f(t) = c_1 e^{a_1 t} + c_2 e^{a_2 t} \tag{7}$$

25. An Example. Consider the problem of solving the linear differential equation

$$\begin{aligned} u'' - 3u' + 2u &= 0 \\ u(0) = 1 \qquad u'(0) &= 0 \end{aligned} \tag{1}$$

Since the characteristic equation is

$$\lambda^2 - 3\lambda + 2 = 0 \tag{2}$$

with roots $\lambda = 1$ and $\lambda = 2$, we see that the solution of (1) will have the form

$$u = c_1 e^{2t} + c_2 e^t \tag{3}$$

where c_1 and c_2 will be determined by the initial conditions in (1). Thus we obtain the linear system

$$\begin{aligned} c_1 + c_2 &= 1 \\ 2c_1 + c_2 &= 0 \end{aligned} \tag{4}$$

yielding

$$c_1 = -1 \qquad c_2 = 2 \tag{5}$$

so that the solution is

$$u = 2e^t - e^{2t} \tag{6}$$

In place of this procedure, let us use the Laplace transform. From (1), for Re $(s) > 2$, we obtain the relation

$$\int_0^\infty u''e^{-st}\,dt - 3\int_0^\infty u'e^{-st}\,dt + 2\int_0^\infty ue^{-st}\,dt = 0 \tag{7}$$

In order to evaluate the first two terms, we integrate by parts, obtaining

$$\begin{aligned}\int_0^\infty u'e^{-st}\,dt &= ue^{-st}\Big]_0^\infty + s\int_0^\infty ue^{-st}\,dt \\ &= -1 + s\int_0^\infty ue^{-st}\,dt \\ \int_0^\infty u''e^{-st}\,dt &= u'e^{-st}\Big]_0^\infty + s\int_0^\infty u'e^{-st}\,dt \\ &= 0 + s\int_0^\infty u'e^{-st}\,dt \\ &= -s + s^2\int_0^\infty ue^{-st}\,dt \end{aligned} \tag{8}$$

upon referring to the relation for $\int_0^\infty u'e^{-st}\,dt$.

Thus, (7) yields the equation

$$(s^2 - 3s + 2)\int_0^\infty ue^{-st}\,dt = s - 3 \tag{9}$$

or

$$\int_0^\infty ue^{-st}\,dt = \frac{s-3}{s^2 - 3s + 2} \tag{10}$$

The rational function on the right side of (10) has a partial fraction decomposition

$$\frac{s-3}{s^2-3s+2} = \frac{a_1}{s-1} + \frac{a_2}{s-2} \tag{11}$$

with

$$\begin{aligned} a_1 &= \lim_{s\to 1} \frac{(s-1)(s-3)}{s^2-3s+2} = \frac{s-3}{s-2}\Big]_{s=1} = 2 \\ a_2 &= \lim_{s\to 2} \frac{(s-2)(s-3)}{s^2-3s+2} = \frac{s-3}{s-1}\Big]_{s=2} = -1 \end{aligned} \tag{12}$$

Hence,

$$\int_0^\infty ue^{-st}\,dt = \frac{2}{s-1} - \frac{1}{s-2} \tag{13}$$

whence $u(t)$ has the form stated in (6).

26. Discussion. If, in place of the second-order system appearing in (5.1), we had studied a tenth-order system, the straightforward approach, based upon undetermined coefficients, would have required the solution of ten simultaneous linear equations in ten unknowns—a formidable problem. The use of the Laplace transform technique avoids this. Actually, this is only a small part of the reason why the Laplace transform plays a fundamental role in analysis.

EXERCISES

1. Show that

$$\int_0^\infty t^n e^{-st} e^{at}\,dt = \frac{n!}{(s-a)^{n+1}}$$

using

(a) Integration by parts

(b) Differentiation with respect to s or a

2. Using this formula, show how to solve linear differential equations whose characteristic equations have multiple roots. In particular, solve

$$u'' - 2u' + u = 0 \qquad u(0) = 1 \qquad u'(0) = 0$$

3. Use the Laplace transform solution of an associated differential equation to obtain a representation for the solution of the system of linear equations $b_k = \sum_{i=1}^{n} e^{\lambda_i k} x_i$, $i = 1, 2, \ldots, N$, where the x_i are the unknowns.

4. How does one treat the corresponding problem when the x_i and N are unknowns?

27. Matrix Case.

Consider the equation

$$\frac{dx}{dt} = Ax \qquad x(0) = c \tag{1}$$

where A is a constant matrix. Taking transforms, we have

$$\int_0^\infty \frac{dx}{dt} e^{-st}\,dt = A \int_0^\infty x e^{-st}\,dt \tag{2}$$

whence, integrating by parts,

$$(A - sI) \int_0^\infty x e^{-st}\,dt = -c \tag{3}$$

Thus, the Laplace transform of the solution is given by

$$\int_0^\infty x e^{-st}\,dt = (sI - A)^{-1} c \tag{4}$$

In order to solve this, we can use the same type of partial fraction decomposition referred to above, namely

$$(sI - A)^{-1} = \sum_{k=1}^{N} A_k (s - \lambda_k)^{-1} \tag{5}$$

We shall assume, for the sake of simplicity, that the characteristic roots of A are distinct. Then

$$A_k = \lim_{s \to \lambda_k} (s - \lambda_k)(sI - A)^{-1} \tag{6}$$

EXERCISES

1. Find an explicit representation for A_k, using the Sylvester interpolation formula of Exercise 34, Miscellaneous Exercises, Chap. 6.

2. What happens if A has multiple characteristic roots?

3. Use the Laplace transform of a linear differential equation of the form $u^{(N)} + a_1u^{(N-1)} + \cdots + a_Nu = 0$, $u(0) = a_0$, $u'(0) = a_1, \ldots, u^{(N-1)}(0) = a_{N-1}$, to obtain a solution of the linear system

$$\sum_{i=1}^{N} \lambda_i^r a_i = a_r \qquad r = 0, 1, \ldots, N-1$$

where $\lambda_i \neq \lambda_j$ for $i \neq j$.

MISCELLANEOUS EXERCISES

1. If $A = I + B$, under what conditions does

$$A^{-1} = I - B + B^2 - \cdots ?$$

2. By considering the equation $x'' + Ax = 0$, where A is positive definite, show that the characteristic roots of a positive definite matrix are positive.

3. Let $A = HU$ be a representation of the complex matrix A where H is a positive definite Hermitian matrix and U is unitary. Show that A is normal if and only if H and U commute.

4. Prove that if A, B and AB are normal, then BA is normal (*N. A. Wiegmann*).

5. A necessary and sufficient condition that the product of two normal matrices be normal is that each commute with the H matrices of the other. By the H matrix of A, we mean the non-negative definite square root of AA^* (*N. A. Wiegmann*).

6. $A^n = I$ for some n if and only if the characteristic roots of A are roots of unity.

7. If $B^k = 0$ for some k, then $|A + B| = |A|$.

8. If $|A + \lambda I| = 0$ has reciprocal roots, A is either orthogonal, or $A^2 = I$ (*Burgatti*).

9. Let P, Q, R, X be matrices of the second order. Then every characteristic root of a solution X of $PX^2 + QX + R = 0$ is a root of $|P\lambda^2 + Q\lambda + R| = 0$ (*Sylvester*).

10. Prove that the result holds for N-dimensional matrices and the equation

$$A_0X^m + A_1X^{m-1} + \cdots + A_{m-1}X + A_m = 0$$

11. The scalar equation $u'' + p(t)u' + q(t)u = 0$ is converted into the Riccati equation $v' + v^2 + p(t)v + q(t) = 0$ by means of the change of variable $u = \exp(\int v\,dt)$ or $v = u'/u$. Show that there is a similar connection between the matrix Riccatian equation $X' = A(t) + B(t)X + XC(t)X$ and a second-order linear differential equation.

12. Every matrix A with $|A| = 1$ can be written in the form $A = BCB^{-1}C^{-1}$ (*Shoda*).

13. If $|X| = |Y| \neq 0$, then two matrices C and D can be found such that

$$X = C^{-1}D^{-1}YCD$$

(*O. Taussky*).

14. Let

$$H_N = \begin{bmatrix} 0 & i\lambda_1 & 0 & & & \\ -i\lambda_1 & 0 & i\lambda_2 & & & \\ & -i\lambda_2 & & & & \\ & & & \ddots & & \\ & & & & 0 & i\lambda_{N-1} \\ & & & & -i\lambda_{N-1} & 0 \end{bmatrix}$$

Show that

$$\lim_{N\to\infty} \frac{1}{N} \log |H_N + \lambda I| = \lim_{N\to\infty} \frac{1}{N} \sum_{i=1}^{N} \log [\lambda + z_n(\lambda)]$$

where $z_n(\lambda)$ represents the continued fraction

$$z_n(\lambda) = \lambda_n^2/\lambda + \lambda_{n+1}^2/\lambda + \cdots$$

Hint: Start with the recurrence relation

$$H_N(\lambda;\lambda_1,\lambda_2, \ldots ,\lambda_{N-1}) = \lambda H_{N-1}(\lambda;\lambda_2,\lambda_3, \ldots ,\lambda_{N-1}) + \lambda_1^2 H_{N-2}(\lambda;\lambda_3,\lambda_4, \ldots ,\lambda_{N-1})$$

for $N \geq 3$, where $H_N(\lambda) = |H_N + \lambda I|$.

15. If A and B are both normal, then if $c_1A + c_2B$ has the characteristic values $c_1\lambda_i + c_2\mu_i$ for all c_1 and c_2, we must have $AB = BA$ (*N. A. Wiegmann-H. Wielandt*).

16. Write

$$e^A e^B = e^{A+B+f(A,B)}$$

Show that $f(A,B) = [A,B]/2 + g(A,B) + h(A,B)$, $[A,B] = AB - BA$, where $g(A,B)$ is a homogeneous polynomial of degree 3 in A and B satisfying the relation $g(A,B) = g(B,A)$, and $h(A,B)$ is a sum of homogeneous polynomials beginning with one of degree 4.

17. Show that

$$f(B,C) + f(A, B + C + f(B,C)) = f(A,B) + f(A + B + f(A,B), C)$$

18. Using this result, show that

$$[A,[B,C]] + [B,[C,A]] + [C,[A,B]] = 0$$

where $[A,B]$ is the Jacobi bracket symbol.

19. A necessary and sufficient condition that $\lim_{n\to\infty} B^n = 0$, for B real or complex, is that there exist a positive definite Hermitian matrix H such that $H - B^*HB$ is positive definite (*P. Stein*).

20. A is diagonalizable if and only if HAH^{-1} is normal for some positive definite Hermitian matrix H (*Mitchell*).

21. Let r be a root of the quadratic equation $r^2 + a_1r + a_2 = 0$. Write $y = x_0 + x_1r$, $yr = x_0r + x_1[-a_1r - a_2] = -a_2x_1 + (x_0 - x_1a_1)r$. Introduce the matrix

$$X = \begin{bmatrix} x_0 & x_1 \\ -a_2 & x_0 - x_1a \end{bmatrix}$$

and write $X \sim x_0 + x_1r$.

If $X \sim x_0 + x_1r$ and $Y \sim y_0 + y_1r$, to what does XY correspond?

22. Determine the characteristic roots of X.

23. Generalize the foregoing results to the case where r is a root of the Nth-order polynomial equation

$$r^N + a_1r^{N-1} + \cdots + a_N = 0$$

24. There exists an orthogonal matrix $B(t)$ such that $y = B(t)z$ transforms $dy/dt = A(t)y$ into $dz/dt = A_1(t)z$ where $A_1(t)$ is semidiagonal (*Diliberto*).

25. $B(t)$ can be chosen to be bounded and nonsingular and $A_1(t)$ diagonal (*Diliberto*).

26. Given one characteristic vector of an $N \times N$ matrix A, can this information be used to reduce the problem of determining all characteristic vectors of A to that of determining all characteristic vectors of an $(N - 1) \times (N - 1)$ matrix?

27. The matrix A is called *circular* if $A^* = A^{-1}$. Show that e^{iR} is circular if R is real.

28. The matrix A has the form e^S where S is real skew-symmetric if and only if A is a real orthogonal matrix with $|A| = 1$ (*Taber*).

29. Every nonsingular matrix can be expressed as the product of

(*a*) A symmetric matrix and an orthogonal matrix, if real

(*b*) A Hermitian matrix and a unitary matrix, if complex

If the first factor is chosen to be positive definite, the factors are uniquely determined. These and many other results are derived by DeBruijn and Szekeres in the paper cited below,[1] using the logarithm of a matrix.

30. Derive the result of Exercise 28 from the canonical representation given above for orthogonal matrices.

31. Let $u_1, u_2, \ldots, u_N$ be a set of linearly independent solutions of the Nth-order linear differential equation

$$u^{(N)} + p_1(t)u^{(N-1)} + \cdots + p_N(t)u = 0$$

The determinant

$$W(u_1,u_2, \ldots ,u_N) = \begin{vmatrix} u_1 & u_2 & & u_N \\ u_1' & u_2' & & u_N' \\ \cdot & \cdot & & \cdot \\ \cdot & \cdot & & \cdot \\ \cdot & \cdot & & \cdot \\ u_1^{(N-1)} & u_2^{(N-1)} & \cdots & u_N^{(N-1)} \end{vmatrix}$$

is called the *Wronskian* of the function $u_1, u_2, \ldots, u_N$. Show that

$$W(t) = W(u_1,u_2, \ldots ,u_N) = W(t_0)e^{-\int_{t_0}^{t} p_1(s)ds}$$

For an extension of the concept of the Wronskian, see A. Ostrowski.[2]

The corresponding determinant associated with the solution of a linear difference equation of the form

$$u(x + N) + p_1(x)u(x + N - 1) + \cdots + p_N(x)u(x) = 0$$

$x = 0, 1, 2, \ldots,$

$$C(u_1, \ldots ,u_N) = \begin{vmatrix} u_1(x) & u_2(x) & \cdots & u_N(x) \\ u_1(x+1) & u_2(x+1) & \cdots & u_N(x+1) \\ \cdot & & & \\ \cdot & & & \\ \cdot & & & \\ u_1(x+N-1) & u_2(x+N-1) & \cdots & u_N(x+N-1) \end{vmatrix}$$

is called a Casorati determinant.[3]

Many further results concerning Wronskians can be found in G. Polya and G. Szego.[4]

[1] N. G. DeBruijn and G. Szekeres, *Nieuw. Arch. Wisk.*, (3), vol. 111, pp. 20–32, 1955.

[2] A. Ostrowski, Über ein Analogon der Wronskischen Determinante bei Funktionen mehrerer Veränderlicher, *Math. Z.*, vol. 4, pp. 223–230, 1919.

[3] For some of its properties, see P. Montel, *Leçons sur les recurrences et leurs applications*, Gauthier-Villars, Paris, 1957.

See also D. M. Krabill, On Extension of Wronskian Matrices, *Bull. Am. Math. Soc.*, vol. 49, pp. 593–601, 1943.

[4] G. Polya and G. Szego, *Aufgaben und Lehrsatze aus der Analysis*, Zweiter Band, p. 113, Dover Publications, New York, 1945.

32. A unitary matrix U can be expressed in the form $U = V^{-1}W^{-1}VW$ with unitary V, W, if and only if $\det U = 1$.

33. The unitary matrices U, V can be expressed in the form $U = W_1W_2W_3$, $V = W_3W_2W_1$ with unitary W_1, W_2, W_3 if and only if $\det U = \det V$.

34. If $AA^* - A^*A$ has all non-negative elements, then actually A is normal.

35. If H_1 and H_2 are Hermitian and one, at least, is positive definite, then H_1H_2 can be transformed into diagonal form (*O. Taussky*).

36. If $\lambda A + \mu B$ can be diagonalized for *all* scalars λ and μ, then $AB = BA$ (*T. S. Motzkin and O. Taussky*).

37. Show that a matrix A, which is similar to a real diagonal matrix D in the sense that $A = TDT^{-1}$, is the product of two Hermitian matrices (*O. Taussky*).

38. The characteristic roots of

$$A = \begin{bmatrix} a & b & \cdots & & & \\ c & a & b & \cdots & & \\ \cdot & c & a & b & \cdots & \\ & & \cdots & & c & a \end{bmatrix}$$

are given by $\lambda_k = a - 2\sqrt{bc}\cos k\theta$, $k = 1, 2, \ldots, N$, where $\theta = \pi/(N + 1)$. Show that $\Delta_N(\lambda) = |A - \lambda I|$ satisfies the difference equation

$$\Delta_N(\lambda) = (a - \lambda)\Delta_{N-1}(\lambda) - bc\Delta_{N-2}(\lambda)$$

39. The characteristic roots of

$$B = \begin{bmatrix} z-b & 2a & b & & & & & \\ 2a & z & 2a & b & & & & \\ b & 2a & z & 2a & b & & & \\ & & & \cdot & & & & \\ & & & \cdot & & & & \\ & & & \cdot & & & & \\ & & & b & 2a & z & 2a & b \\ & & & & b & 2a & z & 2a \\ & & & & & b & 2a & z-b \end{bmatrix}$$

are given by

$$\lambda_k = z - 2b - b^{-1}\{a^2 - (a - 2b\cos k\theta)^2\} \qquad k = 1, 2, \ldots, N$$

(*Rutherford, Todd*)

Obtain the result by relating B to A^2.

(A detailed discussion of the behavior of the characteristic roots and vectors of the matrix $A(\alpha)'A(\alpha)$ may be found in A. Ostrowski.[1])

The matrix $A(\alpha)$ is given by the expression

$$A(\alpha) = \begin{bmatrix} \alpha & 0 & \cdots & & 0 \\ 1 & \alpha & \cdots & & 0 \\ 1 & 1 & \alpha & & 0 \\ \cdot & & & & \\ \cdot & & & & \\ \cdot & & & & \\ 1 & 1 & \cdots & & \alpha \end{bmatrix}$$

[1] A. Ostrowski, On the Spectrum of a One-parametric Family of Matrices, *J. reine angew. Math.*, Bd. 193, pp. 143–160, 1954.

(Another detailed discussion of the properties of a particularly interesting matrix may be found in T. Kato.[1])

40. Denote by $s(A)$ the function $\max_{i,j} |\lambda_i - \lambda_j|$, and by $\|A\|$ the quantity

$$\left(\sum_{i,j} |a_{ij}|^2\right)^{1/2}$$

Show that

$$s(A) \leq \left(2\|A\|^2 - \frac{2}{N} |\text{tr } A|^2\right)^{1/2}$$

and thus that $s(A) \leq 2^{1/2}\|A\|$ (*L. Mirsky*).

41. Let $[A,B] = AB - BA$, as above. If $[A,[A,A^*]] = 0$, then A is normal (*Putnam*).

See T. Kato and O. Taussky.[2]

42. If $\lambda_1, \lambda_2, \ldots, \lambda_N$ are the characteristic values of A, c_1 and c_2 are two nonzero complex numbers, and the characteristic values of $c_1A + c_2A^*$ are $c_1\lambda_i + c_2\bar{\lambda}_{j(i)}$, where $j(i)$ is a permutation of the integers $1, \ldots, N$, then A is normal.

43. Let A and B commute. Then the characteristic values of $f(A,B)$ are $f(\lambda_i,\mu_i)$, where λ_i and μ_i are the characteristic values of A and B arranged in a fixed order, independent of f (*G. Frobenius*).

(Pairs of matrices A and B for which $c_1A + c_2B$ has the characteristic roots $c_1\lambda_i + c_2\mu_i$ are said to possess property L.[3])

44. Prove that A is normal if and only if

$$\text{tr }(A^*A) = \sum_{i=1}^{N} |\lambda_i|^2 \qquad (\textit{I. Schur})$$

45. Use this result to establish Wiegmann's result that BA is normal if A, B and AB are normal (*L. Mirsky*).

46. If A is a Hermitian matrix, show that with $s(A)$ defined as in Exercise 40,

$$s(A) = 2 \sup_{u,v} |(\bar{u},Av)|$$

where the upper bound is taken with respect to all pairs of orthogonal vectors u and v (*L. Mirsky*).

47. If A is Hermitian, then $s(A) \geq 2 \max_{r \neq s} |a_{rs}|$ (*Parker, Mirsky*).

48. If A is normal, then

$$s(A) \leq \sup_{|z|=1} s\left(\frac{zA + \bar{z}A^*}{2}\right) \qquad (\textit{L. Mirsky})^4$$

[1] T. Kato, On the Hilbert Matrix, *Proc. Am. Math. Soc.*, vol. 8, pp. 73–81, 1957.

[2] T. Kato and O. Taussky, Commutators of A and A^*, *J. Washington Acad. Sci.*, vol. 46, pp. 38–40, 1956.

[3] For a discussion and further references, see T. S. Motzkin and O. Taussky, Pairs of Matrices with Property L, *Trans. Am. Math. Soc.*, vol. 73, pp. 108–114, 1952.

[4] For the preceding results, and further ones, see L. Mirsky, Inequalities for Normal and Hermitian Matrices, *Duke Math. J.*, vol. 24, pp. 592–600, 1957.

49. If A and B are normal matrices with characteristic values λ_i and μ_i, respectively, then there exists a suitable rearrangement of the characteristic values so that $\sum_i |\lambda_i - \mu_i|^2 \leq ||A - B||^2$, where $||A||^2 = \sum_{i,j} |a_{ij}|^2$, as above (*Hoffman-Wielandt*).

50. As we know, every matrix A can be written in the form $B + iC$, where B and C are Hermitian. Let the characteristic roots of B be contained between b_1 and b_2 and those of C be between c_1 and c_2. Then if we consider the rectangle in the complex plane with the vertices $b_1 + ic_1$, $b_1 + ic_2$, $b_2 + ic_1$, $b_2 + ic_2$, all the characteristic roots of A are contained within this rectangle (*Bendixson-Hirsch*).

51. By the *domain* D of a matrix A, let us mean the set of complex values assumed by the quadratic form $(x,A\bar{x})$ for values of x satisfying the constraint $(x,\bar{x}) = 1$. Show that the characteristic values of A belong to the domain of A.[1]

52. Let K be the smallest convex domain which includes all characteristic values of A, and let D be a domain of A. If A is normal, K and D are the same.

53. Show, by considering 2×2 matrices, that the result need not be valid if A is not normal.

54. The boundary of D is a convex curve, whether or not A is normal.

55. In addition, D itself is a convex set.[2] For a bibliography of recent results concerning this and related problems, see O. Taussky.[3]

56. Associated with the equation $x' = Ax$, where x is an N-dimensional vector and A an $N \times N$ matrix, is an Nth-order linear differential equation satisfied by each component of x. Suppose that every solution of the corresponding equation associated with $y' = By$ is a solution of the equation associated with $x' = Ax$. What can be said about the relation between the matrices A and B?

57. Consider the equation $dx/dt = Bx$, $x(0) = c$, where B is a constant matrix to be chosen so that the function (x,x) remains constant over time. Show that B will have the required property if and only if it is skew-symmetric. From this, conclude that e^B is orthogonal.

58. Using the fact that the vector differential equation $A\, d^2x/dt^2 + B\, dx/dt + Cx = 0$ can be considered to be equivalent to the system $dx/dt = y$, $A\, dy/dt + By + Cx = 0$, obtain a determinantal relation equivalent to $|A\lambda^2 + B\lambda + C| = 0$ involving only linear terms in λ.

59. Let $x_1, x_2, \ldots, x_m$ be scalars, and $F(x_1,x_2, \ldots ,x_m)$ denote the matrix $A_1x_1 + A_2x_2 + \cdots + A_mx_m$, where A_i are pairwise commutative. Then if $f(x_1,x_2, \ldots ,x_m) = |F(x_1,x_2, \ldots x_m)|$ (the determinant of F), we have $f(X_1,X_2, \ldots ,X_m) = 0$ if the X_i are matrices such that $A_1X_1 + A_2X_2 + \cdots + A_mX_m = 0$. (This generalization of the Hamilton-Cayley theorem is due to H. B. Phillips.)

60. Show that the result remains valid if the linear form $\sum_i A_ix_i$ is replaced by a polynomial in the x_i with matrix coefficients. See Chao Ko and H. C. Lee,[4] where references to earlier work by Ostrowski and Phillips will be found.

[1] For this and the following results see O. Toeplitz, Das algebraische Analogon zu einem Satze von Fejer, *Math. Z.*, vol. 2, pp. 187–197, 1919.

[2] F. Hausdorff, Der Wertvorrat einer Bilinearform, *Math. Z.*, vol. 3, pp. 314–316, 1919.

[3] O. Taussky, *Bibliography on Bounds for Characteristic Roots of Finite Matrices*, National Bureau of Standards, September, 1951.

[4] Chao Ko and H. C. Lee, A Further Generalization of the Hamilton-Cayley Theorem, *J. London Math. Soc.*, vol. 15, pp. 153–158, 1940.

61. If E_i, $i = 1, 2, \ldots, n$, are 4×4 matrices satisfying the relations $E_i^2 = -1$, $E_iE_j = -E_jE_i$, $i \neq j$, then $n \leq 5$ (*Eddington*).

62. If the E_i are restricted to be either real or pure imaginary, then 2 are real and 3 are imaginary (*Eddington*). For a generalization, see M. H. A. Newman.[1]

63. If B commutes with every matrix that commutes with A, then B is a scalar polynomial in A. See J. M. Wedderburn.[2]

64. Let A be a complex matrix with characteristic roots $\lambda_1, \lambda_2, \ldots, \lambda_N$. Then, if we set $\|X\|^2 = \sum_{i,j} |x_{ij}|^2$, we have

$$\operatorname*{Inf}_S \|S^{-1}AS\|^2 = \sum_{i=1}^{N} |\lambda_1|^2$$

where the lower bound is taken with respect to all nonsingular S. The lower bound is attained if and only if A is diagonizable (*L. Mirsky*).

65. Let A, B, and X denote $N \times N$ matrices. Show that a sufficient condition for the existence of at least one solution X of the matrix equation $X^2 - 2AX + B = 0$ is that the characteristic values of

$$R = \begin{pmatrix} A & I \\ A^2 - B & A \end{pmatrix}$$

be distinct (*Treuenfels*).

66. Let $\{X\}$ be a set of $N \times N$ matrices with the property that every real linear combination has only real characteristic values. Then $\lambda_{\max}(X)$, the largest characteristic value, is a convex matrix function, and $\lambda_{\min}(X)$, the smallest characteristic value, is a concave matrix function (*P. D. Lax*).

67. Let X and Y be two matrices all of whose linear combinations have real characteristic values, and suppose that those of X are negative. Then any characteristic root of $X + iY$ has negative real part (*P. D. Lax*).

68. Let $\{X\}$ be a set of $N \times N$ real matrices with the property that every real linear combination has only real characteristic values. Then if X_1 and X_2 are two matrices in the set with the property that $X_1 - X_2$ has non-negative characteristic roots, we have $\lambda_i(X_1) \geq \lambda_i(X_2)$, $i = 1, 2, \ldots, N$, where $\lambda_i(X)$ is the ith characteristic root arranged in order of magnitude (*P. D. Lax*). For proofs of these results based upon the theory of hyperbolic partial differential equations, see P. D. Lax.[3]

For proofs along more elementary lines, see A. F. Weinberger.[4]

69. Let $\{\alpha_i\}$, $\{\beta_i\}$ $(1 \leq i \leq n)$ be complex numbers, each of absolute value 1. Prove: There exist two unitary matrices A, B of order n with the preassigned characteristic roots $\{\alpha_i\}$, $\{\beta_i\}$, respectively, and such that 1 is a characteristic root of AB, if and only if, in the complex plane, the convex hull of the points $\{\alpha_i\}$ and the convex hull of the points $\{\bar{\beta}_i\}$ have a point in common (*Ky Fan*).[5]

[1] M. H. A. Newman, *J. London Math. Soc.*, vol. 7, pp. 93–99, 1932.

[2] J. M. Wedderburn, Lectures on Matrices, *Am. Math. Soc. Colloq. Publ.*, vol. 17, p. 106, 1934.

[3] P. D. Lax, Differential Equations, Difference Equations and Matrix Theory, *Comm. Pure Appl. Math.*, vol. XI, pp. 175–194, 1958.

[4] A. F. Weinberger, Remarks on the Preceding Paper of Lax, *Comm. Pure Appl. Math.*, vol. XI, pp. 195–196, 1958.

[5] This result is analogous to a theorem of H. Wielandt, *Pacific J. Math.*, vol. 5, pp. 633–638, 1955, concerning the characteristic values of the sum of two normal matrices.

70. We know that e^{At} may be written as a polynomial in A, $e^{At} = u_0(t)I + u_1(t)A + \cdots + u_{N-1}(t)A^{N-1}$. Determine differential equations for the $u_i(t)$ using the fact that $d/dt(e^{At}) = Ae^{At}$.

71. Let A be an $N \times N$ matrix and $A(+)$ be formed from A by replacing with zeros all elements of A which are either on or below the principal diagonal. Let $A(-) = A - A(+)$ and suppose that $A(+)$ and $A(-)$ commute. Write $e^A = PQ$, where $P - I$ has nonzero terms only above the diagonal and $Q - I$ has nonzero terms only on or below the diagonal. Then

$$P = e^{A(+)} \qquad Q = e^{A(-)}$$

and the factorization is unique (*G. Baxter, An Operator Identity, Pacific J. Math.*, vol. 8, pp. 649–664, 1958). The analogy between this factorization and the Wiener-Hopf factorization is more than superficial.

72. Consider the linear functional equation $f = e + \lambda T(fg)$, where e is the identity element, λ is a scalar, and T is an operator satisfying a relation of the form $(Tu)(Tv) = T(uT(v)) + T(u)v - \theta uv$, for any two functions u and v with θ a fixed scalar. Show that for small λ

$$f = \exp\left[T\left(\sum_{n=1}^{\infty} n^{-1}(\lambda g)^n\theta^{n-1}\right)\right]$$

The result is due to Baxter. For proofs dependent on differential equations, see Atkinson[1] and Wendel.[2] Operators of the foregoing nature are connected with the Reynolds operator of importance in turbulence theory.

73. Let $|A + \lambda I| = \lambda^n + a_1\lambda^{n-1} + \cdots + a_n$. Show that

$$a_1 = -\operatorname{tr}(A) \qquad a_2 = -\tfrac{1}{2}[a_1 \operatorname{tr}(A) + \operatorname{tr}(A^2)], \ldots$$

and obtain a general recurrence relation connecting a_k with $a_1, a_2, \ldots, a_{k-1}$ (*Bocher*).

74. A matrix A whose main diagonal elements are 0s and 1s, with other elements satisfying the relation $a_{ij} + a_{ji} = l$ is called a *tournament matrix*. Let $\lambda_1, \lambda_2, \ldots, \lambda_N$ be the N characteristic roots of A, with $|\lambda_1| \geq |\lambda_2| \geq \cdots \geq |\lambda_N|$. Then $-\frac{1}{2} \leq \operatorname{Re}(\lambda_1) \leq (n-1)/2$, $|\lambda_1| \leq (n-1)/2$ and $|\lambda_k| \leq (n(n-1)/2k)^{1/2}$ for $k \geq 2$. (*A. Braver and I. C. Gentry, On the Characteristic Roots of Tournament Matrices, Bull. Am. Math. Soc., vol.* 74, pp. 1133–1135, 1968.)

75. If $B = \lim_{n \to \infty} A^n$, what can be said about the characteristic roots of B? (*O. Taussky, Matrices with $C^n \to 0$, J. Algebra*, vol. 1, pp. 5–10, 1964.)

76. Let A be a matrix with complex elements. Show that A is normal (that is, $AA^* = A^*A$), if and only if one of the following holds:

(*a*) $A = B + iC$, where B and C are Hermitian and commute.
(*b*) A has a complete set of orthonormal characteristic vectors.
(*c*) $A = U^*DU$ where U is unitary and D is diagonal.
(*d*) $A = UH$ where U is unitary and H is Hermitian and U and H commute.
(*e*) The characteristic roots of AA^* are $|\lambda_1|^2, \ldots, |\lambda_N|^2$ where $\lambda_1, \ldots, \lambda_N$ are the characteristic roots of A.
(*f*) The characteristic roots of $A + A^*$ are $\lambda_1 + \bar{\lambda}_1, \ldots, \lambda_N + \bar{\lambda}_N$.

The foregoing properties indicate the value of approximating to a given complex matrix B by a normal matrix A in the sense of minimizing a suitable matrix norm

[1] F. V. Atkinson, "Some Aspects of Baxter's Functional Equation," *J. Math. Anal. Appl.*, vol. 6, pp. 1–29, 1963.

[2] J. G. Wendel, "Brief Proof of a Theorem of Baxter," *Math. Scand.*, vol. 11, pp. 107–108, 1962.

$\|B - A\|$. This question was first investigated by Minsky; see Causey[1] and Hoffman and O. Taussky.[2]

77. Consider the matrix differential equation $X' = (P_1(t) + P_2(t + \theta))X$, $X(0) = I$, where $P_1(t + 1) = P_1(t)$, $P_2(t + \lambda) = P_2(t)$, λ is irrational, and θ is a real parameter. Write $X(t,\theta)$ for the solution. Show that $X(t, \theta + \lambda) = X(t,\theta)$, $X(t + 1, \theta) = X(t, \theta + 1)X(1,\theta)$, and hence that $X(t + n, \theta) = X(t, \theta + n)X(1, \theta + n - 1) X(1, \theta + n - 2) \cdots X(1,\theta)$.

78. Does $\prod_{k=0}^{n} X(1, \theta + k)$ possess a limiting behavior as $n \to \infty$? See R. Bellman, A Note on Linear Differential Equations with Quasiperiodic Coefficients, *J. Math. Anal. Appl.*, to appear.

Bibliography and Discussion

§1 to §10. The results and techniques follow

R. Bellman, *Stability Theory of Differential Equations*, Dover Publications, New York, 1969.

S. Lefschetz, Lectures on Differential Equations, *Ann. Math. Studies*, no. 14, 1946.

§3. For a discussion of the problem of determining the characteristic polynomial, see

A. S. Householder and F. L. Bauer, On Certain Methods for Expanding the Characteristic Polynomial, *Numerische Mathematik*, vol. 1, pp. 29–40, 1959.

§6. For some further results, and additional references, see

K. A. Hirsch, A Note on Vandermonde's Determinant, *J. London Math. Soc.*, vol. 24, pp. 144–145, 1949.

§11. For a proof of the Jordan canonical form, see

S. Lefschetz, *Differential Equations, Geometric Theory*, Interscience Publishers, Inc., New York, Appendix I, 1957.

J. H. M. Wedderburn, Lectures on Matrices, *Am. Math. Soc. Colloq. Publs.*, vol. 17, 1934.

§13. The result is given by I. Schur in

I. Schur, Über die characteristischen Wurzeln einer linearen Substitu-

[1] R. L. Causey, *On Closest Normal Matrices*, Computer Sciences Division, Stanford University, Technical Report CS10, June, 1964.

[2] A. J. Hoffman and O. Taussky, On Characterizations of Normal Matrices, *J. Res. Bur. Standards*, vol. 52, pp. 17–19, 1954.

tion mit einer Anwendung auf die Theorie der Integral Gleichungen, *Math. Ann.*, vol. 66, pp. 488–510, 1909.

§14. The concept of a normal matrix appears to have been introduced by Toeplitz,

O. Toeplitz, *Math. Z.*, vol. 2, p. 190, 1919.

§15. This approximate diagonalization was introduced by Perron and used to establish stability theorems. See

O. Perron, *Math. Z.*, vol. 29, pp. 129–160, 1929.

§16. For a discussion of the requisite theorems concerning resultants, see

B. L. Van der Werden, *Moderne Algebra*, Springer-Verlag, Berlin, 1931.

§17. This result was given by Hamilton for quaternions and by Cayley for matrices. On page 18 of Macduffee's book, there is an account of a generalization of the Hamilton-Cayley theorem due to Phillips. For further generalizations, see the paper by Ko and Lee referred to in Exercise 60 of the Miscellaneous Exercises.

§19. This derivation follows the monograph by Lefschetz, cited above. The determination of C is a problem of great difficulty. An excellent survey of results in this field is contained in

V. M. Starzinski, Survey of Works on Conditions of Stability of the Trivial Solution of a System of Linear Differential Equations with Periodic Coefficients, *Trans. Am. Math. Soc.*, series 2, vol. 1, pp. 189–239.

§20. Observe what this result means. Every nonsingular point transformation $y = Ax$ can be considered to arise from a family of continuous transformations furnished by the differential equation $dZ/dt = CZ$, $Z(0) = I$, where $A = e^C$.

§22. Some problems connected with these transformations may be found in the set of Research Problems of the *Bull. Am. Math. Soc.* for 1958.

For the second-order linear differential equation $(pu)' + qu = 0$ there exist a number of interesting representations for the solution. One such, called the *Prüfer transformation,* was generalized by J. H. Barrett to cover the matrix equation $(P(t)X')' + Q(t)X = 0$ (*Proc. Am. Math. Soc.*, *vol.* 8, *pp.* 510–518, 1957). Further extension and refinement were given by

W. T. Reid, A Prufer Transformation for Differential Systems, *Pacific J. Math.*, vol. 8, pp. 575–584, 1958.

For an application to stability theory, see

R. Conti, Sulla t_∞-similitudine tra matrici e l'equivalenza asintotica dei sistemi differenziali lineari, *Riv. Mat. Univ. Parma*, vol. 8, pp. 43–47, 1957.

§23. For a detailed account of the use of biorthogonalization in connection with the basic problems of inversion of matrices and the determination of characteristic roots and vectors, see

M. R. Hestenes, Inversion of Matrices by Biorthogonalization and Related Results, *J. Soc. Ind. Appl. Math.*, vol. 6, pp. 51–90, 1958.

One of the most important problems in the theory of linear differential equations is that of determining the structure of the solutions of linear systems of equations having the form

$$\sum_{j=1}^{N} p_{ij}(D)x_j = 0 \qquad x_i(0) = c_i \qquad i = 1, 2, \ldots, N$$

where D is the operator d/dt.

An equivalent problem in matrix analysis is that of determining the characteristic values and vectors of matrices $A(\lambda)$ whose elements are polynomials in λ.

Since any thorough discussion of this problem, which we have merely touched upon in treating $A - \lambda I$, would take us too far afield in this volume, we would like to refer the reader to

R. A. Frazer, W. J. Duncan, and A. R. Collar, *Elementary Matrices*, The Macmillan Company, New York, 1946,

where many applications are given to the flutter problem, etc.

A consideration of the analytic properties of these matrices brings us in contact with the general theory of matrices whose elements are functions of a complex variable. These questions are not only of importance in themselves, but have important ramifications in modern physics and prediction theory. See

P. Masani, The Laurent Factorization of Operator-valued Functions, *Proc. London Math. Soc.*, vol. 6, pp. 59–69, 1956.

P. Masani and N. Wiener, The Prediction Theory of Multivariate

Stochastic Processes, *Acta Math.*, *I*, vol. 98, pp. 111–150, 1957; II, vol. 99, pp. 96–137, 1958.

H. Helson and D. Lowdenslager, Prediction Theory and Fourier Series in Several Variables, *Acta Math.*, vol. 99, 1958.

Results on the structure of matrices with quaternion elements may be found in

N. A. Wiegmann, The Structure of Unitary and Orthogonal Quaternion Matrices, *Illinois J. Math.*, vol. 2, pp. 402–407, 1958,

where further references may be found.

Some recent papers of interest concerning the location of characteristic roots are

J. L. Brenner, New Root-location Theorems for Partitioned Matrices, *SIAM J. Appl. Math.*, vol. 16, pp. 889–896, 1968.

A. S. Householder, Norms and the Location of Roots of Matrices, *Bull. Am. Math. Soc.*, vol. 74, pp. 816–830, 1968.

J. L. Brenner, Gersgorin Theorems by Householder's Proof, *Bull. Am. Math. Soc.*, vol. 74, pp. 625–627, 1968.

D. G. Feingold and R. S. Varga, Block Diagonally Dominant Matrices and Generalizations of the Gersgorin Circle Theorem, *Pacific J. Math.*, vol. 12, pp. 1241–1250, 1962.

R. S. Varga, Minimal Gerschgorin Sets, *Pacific J. Math.*, vol. 15, pp. 719–729, 1965.

B. W. Levinger and R. S. Varga, Minimal Gerschgorin Sets, II, *Pacific J. Math.*, vol. 17, pp. 199–210, 1966.

§24. For discussions of analytic and computational aspects of the Laplace transform, see

R. Bellman and K. L. Cooke, *Differential Difference Equations*, Academic Press Inc., New York, 1963.

R. Bellman, R. E. Kalaba and J. Lockett, *Numerical Inversion of the Laplace Transform*, American Elsevier, New York, 1966.

12

Symmetric Functions, Kronecker Products and Circulants

1. Introduction. In this chapter, we shall show that some problems of apparently minor import, connected with various symmetric functions of the characteristic roots, lead to a quite important concept in matrix theory, the Kronecker product of two matrices. As a limiting case, we obtain the Kronecker sum, a matrix function which plays a basic role in the theory of the matrix equation

$$AX + XB = C \tag{1}$$

We have already met this equation in an earlier chapter, and we shall encounter it again in connection with stability theory. Furthermore, the Kronecker product will crop up in a subsequent chapter devoted to stochastic matrices.

Following this, we shall construct another class of compound matrices which arise from the consideration of certain skew-symmetric functions. Although these are closely tied in with the geometry of N-dimensional Euclidean space, we shall not enter into these matters here.

Finally, again motivated by conditions of symmetry, we shall discuss circulants.

Throughout the chapter, we shall employ the same basic device to determine the characteristic roots and vectors of the matrices under discussion, namely, that of viewing a matrix as an equivalent of a transformation of one set of quantities into another. A number of problems in the exercises will further emphasize this point of view.

2. Powers of Characteristic Roots. A very simple set of symmetric functions of the characteristic roots are the power sums

$$p_k = \sum_{i=1}^{N} \lambda_i^k \tag{1}$$

$k = 1, 2, \ldots .$

Suppose that we wish to determine $\sum_{i=1}^{N} \lambda_i^2$. Since

$$\sum_{i=1}^{N} \lambda_i^2 = \left(\sum_{i=1}^{N} \lambda_i\right)^2 - 2 \sum_{i \neq j} \lambda_i \lambda_j \tag{2}$$

we can determine the sum of the squares of the characteristic roots in terms of two of the coefficients of the characteristic polynomial. Moreover, since $\sum_{i=1}^{N} \lambda_i^k$, $k = 1, 2, \ldots$, is a symmetric function of the characteristic roots, we know, in advance, that it can be written as a polynomial in the elementary symmetric functions.

For many purposes, however, this is not the most convenient representation. We wish to demonstrate

Theorem 1. *For $k = 1, 2, \ldots$, we have*

$$\sum_{i=1}^{N} \lambda_i^k = \operatorname{tr}(A^k) \tag{3}$$

Proof. If A has distinct characteristic roots, the diagonal representation

$$A = T \begin{bmatrix} \lambda_1 & & & & \\ & & & 0 & \\ & \lambda_2 & & & \\ & & \cdot & & \\ & & & \cdot & \\ & 0 & & & \cdot \\ & & & & \lambda_N \end{bmatrix} T^{-1} \tag{4}$$

which yields the representation for A^k,

$$A^k = T \begin{bmatrix} \lambda_1^k & & & & \\ & \lambda_2^k & & & \\ & & & 0 & \\ & & \cdot & & \\ & & & \cdot & \\ & 0 & & & \cdot \\ & & & & \lambda_N^k \end{bmatrix} T^{-1} \tag{5}$$

makes the result evident. It is easily seen that an appeal to continuity along by now familiar lines yields (3) for general matrices.

If we do not like this route, we can employ the triangular representation

$$A = T \begin{bmatrix} \lambda_1 & & \cdots & \\ & \lambda_2 & & \cdot \\ & & & \cdot \\ & & \cdots & \cdot \\ & 0 & & \\ & & & \lambda_N \end{bmatrix} T^{-1} \tag{6}$$

where the elements above the main diagonal are not necessarily zero and the λ_1 are not necessarily distinct.

It is easily seen that

$$A^k = T \begin{bmatrix} \lambda_1^k & & \cdots & \\ & \lambda_2^k & & \cdot \\ & & & \cdot \\ & & & \cdot \\ & & \cdots & \\ & 0 & & \\ & & & \lambda_N^k \end{bmatrix} T^{-1} \tag{7}$$

This representation yields an alternative proof of (3).

3. Polynomials and Characteristic Equations. Although we know that every characteristic equation produces a polynomial, it is not clear that every polynomial can be written as the characteristic polynomial of a matrix.

Theorem 2. *The matrix*

$$A = \begin{bmatrix} -a_1 & -a_2 & \cdots & -a_{n-1} & -a_n \\ 1 & 0 & \cdots & 0 & 0 \\ 0 & 1 & \cdots & 0 & 0 \\ \cdot & \cdot & & \cdot & \cdot \\ \cdot & \cdot & & \cdot & \cdot \\ \cdot & \cdot & & \cdot & \cdot \\ 0 & 0 & \cdots & 1 & 0 \end{bmatrix} \tag{1}$$

has the characteristic equation

$$|A - \lambda I| = 0 = \lambda^n + a_1\lambda^{n-1} + \cdots + a_n \tag{2}$$

We leave the proof as an exercise.

EXERCISE

1. Using the result given above, determine the sum of the cubes of the roots of the equation $\lambda^n + a_1\lambda^{n-1} + \cdots + a_n = 0$.

4. Symmetric Functions. Although we have solved the problem of determining matrices which have as characteristic roots the quantities $\lambda_1^k, \lambda_2^k, \ldots, \lambda_N^k$, we do not as yet possess matrices whose characteristic roots are prescribed functions of the characteristic roots, such as $\lambda_i\lambda_j$, $i, j = 1, 2, \ldots, N$.

In order to solve this problem, we begin with the apparently more difficult problem of determining a matrix whose characteristic roots are $\lambda_i\mu_j$, where λ_i are the characteristic roots of A and μ_j those of B. Specializing $\lambda_i = \mu_i$, we obtain the solution of the original problem.

Let us discuss the 2×2 case where the algebra is more transparent. Start with the two sets of equations

$$\begin{aligned} \lambda_1 x_1 &= a_{11}x_1 + a_{12}x_2 & \mu_1 y_1 &= b_{11}y_1 + b_{12}y_2 \\ \lambda_1 x_2 &= a_{21}x_1 + a_{22}x_2 & \mu_1 y_2 &= b_{21}y_1 + b_{22}y_2 \end{aligned} \tag{1}$$

and perform the following multiplications:

$$\begin{aligned} \lambda_1\mu_1 x_1y_1 &= a_{11}b_{11}x_1y_1 + a_{11}b_{12}x_1y_2 + a_{12}b_{11}x_2y_1 + a_{12}b_{12}x_2y_2 \\ \lambda_1\mu_1 x_1y_2 &= a_{11}b_{21}x_1y_1 + a_{11}b_{22}x_1y_2 + a_{12}b_{21}x_2y_1 + a_{12}b_{22}x_2y_2 \\ \lambda_1\mu_1 x_2y_1 &= a_{21}b_{11}x_1y_1 + a_{21}b_{12}x_1y_2 + a_{22}b_{11}x_2y_1 + a_{22}b_{12}x_2y_2 \\ \lambda_1\mu_1 x_2y_2 &= a_{21}b_{21}x_1y_1 + a_{21}b_{22}x_1y_2 + a_{22}b_{21}x_2y_1 + a_{22}b_{22}x_2y_2 \end{aligned} \tag{2}$$

We note that the four quantities x_1y_1, x_1y_2, x_2y_1, x_2y_2, occur on the right and on the left in (2). Hence if we introduce the four-dimensional vector

$$z = \begin{bmatrix} x_1y_1 \\ x_1y_2 \\ x_2y_1 \\ x_2y_2 \end{bmatrix} \tag{3}$$

and the 4×4 matrix

$$C = \begin{bmatrix} a_{11}b_{11} & a_{11}b_{12} & a_{12}b_{11} & a_{12}b_{12} \\ a_{11}b_{21} & a_{11}b_{22} & a_{12}b_{21} & a_{12}b_{22} \\ a_{21}b_{11} & a_{21}b_{12} & a_{22}b_{11} & a_{22}b_{12} \\ a_{21}b_{21} & a_{22}b_{22} & a_{22}b_{21} & a_{22}b_{22} \end{bmatrix} \tag{4}$$

we may write the equations of (2) in the form

$$\lambda_1\mu_1 z = Cz \tag{5}$$

It follows that z is a characteristic vector of C with the associated characteristic value $\lambda_1\mu_1$. In precisely the same way, we see that C has $\lambda_1\mu_2$, $\lambda_2\mu_1$, and $\lambda_2\mu_2$ as characteristic roots with associated characteristic vectors formed in the same fashion as in (3).

We thus have solved the problem of determining a matrix whose characteristic roots are $\lambda_i\mu_j$, $i, j = 1, 2$.

5. Kronecker Products. Since even the 2×2 case appears to introduce an unpleasant amount of calculation, let us see whether or not we can introduce a better notation. Referring to (4.4), we note that there is a certain regularity to the structure of C. Closer observation shows that C may be written as a compound matrix

$$C = \begin{bmatrix} a_{11}B & a_{12}B \\ a_{21}B & a_{22}B \end{bmatrix} \tag{1}$$

or, even more simply,

$$C = (a_{ij}B) \tag{2}$$

Once this representation has been obtained, it is easy to see that it is independent of the dimensions of A and B. Hence

Definition. Let A be an M-dimensional matrix and B an N-dimensional matrix. The MN-dimensional matrix defined by (2) *is called the Kronecker product of A and B and written*

$$A \times B = (a_{ij}B) \tag{3}$$

The above argumentation readily yields Theorem 3.

Theorem 3. *The characteristic roots of $A \times B$ are $\lambda_i\mu_j$ where λ_i are the characteristic roots of A and μ_j the characteristic roots of B.*

The characteristic vectors have the form

$$z_{ij} = \begin{bmatrix} x_1{}^i y^j \\ x_2{}^i y^j \\ \cdot \\ \cdot \\ \cdot \\ x_M{}^i y^j \end{bmatrix} \tag{4}$$

Here by $x_k{}^i$, $k = 1, 2, \ldots, M$, we mean the components of the characteristic vector x^i of A, while y^j is a characteristic vector of B.

EXERCISES

1. Show that $\operatorname{tr}(A \times B) = (\operatorname{tr} A)(\operatorname{tr} B)$.

2. Determine $|A \times B|$.

3. Show that

$$I \times B = \begin{bmatrix} B & & & & 0 \\ & B & & & \\ & & \cdot & & \\ & & & \cdot & \\ & & & & \cdot \\ 0 & & & & B \end{bmatrix}$$

6. Algebra of Kronecker Products. In order to justify the name "Kronecker product" and the notation we have used, we must show that $A \times B$ possesses a number of the properties of a product.

We shall leave as exercises for the reader proofs of the following results:

$$A \times B \times C = (A \times B) \times C = A \times (B \times C) \tag{1a}$$
$$(A + B) \times (C + D) = A \times C + A \times D + B \times C + B \times D \tag{1b}$$
$$(A \times B)(C \times D) = (AC) \times (BD) \tag{1c}$$

7. Kronecker Powers—I. We shall also consider the *Kronecker power* and write

$$\begin{aligned} A^{[2]} &= A \times A \\ A^{[k+1]} &= A \times A^{[k]} \end{aligned} \tag{1}$$

If A and B do not commute, $(AB)^k \neq A^k B^k$ in general; and never if $k = 2$. It is, however, true that

$$(AB)^{[k]} = A^{[k]} B^{[k]} \tag{2}$$

for all A and B.

This important property removes many of the difficulties due to noncommutativity, and will be used for this purpose in a later chapter devoted to stochastic matrices.

EXERCISE

1. Prove that $A^{[k+l]} = A^{[k]} \times A^{[l]}$.

8. Kronecker Powers—II. If we are interested only in Kronecker powers of a particular matrix, rather than general products, we can define matrices with these properties which are of much smaller dimension. Starting with the equations

$$\begin{aligned} \lambda_1 x_1 &= a_{11} x_1 + a_{12} x_2 \\ \lambda_1 x_2 &= a_{21} x_1 + a_{22} x_2 \end{aligned} \tag{1}$$

we form the products

$$\begin{aligned} \lambda_1^2 x_1^2 &= a_{11}^2 x_1^2 + 2a_{11} a_{12} x_1 x_2 + a_{12}^2 x_2^2 \\ \lambda_1^2 x_1 x_2 &= a_{11} a_{21} x_1^2 + (a_{11} a_{22} + a_{12} a_{21}) x_1 x_2 + a_{12} a_{22} x_2^2 \\ \lambda_1^2 x_2^2 &= a_{21}^2 x_1^2 + 2a_{21} a_{22} x_1 x_2 + a_{22}^2 x_2^2 \end{aligned} \tag{2}$$

It is clear then that the matrix

$$A_{[2]} = \begin{bmatrix} a_{11}^2 & 2a_{11}a_{12} & a_{12}^2 \\ a_{11}a_{21} & (a_{11}a_{22} + a_{12}a_{21}) & a_{12}a_{22} \\ a_{21}^2 & 2a_{21}a_{22} & a_{22}^2 \end{bmatrix} \tag{3}$$

possesses the characteristic roots λ_1^2, $\lambda_1\lambda_2$, and λ_2^2. The corresponding characteristic vectors are readily obtained.

To obtain the expression for $A_{[k]}$, we proceed in a more systematic fashion.

9. Kronecker Products—III. Consider the equations, for a fixed integer k,

$$
\begin{aligned}
(a_{11}x_1 + a_{12}x_2)^k &= a_{11}{}^k x_1{}^k + k a_{11}{}^{k-1} a_{12} x_1{}^{k-1} x_2 + \cdots \\
(a_{11}x_1 + a_{12}x_2)^{k-1}(a_{21}x_1 + a_{22}x_2) &= a_{11}{}^{k-1} a_{21} x_1{}^k \\
&\quad + ((k-1)a_{11}{}^{k-2} a_{22} a_{21} + a_{11}{}^{k-1} a_{22}) x_1{}^{k-1} x_2 \\
&\ \ \vdots \\
(a_{11}x_1 + a_{12}x_2)^{k-i}(a_{21}x_1 + a_{22}x_2)^i &= a_{11}{}^{k-i} a_{21}{}^i x_1{}^k + \cdots
\end{aligned}
\tag{1}
$$

where $i = 0, 1, 2, \ldots, k$.

The matrix $A_{[k]}$ is then defined by the tableau of coefficients of the terms $x_1{}^{k-i} x_2{}^i$, $i = 0, 1, 2, \ldots, k$,

$$
A_{[k]} = \begin{bmatrix}
a_{11}{}^k & k a_{11}{}^{k-1} a_{12} & \cdots \\
a_{11}{}^{k-1} a_{21} & ((k-1)a_{11}{}^{k-2} a_{12} a_{21} + a_{11}{}^{k-1} a_{22}) & \cdots \\
\cdot & & \\
\cdot & & \\
\cdot & &
\end{bmatrix}
\tag{2}
$$

This representation holds only for 2×2 matrices. The representation of $A_{[k]}$ for matrices of dimension N can be obtained similarly.

EXERCISES

1. Show that $(AB)_{[k]} = A_{[k]} B_{[k]}$.
2. Show that $A_{[k+l]} = A_{[k]} A_{[l]}$.

10. Kronecker Logarithm. Let us now derive an infinite matrix which we can think of as a Kronecker logarithm. We begin by forming a generalized kth power for nonintegral k. To do this, we consider the infinite sequence of expression $(a_{11}x_1 + a_{12}x_2)^{k-i}(a_{21}x_1 + a_{22}x_2)^i$, $i = 0, 1, 2, \ldots$, for k nonintegral.

This yields, in place of the finite tableau of (8.2), an infinite array, since the binomial series $(a_{11}x_1 + a_{12}x_2)^{k-i}$ is now not finite.

The change of variable

$$
\begin{bmatrix} x_1 \\ x_2 \end{bmatrix} = B \begin{bmatrix} x_1' \\ x_2' \end{bmatrix}
\tag{1}
$$

shows that

$$
(AB)_{[k]} = A_{[k]} B_{[k]}
\tag{2}
$$

holds for arbitrary values of k.

Let us now expand each element in $A_{[k]}$ about $k = 0$. Write

$$A_{[k]} = A_{[0]} + kL(A) + \cdots \tag{3}$$

This defines an infinite matrix $L(A)$, which we can think of as a Kronecker logarithm. Substituting in (2), we see that

$$L(AB) = A_{[0]}L(B) + L(A)B_{[0]} \tag{4}$$

an analogue of the functional equation for the scalar logarithm. Observe that $A_{[0]}$ is not an identity matrix.

EXERCISE

1. Determine the ijth element in $L(A)$, for $i = 0, 1, j = 0, 1$.

11. Kronecker Sum—I. Let us now turn to the problem of determining matrices which possess the characteristic values $\lambda_i + \mu_j$. Following a method used before, we shall derive the additive result from the multiplicative case. Consider, for a parameter ϵ, the relation

$$\begin{aligned}(I_M + \epsilon A) \times (I_N + \epsilon B)& \\ &= I_M \times I_N + \epsilon(I_M \times B + A \times I_N) + \epsilon^2 A \times B\end{aligned} \tag{1}$$

Since the characteristic roots of $(I_M + \epsilon A) \times (I_N + \epsilon B)$ are

$$(1 + \epsilon\lambda_i)(1 + \epsilon\mu_j) = 1 + \epsilon(\lambda_i + \mu_j) + \epsilon^2\lambda_i\mu_j$$

we see that the characteristic roots of $I_M \times B + A \times I_N$ must be $\lambda_i + \mu_j$.

EXERCISE

1. Determine the characteristic vectors of $I_M \times B + A \times I_N$.

12. Kronecker Sum—II. The matrix $I_M \times B + A \times I_N$ can also be obtained by use of differential equations.

Consider the equations

$$\begin{aligned}\frac{dx_1}{dt} &= a_{11}x_1 + a_{12}x_2 \qquad & \frac{dy_1}{dt} &= b_{11}y_1 + b_{12}y_2 \\ \frac{dx_2}{dt} &= a_{21}x_1 + a_{22}x_2 \qquad & \frac{dy_2}{dt} &= b_{21}y_1 + b_{22}y_2\end{aligned} \tag{1}$$

Let us now compute the derivatives of the four products x_1y_1, x_1y_2, x_2y_1, x_2y_2. We have

$$\begin{aligned}\frac{d}{dt}(x_1y_1) &= (a_{11}x_1 + a_{12}x_2)y_1 + x_1(b_{11}y_1 + b_{12}y_2) \\ &= (a_{11} + b_{11})x_1y_1 + b_{12}x_1y_2 + a_{12}x_2y_1\end{aligned} \tag{2}$$

and so on. It is easy to see that the matrix we obtain is precisely $A \times I_N + I_M \times B$.

13. The Equation $AX + XB = C$, the Lyapunov Equation. In Chap. 11, we showed that the equation

$$AX + XB = C \tag{1}$$

possessed the unique solution

$$X = - \int_0^\infty e^{At} C e^{Bt} \, dt \tag{2}$$

provided that the integral on the right exists for all C.

Let us now complete this result.

Consider the case where A, B, C, and X are 2×2 matrices. The equations to determine the unknown components x_{ij} are

$$\begin{aligned}
a_{11}x_{11} + a_{12}x_{21} + x_{11}b_{11} + x_{12}b_{21} &= c_{11} \\
a_{11}x_{12} + a_{12}x_{22} + x_{11}b_{12} + x_{12}b_{22} &= c_{12} \\
a_{21}x_{11} + a_{22}x_{21} + x_{21}b_{11} + x_{22}b_{21} &= c_{21} \\
a_{21}x_{12} + a_{22}x_{22} + x_{21}b_{12} + x_{22}b_{22} &= c_{22}
\end{aligned} \tag{3}$$

The matrix of coefficients is

$$\begin{bmatrix}
a_{11} + b_{11} & b_{21} & a_{12} & 0 \\
b_{12} & a_{11} + b_{22} & 0 & a_{12} \\
a_{21} & 0 & a_{22} + b_{11} & b_{21} \\
0 & a_{21} & b_{12} & a_{22} + b_{22}
\end{bmatrix} \tag{4}$$

which we recognize as the matrix $A \times I + I \times B'$.

The characteristic roots are thus $\lambda_i + \mu_j$, since B and B' have the same characteristic roots. It is easy to see that the same results generalize to matrices of arbitrary dimension so that we have established Theorem 4.

Theorem 4. *A necessary and sufficient condition that* (1) *have a solution for all C is that $\lambda_i + \mu_j \neq 0$ where λ_i are the characteristic roots of A and μ_i the characteristic roots of B.*

EXERCISES

1. Prove that a necessary and sufficient condition that $AX + XA' = C$ have a unique solution for all C is that $\lambda_i + \lambda_j \neq 0$.

2. Prove the foregoing result in the following steps.

(*a*) Let T be a matrix reducing A to triangular form,

$$T^{-1}AT = B = \begin{bmatrix}
b_{11} & b_{12} & \cdots & b_{1N} \\
 & b_{22} & \cdots & b_{2N} \\
 & & \ddots & \\
0 & & & b_{NN}
\end{bmatrix}$$

(b) Then $AX + XA' = C$ becomes $B'(T'XT) + (T'XT)B = T'CT$.
(c) Let $Y = T'XT$ and consider the linear equations for the elements of Y. Show that the determinant of the coefficients is $\prod_{i,j=1}^{N} (b_{ii} + b_{jj})$ and thus derive the stated results (*Hahn*).

14. An Alternate Route. We observed in the foregoing sections how Kronecker products arose from the consideration of various symmetric functions of the roots of two distinct matrices, A and B, and how a particular Kronecker power could be formed from a single matrix A.

Let us now consider a different type of matrix "power" which is formed if we consider certain skew-symmetric functions, namely, determinants. There is strong geometric motivation for what at first sight may seem to be quite formal manipulation. The reader who is interested in the geometric background will find references to books by Bocher and Klein listed in the Bibliography at the end of the chapter.

To emphasize the basic idea with the arithmetic and algebraic level kept at a minimum, we shall consider a 3×3 matrix and a set of 2×2 determinants formed from the characteristic vectors. In principle, it is clear how the procedure can be generalized to treat $R \times R$ determinants associated with the characteristic vectors of $N \times N$ matrices. In practice, to carry out the program would require too painstaking a digression into the field of determinants. Consequently, we shall leave the details, by no means trivial, to the interested reader.

Consider the matrix

$$A = \begin{bmatrix} a_1 & a_2 & a_3 \\ b_1 & b_2 & b_3 \\ c_1 & c_2 & c_3 \end{bmatrix} \tag{1}$$

where we have momentarily departed from our usual notation to simplify keeping track of various terms, whose characteristic roots are λ_1, λ_2, λ_3, with associated characteristic vectors x^1, x^2, x^3.

For any two characteristic vectors, x^i and x^j, $i \neq j$, we wish to obtain a set of linear equations for the 2×2 determinants y_1, y_2, y_3 given by the relations

$$\lambda_i\lambda_j y_1 = \begin{vmatrix} \lambda_i x_1{}^i & \lambda_j x_1{}^j \\ \lambda_i x_2{}^i & \lambda_j x_2{}^j \end{vmatrix} \qquad \lambda_i\lambda_j y_2 = \begin{vmatrix} \lambda_i x_2{}^i & \lambda_j x_2{}^j \\ \lambda_i x_3{}^i & \lambda_j x_3{}^j \end{vmatrix}$$
$$\lambda_i\lambda_j y_3 = \begin{vmatrix} \lambda_i x_3{}^i & \lambda_j x_3{}^j \\ \lambda_i x_1{}^i & \lambda_j x_1{}^j \end{vmatrix} \tag{2}$$

Since

$$\begin{aligned} \lambda_i x_1{}^i &= a_1 x_1{}^i + a_2 x_2{}^i + a_3 x_3{}^i \\ \lambda_i x_2{}^i &= b_1 x_1{}^i + b_2 x_2{}^i + b_3 x_3{}^i \\ \lambda_i x_3{}^i &= c_1 x_1{}^i + c_2 x_2{}^i + c_3 x_3{}^i \end{aligned} \tag{3}$$

with a similar equation for the components of x^j, we see that

$$\lambda_i\lambda_j y_1 = \begin{vmatrix} a_1x_1^i + a_2x_2^i + a_3x_3^i & a_1x_1^j + a_2x_2^j + a_3x_3^j \\ b_1x_1^i + b_2x_2^i + b_3x_3^i & b_1x_1^j + b_2x_2^j + b_3x_3^j \end{vmatrix} \tag{4}$$

If a is the vector whose components are the a_i, and b the vector whose components are the b_i, we see that (4) is equivalent to the equation

$$\lambda_i\lambda_j y_1 = \begin{vmatrix} (a,x^i) & (a,x^j) \\ (b,x^i) & (b,x^j) \end{vmatrix} \tag{5}$$

What we want at this point is a generalization of the expansion for the Gramian given in Sec. 5 of Chap. 4. Fortunately, it exists. It is easy to verify that

$$\begin{vmatrix} (a,x^i) & (a,x^j) \\ (b,x^i) & (b,x^j) \end{vmatrix} = \begin{vmatrix} a_1 & b_1 \\ a_2 & b_2 \end{vmatrix}\begin{vmatrix} x_1^i & x_1^j \\ x_2^i & x_2^j \end{vmatrix} + \begin{vmatrix} a_2 & b_1 \\ a_3 & b_3 \end{vmatrix}\begin{vmatrix} x_2^i & x_2^j \\ x_3^i & x_3^j \end{vmatrix} + \begin{vmatrix} a_3 & b_3 \\ a_1 & b_1 \end{vmatrix}\begin{vmatrix} x_3^i & x_3^j \\ x_1^i & x_1^j \end{vmatrix} \tag{6}$$

From this it follows immediately from consideration of symmetry that

$$\lambda_i\lambda_j \begin{bmatrix} y_1 \\ y_2 \\ y_3 \end{bmatrix} = \begin{bmatrix} g_{12}(a,b) & g_{23}(a,b) & g_{31}(a,b) \\ g_{12}(a,c) & g_{23}(a,c) & g_{31}(a,c) \\ g_{12}(b,c) & g_{23}(b,c) & g_{31}(b,c) \end{bmatrix} \begin{bmatrix} y_1 \\ y_2 \\ y_3 \end{bmatrix} \tag{7}$$

where

$$g_{ij}(x,y) = \begin{vmatrix} x_i & y_i \\ x_j & y_j \end{vmatrix} \tag{8}$$

Our first observation is that $\lambda_i\lambda_j$ is a characteristic root and y an associated characteristic vector of the matrix occurring in (7). Again, since the matrix is independent of i and j, we see that the three characteristic roots must be $\lambda_1\lambda_2$, $\lambda_1\lambda_3$, and $\lambda_2\lambda_3$.

We now have a means of obtaining a matrix whose characteristic roots are $\lambda_i\lambda_j$, $i \neq j$.

EXERCISES

1. Reverting to the usual notation for the elements of A, write out the representation for the ijth element of the matrix appearing in (7).

2. Let A be an $N \times N$ matrix. Write out the formula for the elements of the $N(N-1)/2$-dimensional matrix, obtained in the foregoing fashion, whose roots are $\lambda_i\lambda_j$, $i \neq j$.

3. For the case where A is symmetric, show that we can obtain the inequality

$$\lambda_1\lambda_2 \geq \begin{vmatrix} a_{11} & a_{12} \\ a_{21} & a_{22} \end{vmatrix}$$

from the preceding results, and many similar inequalities.

4. Let A be an $N \times N$ matrix and let r be an integer between 1 and N. Denote by S_r the ensemble of all sets of r distinct integers chosen from the integers $1, 2, \ldots, N$.

Let s and t be two elements of S_r and A_{st} denote the matrix formed from A by deleting all rows whose indices do not belong to s and all columns whose elements do not belong to t.

Let the elements of S_r be enumerated in some fixed order, say in numerical value, $s_1, s_2, \ldots, s_M$. The $M \times M$ matrix, where $M = N!/r!(N-r)!$,

$$C_r(A) = (|A_{s_is_j}|)$$

is called the rth *compound* or rth *adjugate* of A.

Establish the following results.

(a) $C_r(AB) = C_r(A)C_r(B)$

(b) $C_r(A') = C_r(A)'$

(c) $C_r(A^{-1}) = C_r(A)^{-1}$

(d) $|C_r(A)| = |A|^k \qquad k = (N-1)!/(r = 1)!(N-r)!$

(e) The characteristic roots of $C_r(A)$ are the expressions $\mu_1, \mu_2, \ldots, \mu_M$, where $\mu_1 + \mu_2 + \cdots + \mu_M$ is the rth elementary symmetric function of $\lambda_1, \lambda_2, \ldots, \lambda_N$, e.g., $\mu_1 = \lambda_1\lambda_2 \cdots \lambda_R$, $\mu_2 = \mu_1\mu_2 \cdots \lambda_{R-1}\lambda_{R+1}, \ldots$.[1]

For a discussion of when a given matrix is a compound of another, see D. E. Rutherford, Compound Matrices, *Koninkl. Ned. Akad. Wetenschap. Amsterdam, Proc. Sect. Sci., ser. A*, vol. 54, pp. 16–22, 1951.

Finally, for the connection with invariant theory, see R. Weitzenbuck, *Invariantentheorie*, Groningen, 1923.

15. Circulants. Matrices of the form

$$C = \begin{bmatrix} c_0 & c_1 & & \cdots & c_{N-1} \\ c_{N-1} & c_0 & & \cdots & c_{N-2} \\ c_{N-2} & c_{N-1} & c_0 & \cdots & c_{N-3} \\ \cdot & & & & \cdot \\ \cdot & & & & \cdot \\ \cdot & & & & \cdot \\ c_1 & c_2 & & \cdots & c_0 \end{bmatrix} \tag{1}$$

occur in a variety of investigations. Let us determine the characteristic roots and vectors.

Let r_1 be a root of the scalar equation $r^N = 1$, and set

$$y_1 = c_0 + c_1r_1 + \cdots + c_{N-1}r_1^{N-1} \tag{2}$$

Then we see that y_1 satisfies the following system of equations:

$$\begin{aligned} y_1 &= c_0 + c_1r_1 + \cdots + c_{N-1}r_1^{N-1} \\ r_1y_1 &= c_{N-1} + c_0r_1 + \cdots + c_{N-2}r_1^{N-2} \\ &\cdot \\ &\cdot \\ &\cdot \\ r_1^{N-1}y_1 &= c_1 + c_2r_1 + \cdots + c_0r_1^{N-1} \end{aligned} \tag{3}$$

[1] See, for further results and references, H. J. Ryser, Inequalities of Compound and Induced Matrices with Applications to Combinatorial Analysis, *Illinois J. Math.*, vol. 2, pp. 240–253, 1958.

It follows that y_1 is a characteristic root of C with associated characteristic vector

$$x^1 = \begin{bmatrix} 1 \\ r_1 \\ \cdot \\ \cdot \\ \cdot \\ r_1^{N-1} \end{bmatrix} \tag{4}$$

Since the equation $r^N = 1$ has N distinct roots, we see that we obtain N distinct characteristic vectors. Consequently, we have the complete set of characteristic roots and vectors in this way.

EXERCISES

1. Use the scalar equation $r^5 = a_1 r + 1$ in a similar fashion to obtain the characteristic roots and vectors of the matrix

$$\begin{bmatrix} c_0 & c_1 & c_2 & c_3 & c_4 \\ c_1 & c_2 + c_1 a & c_3 + c_2 a & c_4 + c_3 a & c_0 + c_4 a \\ c_2 & c_3 + c_2 a & c_4 + c_3 a & c_0 + c_4 a & c_1 \\ c_3 & c_4 + c_3 a & c_0 + c_4 a & c_1 & c_2 \\ c_4 & c_0 + c_4 a & c_1 & c_2 & c_3 \end{bmatrix}$$

2. Generalize, using the defining equation $r^N = b_1 r^{N-1} + \cdots + b_N$.

MISCELLANEOUS EXERCISES

1. Let $f(X)$ be a function of the N^2 variables x_{ij} possessing a power series development in these variables about zero. Show that we may write

$$f(X) = \sum_{k=0}^{\infty} \operatorname{tr} (X^{[k]} C_k)$$

2. Is it true that

$$e^{\operatorname{tr} X} = \sum_{k=0}^{\infty} \frac{\operatorname{tr} (X^{[k]})}{k!} \; ?$$

3. If A and B are positive definite, then $A \times B$ is positive definite.

4. If A and B are symmetric matrices with $A \geq B$, then $A^{[n]} \geq B^{[n]}$, for $n = 1, 2, \ldots$.

5. If r satisfies the equation in Exercise 23 of Chap. 11 and s the equation $s^M + b_1 s^{M-1} + \cdots + b_M = 0$, how does one form an equation whose roots are $r_i s_j$?

6. Let $1, \alpha_1, \ldots, \alpha_{N-1}$ be the elements of a finite group. Write $x = x_0 + x_1\alpha_1 + \cdots + x_{N-1}\alpha_{N-1}$ where the x_i are scalars. Consider the products $\alpha_i x = \alpha_i(0)x_0 + \alpha_i(1)x_1 + \cdots + \alpha_i(N-1)x_{N-1}$, where the elements $\alpha_i(j)$ are the α_i in some order, or $\alpha_i x = x_0(i) + x_1(i)\alpha_1 + \cdots + x_{N-1}(i)\alpha_{N-1}$, where the $x_i(j)$ are the x_i

in some order. Introduce the matrix $X = (x_i(j))$ and write $X \sim x$. If $X \sim x$ $Y \sim y$, does $XY \sim xy$?

7. Let $1, \alpha_1, \ldots, \alpha_{N-1}$ be the elements of a finite group G, and $1, \beta_1, \ldots, \beta_{M-1}$ the elements of a finite group H. Consider the *direct product* of G and H defined as the group of order MN whose elements are $\alpha_i\beta_j$. Using the procedure outlined in the preceding exercise, form a matrix corresponding to

$$\left(\sum_{i=0}^{N-1} x_i\alpha_i\right)\left(\sum_{j=0}^{M-1} y_j\beta_j\right)$$

What is the connection between this matrix and the matrices

$$X \sim \sum_{i=0}^{N-1} x_i\alpha_i \qquad Y \sim \sum_{j=0}^{M-1} y_j\beta_j?$$

For an interesting relation between the concept of group matrices and normality, see O. Taussky.[1]

8. The equation $AX - XA = \lambda X$ possesses nontrivial solutions for X if and only if $\lambda = \lambda_i - \lambda_j$ where $\lambda_1, \lambda_2, \ldots, \lambda_N$ are the characteristic values of A (*Lappo-Danilevsky*).

9. Let F be a matrix of order N^2, portioned into an array of N submatrices f_{ij}, each of order N, such that each f_{ij} is a rational function, $f_{ij}(A)$, of a fixed matrix A of order N. If the characteristic values of A are $\lambda_1, \lambda_2, \ldots, \lambda_N$, then those of F are given by the characteristic values of the N matrices, $(f_{ij}(\lambda_k))$, $k = 1, 2, \ldots, N$, each of order N. See J. Williamson[2] and S. N. Afriat.[3] For an application to the solution of partial differential equations by numerical techniques, see J. Todd[4] and A. N. Lowan.[5]

10. Let $f(\theta)$ be a real function of θ for $-\pi \leq \theta \leq \pi$, and form the Fourier coefficients of $f(\theta)$,

$$c_n = \frac{1}{2\pi}\int_{-\pi}^{\pi} f(\theta)e^{-in\theta}\,d\theta \qquad n = 0, \pm 1, \pm 2, \ldots$$

The finite matrices

$$T_N = (c_{k-l}) \qquad k, l = 0, 1, 2, \ldots, N$$

are called *Toeplitz matrices* of order N. Show that T_N is Hermitian.

11. Show that if we denote by $\lambda_1^{(N)}, \lambda_2^{(N)}, \ldots, \lambda_{N+1}^{(N)}$, the $N + 1$ characteristic values of T_N, then

$$\lim_{N\to\infty} \frac{\lambda_1^{(N)} + \lambda_2^{(N)} + \cdots + \lambda_{N+1}^{(N)}}{N+1} = \frac{1}{2\pi}\int_{-\pi}^{\pi} f(\theta)\,d\theta$$

[1] O. Taussky, A Note on Group Matrices, *Proc. Am. Math. Soc.*, vol. 6, pp. 984–986, 1955.

[2] J. Williamson, The Latent Roots of a Matrix of Special Type, *Bull. Am. Math. Soc.*, vol. 37, pp. 585–590, 1931.

[3] S. N. Afriat, Composite Matrices, *Quart. J. Math., Oxford* 2d, ser. 5, pp. 81–98, 1954.

[4] J. Todd, The Condition of Certain Matrices, III, *J. Research Natl. Bur. Standards*, vol. 60, pp. 1–7, 1958.

[5] A. N. Lowan, The Operator Approach to Problems of Stability and Convergence of Solutions of Difference Equations, *Scripta Math.*, no. 8, 1957.

12. Show that

$$\lim_{N\to\infty} \frac{(\lambda_1^{(N)})^2 + (\lambda_2^{(N)})^2 + \cdots + (\lambda_{N+1}^{(N)})^2}{N+1} = \frac{1}{2\pi}\int_{-\pi}^{\pi} f^2(\theta)\,d\theta$$

(These last two results are particular results of a general result established by Szego in 1917. A discussion of the most recent results, plus a large number of references to the ways in which these matrices enter in various parts of analysis, will be found in Kac, Murdock, and Szego.[1])

Generalizations of the Toeplitz matrix have been considered in U. Grenander.[2] See also H. Widom.[3]

13. Having defined what we mean by the Kronecker sum of two matrices, $A \oplus B$, define a Kronecker derivative of a variable matrix $X(t)$ as follows:

$$\frac{\delta X}{\delta t} = \lim_{h\to 0} \frac{X(t+h) \oplus (-X(t))}{h}$$

14. Consider the differential equation

$$\frac{\delta Y}{\delta t} = A \times Y \qquad Y(0) = I$$

Does it have a solution, and if so, what is it?

15. Establish the Schur result that $C = (a_{ij}b_{ij})$ is positive definite if A and B are by considering C as a suitable submatrix of the Kronecker product $A \times B$ (*Marcus*).

16. Let $a^{(i)}$ be the ith column of A. Introduce the "stacking operator," $S(A)$, which transforms a matrix into a vector

$$S(A) = \begin{bmatrix} a^{(1)} \\ a^{(2)} \\ \cdot \\ \cdot \\ \cdot \\ a^{(N)} \end{bmatrix}$$

Prove that $S(PAQ) = (Q' \otimes P)S(A)$. See D. Nissen, A Note on the Variance of a Matrix, *Econometrica,* vol. 36, pp. 603–604, 1968.

17. Consider the Selberg quadratic form (see Appendix B), $Q(x) = \sum_{k=1}^{N} \Big(\sum_{v/a_k} x_v\Big)^2$, where a_k, $k = 1, 2, \ldots, N$, is a set of positive integers, $x_1 = 1$, and the remaining x_v are real. Consider the extended form $Q_r(x,z) = \sum_{k=1}^{N} \Big(z_k + \sum_{v/a_k} x_v\Big)^2$ and write $f_r(z) = \min_x Q_r(x,z)$, where the minimization is over $x_k, x_{k+1}, \ldots, x_N$. Obtain a recurrence relation connecting $f_r(z)$ with $f_{r+1}(z)$.

[1] M. Kac, W. L. Murdock, and G. Szego, On the Eigenvalues of Certain Hermitian Forms, *J. Rat. Mech. Analysis,* vol. 2, pp. 767–800, 1953.

[2] U. Grenander, *Trans. Am. Math. Soc.,* 1958.

[3] H. Widom, On the Eigenvalues of Certain Hermitian Operators, *Trans. Am. Math. Soc.,* vol. 88, 491–522, 1958.

18. Use the fact that $f_r(z)$ is quadratic in z to simplify this recurrence relation; see Chap. 9.

19. Discuss the use of the recurrence relation for analytic and computational purposes. See R. Bellman, Dynamic Programming and the Quadratic Form of Selberg, *J. Math. Anal. Appl.*, vol. 15, pp. 30–32, 1966.

20. What conditions must b and c satisfy so that $Xb = c$ where X is positive definite? (*Wimmer*)

21. Establish the Schur result concerning $(a_{ij}b_{ij})$ using Kronecker products and a suitable submatrix (*Marcus*).

Bibliography and Discussion

§1. The direct product of matrices arises from the algebraic concept of the direct product of groups. For a discussion and treatment from this point of view, see

C. C. MacDuffee, *The Theory of Matrices,* Chelsea Publishing Co., chapter VII, New York, 1946.

F. D. Murnaghan, *The Theory of Group Representation,* Johns Hopkins Press, Baltimore, 1938.

We are interested here in showing how this concept arises from two different sets of problem areas, one discussed here and one in Chap. 15. See also

E. Cartan, *Oeuvres Complètes,* part 1, vol. 2, pp. 1045–1080, where the concept of a *zonal harmonic* of a positive definite matrix is introduced.

§4. This procedure is taken from

R. Bellman, Limit Theorems for Non-commutative Operations—I, *Duke Math. J.*, vol. 21, pp. 491–500, 1954.

§5. Some papers concerning the application and utilization of the Kronecker product are

F. Stenger, Kronecker Product Extensions of Linear Operators, *SIAM J. Numer. Anal.*, vol. 5, pp. 422–435, 1968.

J. H. Pollard, On the Use of the Direct Matrix Product in Analyzing Certain Stochastic Population Models, *Biometrika,* vol. 53, pp. 397–415, 1966.

§10. The Kronecker logarithm is introduced in the paper cited above.

§11. The result of Theorem 4 is given in the book by MacDuffee referred to above. It was independently derived by W. Hahn, using the procedure outlined in the exercises; see

W. Hahn, Eine Bemerkung zur zweiten Methode von Lyapunov, *Math. Nachr.*, 14 Band., Heft 4/6, pp. 349–354, 1956.

The result was obtained by R. Bellman, as outlined in the text, unaware of the discussion in the book by MacDuffee, cited above,

R. Bellman, Kronecker Products and the Second Method of Lyapunov, *Math. Nachr.*, 1959.

§13. A great deal of work has been done on the corresponding operator equations, partly for their own sake and partly because of their central position in quantum mechanics. See

M. Rosenbloom, On the Operator Equation $BX - XA = C$, *Duke Math. J.*, vol. 23, 1956.

E. Heinz, *Math. Ann.*, vol. 123, pp. 415–438, 1951.

D. E. Rutherford, *Koninkl. Ned. Akad. Wetenschap., Proc.*, ser. A, vol. 35, pp. 54–59, 1932.

§14. The motivation for this section lies within the theory of invariants of linear transformations and the geometric interconnections. See

M. Bocher, *Introduction to Higher Algebra,* The Macmillan Company, New York, reprinted, 1947.

F. Klein, *Elementary Mathematics from an Advanced Standpoint, Geometry,* Dover Publications, New York.

H. Weyl, *The Classical Groups,* Princeton University Press, Princeton, N.J., 1946.

The 2×2 determinants are the Plucker coordinates. The basic idea is that every linear transformation possesses a large set of associated *induced* transformations that permit us to derive certain properties of the original transformation in a quite simple fashion. As indicated in the text, we have not penetrated into this area in any depth because of a desire to avoid a certain amount of determinantal manipulation.

See also

H. Schwerdtfeger, Skew-symmetric Matrices and Projective Geometry, *Am. Math. Monthly,* vol. LI, pp. 137–148, 1944.

It will be clear from the exercises that once this determinantal groundwork has been laid, we have a new way of obtaining a number of the inequalities of Chap. 8.

For an elegant probabilistic interpretation of Plucker coordinates, and the analogous multidimensional sets, see Exercises 15 and 16 of Chap. 15, where some results of Karlin and MacGregor are given.

§15. Circulants play an important role in many mathematical-physical theories. See the paper

T. H. Berlin and M. Kac, The Spherical Model of a Ferromagnet, *Phys. Rev.*, vol. 86, pp. 821–835, 1952.

for an evaluation of some interesting circulants and further references to work by Onsager, et al., connected with the direct product of matrices and groups.

Treatments of the theory of compound matrices, and further references may be found in

H. J. Ryser, Inequalities of Compound and Induced Matrices with Applications to Combinatorial Analysis, *Illinois J. Math.*, vol. 2, pp. 240–253, 1958.

N. G. deBruijn, Inequalities Concerning Minors and Eigenvalues, *Nieuw. Arch. Wisk.* (3), vol. 4, pp. 18–35, 1956.

D. E. Littlewood, *The Theory of Group Characters and Matrix Representations of Groups*, Oxford, New York, 1950.

I. Schur, *Über eine Klasse von Matrizen die sich einer gegebenen Matrix zuordnen lassen*, Dissertation, Berlin, 1901.

J. H. M. Wedderburn, Lectures on Matrices, *Am. Math. Soc. Colloq. Publ.*, vol. 17, 1934.

C. C. MacDuffee, *The Theory of Matrices*, Chelsea Publishing Company, New York, 1946.

O. Toeplitz, Das algebraische Analogon zu einem Satze von Fejer, *Math. Z.*, vol. 2, pp. 187–197, 1918.

M. Marcus, B. N. Moyls, and R. Westwick, Extremal Properties of Hermitian Matrices II, *Canad. J. Math.*, 1959.

13

Stability Theory

1. Introduction. A problem of great importance is that of determining the behavior of a physical system in the neighborhood of an equilibrium state. If the system returns to this state after being subjected to small disturbances, it is called *stable;* if not, it is called *unstable.*

Although physical systems can often be tested for this property, in many cases this experimental procedure is both too expensive and too time-consuming. Consequently, when designing a system we would like to have mathematical criteria for stability available.

It was pointed out in Chap. 11 that a linear equation of the form

$$\frac{dx}{dt} = Ax \qquad x(0) = c \tag{1}$$

can often be used to study the behavior of a system in the vicinity of an equilibrium position, which in this case is $x = 0$.

Consequently, we shall begin by determining a necessary and sufficient condition that the solution of (1) approach zero as $t \to \infty$. The actual economic, engineering, or physical problem is, however, more complicated, since the equation describing the process is not (1), but nonlinear of the form

$$\frac{dy}{dt} = Ay + g(y) \qquad y(0) = c \tag{2}$$

The question then is whether or not criteria derived for linear systems are of any help in deciding the stability of nonlinear systems. It turns out that under quite reasonable conditions, the two are equivalent. This is the substance of the classical work of Poincaré and Lyapunov. However, we shall not delve into these more recondite matters here, restricting ourselves solely to the consideration of the more tractable linear equations.

2. A Necessary and Sufficient Condition for Stability. Let us begin by demonstrating the fundamental result in these investigations.

Theorem 1. *A necessary and sufficient condition that the solution of*

$$\frac{dx}{dt} = Ax \qquad x(0) = c \tag{1}$$

regardless of the value of c, approach zero as $t \to \infty$, is that all the characteristic roots of A have negative real parts.

Proof. If A has distinct characteristic roots, then the representation

$$e^{At} = T \begin{bmatrix} e^{\lambda_1 t} & & & & 0 \\ & e^{\lambda_2 t} & & & \\ & & \cdot & & \\ & & & \cdot & \\ & & & & \cdot \\ 0 & & & & e^{\lambda_N t} \end{bmatrix} T^{-1} \tag{2}$$

establishes the result. We cannot make an immediate appeal to continuity to obtain the result for general matrices, but we can proceed in the following way.

In place of reducing A to diagonal form, let us transform it into triangular form by means of a similarity transformation, $T^{-1}AT = B$. The system of equations in (1) takes the form

$$\frac{dz}{dt} = Bz \qquad z(0) = c' \tag{3}$$

where B is a triangular matrix upon the substitution $x = Tz$. Written out in terms of the components, we have

$$\begin{aligned} \frac{dz_1}{dt} &= b_{11}z_1 + b_{12}z_2 + \cdots + b_{1N}z_N & z_1(0) &= c_1' \\ \frac{dz_2}{dt} &= \qquad\quad b_{22}z_2 + \cdots + b_{2N}z_N & z_2(0) &= c_2' \\ &\cdot \\ &\cdot \\ &\cdot \\ \frac{dz_N}{dt} &= \qquad\qquad\qquad\qquad b_{NN}z_N & z_N(0) &= c_N' \end{aligned} \tag{4}$$

Since the b_{ii} are the characteristic roots of A, we have, by assumption, Re $(b_{ii}) < 0$ for $i = 1, 2, \ldots, N$.

Solving for z_N,

$$z_N = c_N' e^{b_{NN}t} \tag{5}$$

we see that $z_N \to 0$ as $t \to \infty$.

In order to show that all $z_i \to 0$ as $t \to \infty$, we proceed inductively based upon the following result.

If $v(t) \to 0$ as $t \to \infty$, then $u(t)$ as determined by

$$\frac{du}{dt} = b_1u + v(t) \qquad u(0) = a_1 \tag{6}$$

approaches zero as $t \to \infty$, provided that $\text{Re}\,(b_1) < 0$.

Since

$$u(t) = a_1e^{b_1t} + e^{b_1t}\int_0^t e^{-b_1s}v(s)\,ds \tag{7}$$

it is easy to see that the stated result is valid.

Starting with the result for z_N based upon (5), we obtain successively the corresponding results for $z_{N-1}, \ldots, z_1$.

EXERCISE

1. Prove Theorem 1 using the Jordan canonical form.

3. Stability Matrices. To avoid wearisome repetition, let us introduce a new term.

Definition. A matrix A will be called a stability matrix if all of its characteristic roots have negative real parts.

EXERCISES

1. Derive a necessary and sufficient condition that a real matrix be a stability matrix in terms of the matrix $(\text{tr}\,(A^{i+j}))$ (*Bass*).
2. What is the corresponding condition if A is complex?

4. A Method of Lyapunov. Let us now see how we can use quadratic forms to discuss questions of asymptotic behavior of the solutions of linear differential equations. This method was devised by Lyapunov and is of great importance in the modern study of the stability of solutions of nonlinear functional equations of all types.

Consider the equation

$$\frac{dx}{dt} = Ax \qquad x(0) = c \tag{1}$$

where c and A are taken to be real, and the quadratic form

$$u = (x,Yx) \tag{2}$$

where Y is a symmetric constant matrix as yet undetermined. We have

$$\begin{aligned}\frac{du}{dt} &= (x',Yx) + (x,Yx') \\ &= (Ax,Yx) + (x,YAx) \\ &= (x,(A'Y + YA)x)\end{aligned} \tag{3}$$

Suppose that we can determine Y so that

$$A'Y + YA = -I \tag{4}$$

with the further condition that Y be positive definite. Then the relation in (3) becomes

$$\frac{du}{dt} = -(x,x) \tag{5}$$

which yields

$$\frac{du}{dt} \leq -\frac{1}{\lambda_N} u \tag{6}$$

where λ_N is the largest characteristic root of Y. From (6) we have $u \leq u(0)e^{-t/\lambda_N}$. Hence $u \to 0$ as $t \to \infty$. It follows from the positive definite property of Y that each component of x must approach zero as $t \to \infty$.

We know, however, from the result of Sec. 13 of Chap. 12 that if A is a stability matrix, we can determine the symmetric matrix Y uniquely from (4). Since Y has the representation

$$Y = \int_0^\infty e^{A't_1}e^{At_1}\,dt_1 \tag{7}$$

we have

$$\begin{aligned}(x,Yx) &= \int_0^\infty (x,e^{A't_1}e^{At_1}x)\,dt_1 \\ &= \int_0^\infty (e^{At_1}x,e^{At_1}x)\,dt_1\end{aligned} \tag{8}$$

It follows that Y is positive definite, since e^{At} is never singular.

EXERCISES

1. Consider the equation $dx/dt = Ax + g(x)$, $x(0) = c$, where
(a) A is a stability matrix,
(b) $\|g(x)\|/\|x\| \to 0$ as $\|x\| \to 0$,
(c) $\|c\|$ is sufficiently small.

Let Y be the matrix determined above. Prove that if x satisfies the foregoing nonlinear equation and the preceding conditions are satisfied, then

$$\frac{d}{dt}(x,Yx) \leq -r_1(x,Yx)$$

where r_1 is a positive constant. Hence, show that $x \to 0$ as $t \to \infty$.

2. Extend the foregoing argument to treat the case of complex A.

5. Mean-square Deviation. Suppose that A is a stability matrix and that we wish to calculate

$$J = \int_0^\infty (x,Bx)\,dt \tag{1}$$

where x is a solution of (4.1).

It is interesting to note that J can be calculated as a rational function of the elements of A without the necessity of solving the linear differential equation for x. In particular, it is not necessary to calculate the characteristic roots of A.

Let us determine a constant matrix Y such that

$$(x,Bx) = \frac{d}{dt}(x,Yx) \tag{2}$$

We see that

$$B = A'Y + YA \tag{3}$$

With this determination of Y, the value of J is given by

$$J = -(c,Yc) \tag{4}$$

Since A is a stability matrix, (3) has a unique solution which can be found using determinants.

6. Effective Tests for Stability. The problem of determining when a given matrix is a stability matrix is a formidable one, and at the present time there is no simple solution. What complicates the problem is that we are not so much interested in resolving the problem for a particular matrix A as we are in deriving conditions which enable us to state when various members of a class of matrices, $A(\mu)$, are stability matrices. Questions of this type arise constantly in the design of control mechanisms, in the field of mathematical economics, and in the study of computational algorithms.

Once the characteristic polynomial of A has been calculated, there are a variety of criteria which can be applied to determine whether or not all the roots have negative real parts. Perhaps the most useful of these are the criteria of Hurwitz.

Consider the equation

$$|\lambda I - A| = \lambda^N + a_1\lambda^{N-1} + \cdots + a_{N-1}\lambda + a_N = 0 \tag{1}$$

and the associated infinite array

$$\begin{matrix} a_1 & 1 & 0 & 0 & 0 & 0 & \cdots \\ a_3 & a_2 & a_1 & 1 & 0 & 0 & \cdots \\ a_5 & a_4 & a_3 & a_2 & a_1 & 0 & \cdots \\ a_7 & a_6 & a_5 & a_4 & a_3 & a_2 & \cdots \\ \cdot & \cdot & \cdot & \cdot & \cdot & \cdot & \\ \cdot & \cdot & \cdot & \cdot & \cdot & \cdot & \\ \cdot & \cdot & \cdot & \cdot & \cdot & \cdot & \end{matrix} \tag{2}$$

where a_k is taken to be zero for $k > N$.

A necessary and sufficient condition that all the roots of (1) have negative real parts is that the sequence of determinants

$$h_1 = |a_1| \qquad h_2 = \begin{vmatrix} a_1 & 1 \\ a_3 & a_2 \end{vmatrix} \qquad h_3 = \begin{vmatrix} a_1 & 1 & 0 \\ a_3 & a_2 & a_1 \\ a_5 & a_4 & a_3 \end{vmatrix} \tag{3}$$

formed from the preceding array, be positive.

There are no simple direct proofs of this result, although there are a number of elegant proofs. We shall indicate in the following section one line that can be followed, and in Appendix C discuss briefly the chain of ideas, originating in Hermite, giving rise to Hurwitz's proof. Both of these depend upon quadratic forms. References to other types of proof will be found at the end of the chapter.

The reason why this result is not particularly useful in dealing with stability matrices is that it requires the evaluation of $|\lambda I - A|$, something we wish strenuously to avoid if the dimension of A is high.

EXERCISES

1. Using the foregoing criteria, show that a necessary and sufficient condition that $\lambda^2 + a_1\lambda + a_2$ be a *stability polynomial* is that $a_1 > 0$, $a_2 > 0$. By this we mean that the roots of the polynomial have negative real parts.

2. For $\lambda^3 + a_1\lambda^2 + a_2\lambda + a_3$ show that corresponding conditions are

$$a_1, a_2, a_3 > 0 \qquad a_1a_2 > a_3$$

3. For $\lambda^4 + a_1\lambda^3 + a_2\lambda^2 + a_3\lambda + a_4$ show that the conditions are a_1, $a_4 > 0$, $a_1a_2 > a_3$, $a_3(a_1a_2 - a_3) > a_1^2a_4$.

7. A Necessary and Sufficient Condition for Stability Matrices. Let us now show that the results of Sec. 4 yield

Theorem 2. *Let Y be determined by the relation*

$$A'Y + YA = -I \tag{1}$$

Then a necessary and sufficient condition that the real matrix A be a stability matrix is that Y be positive definite.

Proof. Referring to Sec. 5, we see that

$$\int_0^T (x,x)\,dt = (x(0),Yx(0)) - (x(T),Yx(T)) \tag{2}$$

or

$$(x(T),Yx(T)) + \int_0^T (x,x)\,dt = (x(0),Yx(0)) \tag{3}$$

Here x is a solution of the equation $dx/dt = Ax$.

If Y is positive definite, $\int_0^T (x,x)\,dt$ is uniformly bounded, which means that $x(t) \to 0$ as $t \to \infty$, whence A is a stability matrix. We have already established the fact that Y is positive definite if A is a stability matrix.

8. Differential Equations and Characteristic Values. Consider the differential equation

$$Ax'' + 2Bx' + Cx = 0 \qquad x(0) = c^1 \qquad x'(0) = c^2 \tag{1}$$

which, if considered to arise from the study of electric circuits possessing capacitances, inductances, and resistances, is such that A, B, and C are non-negative definite.

On these grounds, it is intuitively clear that the following result holds.

Theorem 3. *If A, B, and C are non-negative definite, and either A or C positive definite, then*

$$|\lambda^2 A + 2\lambda B + C| = 0 \tag{2}$$

has no roots with positive real parts.

If A and C are non-negative definite and B is positive definite, then the only root with zero real part is $\lambda = 0$.

Proof. Let us give a proof which makes use of the physical background of the statement. In this case, we shall use energy considerations. Starting with (1), let us write

$$(x',Ax'') + 2(x',Bx') + (x',Cx) = 0 \tag{3}$$

Thus, for any $s > 0$,

$$\int_0^s [(x',Ax'') + 2(x',Bx') + (x',Cx)]\, dt = 0 \tag{4}$$

or

$$(x',Ax')\Big]_0^s + 4\int_0^s (x',Bx')\, dt + (x,Cx)\Big]_0^s = 0 \tag{5}$$

This is equivalent to the equation

$$(x'(s),Ax'(s)) + 4\int_0^s (x',Bx')\, dt + (x(s),Cx(s)) = c_3 \tag{6}$$

where $c_3 = (c^2,Ac^2) + (c^1,Cc^1)$.

If λ is a root of (2), then (1) has a solution of the form $e^{\lambda t}c$. If λ is real, c is real. If λ is complex, $\lambda = r_1 + ir_2$, then the real part of $e^{\lambda t}c$, which has the form $e^{r_1 t}(a^1 \cos r_2 t + a^2 \sin r_2 t)$, is also a solution. Substituting in (6), we see that

$$e^{2r_1 s}(b^1,Ab^1) + 4\int_0^s e^{2r_1 t}(b^2(t),Bb^2(t))\, dt + e^{2r_1 s}(b^3,Cb^3) = c_3 \tag{7}$$

where b^1 and b^3 are constant vectors and b^2 is a variable vector given by $(a^1 r_1 + a^2 r_2) \cos r_2 t + (a^2 r_1 - a^1 r_2) \sin r_2 t$.

If A or C is positive definite, with $B \geq 0$, we see that $r_1 > 0$ leads to a contradiction as $s \to \infty$.

If A, $C \geq 0$, then B positive definite requires that $r_1 \leq 0$. Furthermore, since $a^1 \cos r_2 t + a^2 \sin r_2 t$ is periodic, we see that the integral $\int_0^s (b^2(t),Bb^2(t))\, dt$ diverges to plus infinity as $s \to \infty$, unless $r_2 = 0$, if $r_1 = 0$.

We have presented this argument in some detail since it can be extended to treat similar questions for equations in which the coefficients A, B, and C are variable.

EXERCISES

1. Following Anke,[1] let us use the foregoing techniques to evaluate $\int_0^\infty u^2\,dt$, given that u is a solution of $u''' + a_2u'' + a_1u' + a_0u = 0$, all of whose solutions tend to zero. Let $u(0) = c_0$, $u'(0) = c_1$, $u''(0) = c_2$. Establish the results

$$\int_0^\infty u'''u\,dt = u''u\Big|_0^\infty - \int_0^\infty u''u'\,dt = -c_0c_2 + c_1^2/2$$
$$\int_0^\infty u''u\,dt = u'u\Big|_0^\infty - \int_0^\infty u'^2\,dt = -c_0c_1 - \int_0^\infty u'^2\,dt$$
$$\int_0^\infty u'u\,dt = -c_0^2/2$$

2. Derive from the equations

$$\int_0^\infty u^{(i)}(u''' + a_2u'' + a_1u' + a_0u)\,dt = 0 \qquad i = 0,\ 1,\ 2$$

and the results of Exercise 1, a set of linear equations for the quantities $\int_0^\infty u^2\,dt$, $\int_0^\infty (u')^2\,dt$, $\int_0^\infty (u'')^2\,dt$.

3. Using these linear equations, express $\int_0^\infty u^2\,dt$ as a quadratic form in c_0, c_1, c_2.

4. Using this quadratic form, obtain a set of necessary and sufficient conditions that $r^3 + a_2r^2 + a_1r + a_0$ be a stability polynomial.

5. Show that these conditions are equivalent to the Hurwitz conditions given in Exercise 2 of Sec. 6.

9. Effective Tests for Stability Matrices. As indicated in the foregoing sections, there exist reasonably effective techniques for determining when the roots of a given polynomial have negative real parts. Since, however, the task of determining the characteristic polynomial of a matrix of large dimension is a formidable one, we cannot feel that we have a satisfactory solution to the problem of determining when a given matrix is a stability matrix.

In some special cases, nonetheless, very elegant criteria exist. Thus:

Theorem 4. *If A has the form*

$$A = \begin{bmatrix} a_1 + b_1 & a_2 & \cdots & & & & \\ -1 & b_2 & a_3 & \cdots & & & \\ \cdot & -1 & b_3 & a_4 & \cdots & & \\ & & & & -1 & b_{N-1} & a_N \\ & & \cdots & & \cdot & -1 & b_N \end{bmatrix} \tag{1}$$

[1] *Z. angew. Math. u. Phys.*, vol. VI, pp. 327–332, 1955.

where the dots indicate that all other terms are zero, the a_i are real, and the b_i are zero or pure imaginary, then the number of positive terms in the sequence of products $a_1, a_1a_2, \ldots, a_1a_2 \ldots a_{N-1}a_N$ is the number of characteristic roots of A with positive real part.

For the proof, which is carried out by means of the theory of Sturmian sequences, we refer to the paper by Schwarz given in the Bibliography at the end of the chapter. In this paper it is also shown that any matrix B with complex elements can be transformed into the form appearing in (1) by means of a transformation of the type $A = T^{-1}BT$.

As we shall see in Chap. 16, the theory of positive matrices gives us a foothold on the stability question when A is a matrix all of whose off-diagonal elements are non-negative.

MISCELLANEOUS EXERCISES

1. Obtain the solution of the scalar equation $u'' + u = f_1(t)$ by writing $u' = v$, $v' = -u + f_1(t)$ and determining the elements of e^{At} where

$$A = \begin{bmatrix} 0 & 1 \\ -1 & 0 \end{bmatrix}$$

2. Let the characteristic roots of A be distinct and $B(t) \to 0$ as $t \to \infty$. Then the characteristic roots of $A + B(t)$, which we shall call $\lambda_i(t)$, are distinct for $t \geq t_0$.

3. If, in addition, $\int^\infty \|B(t)\|\, dt < \infty$, there exists a matrix $T(t)$ having the property that the change of variable $y = Tz$ converts $y' = (A + B(t))y$ into $z' = (L(t) + C(t))z$ where

$$L(t) = \begin{bmatrix} \lambda_1(t) & & & 0 \\ & \lambda_2(t) & & \\ & & \ddots & \\ 0 & & & \lambda_N(t) \end{bmatrix}$$

and $\int^\infty \|C(t)\|\, dt < \infty$.

4. If $\sum_{i,j=1}^{N} \int_0^\infty |a_{ij}(t) + a_{ji}(t)|\, dt < \infty$, all solutions of $y' = A(t)y$ are bounded as $t \to \infty$.

5. There exists an orthogonal matrix $B(t)$ such that the transformation $y = B(t)z$ converts $y' = A(t)y$ into $z' = C(t)z$ where $C(t)$ is semidiagonal (*Diliberto*).

6. There exists a bounded nonsingular matrix $B(t)$ with the property that $C(t)$ is diagonal.

7. The usual rule for differentiation of a product of two functions u and v has the form $d(uv)/dt = u\, dv/dt + (du/dt)v$. Consider the linear matrix function $d(X)$ defined by the relation

$$d(X) = AX - XA$$

where A is a fixed matrix. Show that

$$d(XY) = Xd(Y) + d(X)Y$$

8. Obtain a representation for $d(X_1, X_2 \cdots X_N)$ and thus for $d(X^N)$.

9. Let $d_A(X) = AX - XA$, $d_B(X) = BX - XB$. Show that

$$d_A(d_B(X)) = d_B(d_A(X))$$

10. When does the equation $d(X) = \lambda X$, λ a scalar, have a solution, and what is it? For an extensive and intensive discussion of these and related questions, see J. A. Lappo-Danilevsky.[1]

11. When do the equations

$$\begin{aligned} d(X_1) &= a_{11}X_1 + a_{12}X_2 \\ d(X_2) &= a_{21}X_1 + a_{22}X_2 \end{aligned}$$

have solutions, and what are they?

12. Since $d(X)$ is an analogue of a derivative, are there analogues of a Taylor series expansion?

13. Given a real matrix A, can we always find a diagonal matrix B with elements ± 1 such that all solutions of $Bdx/dt = Ax$ approach zero as $t \to \infty$ (*Brock*)?

14. Let A be nonsingular. Then a permutation matrix P and a diagonal matrix D can be chosen so that DPA is a stability matrix with distinct characteristic roots (*Folkman*).

15. What is the connection between the foregoing results and the idea of solving $Ax = b$ by use of the limit of the solution of $y' = Ay - b$ as $t \to \infty$? (Unfortunately, the result of Exercise 14 is an existence proof. No constructive way of choosing D and P is known.)

16. If C has all characteristic roots less than one in absolute value, there exists a positive definite G such that $G - CGC^*$ is positive definite (*Stein*).

17. Establish Stein's result using difference equations. For extensions of the results in Exercises 14–17 and additional references, see O. Taussky, On Stable Matrices, *Programmation en mathématiques numériques*, Besançon, pp. 75–88, 1966.

18. Let A be an $N \times N$ complex matrix with characteristic roots λ_i, $\lambda_i + \bar{\lambda}_k \neq 0$. Then the $N \times N$ matrix $G = G^*$, the solution of $AG + GA^* = I$ is nonsingular and has as many positive characteristic roots as there are λ_i with positive real parts. (*O. Taussky, A Generalization of a Theorem of Lyapunov, J. Soc. Ind. Appl. Math.*, vol. 9, pp. 640–643, 1961.)

19. Given the equation $\lambda^n + a_1\lambda^{n-1} + \cdots + a_n = 0$, with the roots $\lambda_1, \lambda_2, \ldots, \lambda_n$, find the equation of degree n^2 whose roots are $\lambda_i + \lambda_j$, $i, j = 1, 2, \ldots, n$.

20. Find the equation of degree $n(n+1)/2$ whose roots are $\lambda_i + \lambda_j$, $i = 1, 2, \ldots, n$, $j = 1, 2, \ldots, i$, and the equation of degree $n(n-1)/2$ whose roots are $\lambda_i + \lambda_j$, $i = 2, 3, \ldots, n$; $j = 1, 2, \ldots, i - 1$.

21. Use these equations to determine necessary and sufficient conditions that all of the roots of the original equation have negative real parts (*Clifford-Routh*). See Fuller,[2] where many additional results will be found, and Barnett and Storey[3] and the survey paper by O. Taussky cited above.

[1] J. A. Lappo-Danilevsky, *Mémoires sur la théorie des systèmes des équations différentielles linéaires*, vol. 1, Chelsea Publishing Co., New York, 1953.

[2] A. T. Fuller, Conditions for a Matrix to Have Only Characteristic Roots with Negative Real Parts, *J. Math. Anal. Appl.*, vol. 23, pp. 71–98, 1968.

[3] S. Barnett and C. Storey, Analysis and Synthesis of Stability Matrices, *J. Diff. Eq.*, vol. 3, pp. 414–422, 1967.

Bibliography and Discussion

§1. We are using the word "stability" here in the narrow sense that all solutions of $dx/dt = Ax$ tend to the null solution as $t \to \infty$. For a wider definition and further results, see

R. Bellman, *Stability Theory of Differential Equations*, Dover Publications, New York, 1969.

The problem of determining the asymptotic behavior of the solutions of $dx/dt = A(t)x$ in terms of the nature of the characteristic roots of $A(t)$ as functions of t is much more complex. See the book cited above, and

E. A. Coddington and N. Levinson, *Theory of Ordinary Differential Equations*, McGraw-Hill Book Company, Inc., New York, 1955.

§2. An extensive generalization of this result is the theorem of Poincaré-Lyapunov which asserts that all solutions of $dx/dt = Ax + g(x)$ for which $\|x(0)\|$ is sufficiently small tend to zero as $t \to \infty$, provided that A is a stability matrix and $g(x)$ is nonlinear, that is, $\|g(x)\|/\|x\| \to 0$ as $\|x\| \to 0$. See the books cited above for several proofs, one of which is sketched in Exercise 1 of Sec. 4. See also

R. Bellman, *Methods of Nonlinear Analysis*, Academic Press, Inc., New York, 1970.

§4. This method plays a fundamental role in the modern theory of the stability of functional equations. It was first presented by Lyapunov in his memoir published in 1897 and reprinted in 1947,

A. Lyapunov, *Probleme general de la stabilite du mouvement, Ann. Math. Studies*, no 17, 1947.

For a report on its current applications, see

H. Antosiewicz, *A Survey of Lyapunov's Second Method*, Applied Mathematics Division, National Bureau of Standards, 1956.

H. Antosiewicz, Some Implications of Lyapunov's Conditions for Stability, *J. Rat. Mech. Analysis*, vol. 3, pp. 447–457, 1954.

§5. We follow the results contained in

R. Bellman, Notes on Matrix Theory—X: A Problem in Control, *Quart. Appl. Math.*, vol. 14, pp. 417–419, 1957.

K. Anke, Eine neue Berechnungsmethode der quadratischen Regelflache, *Z. angew. Math. u. Phys.*, vol. 6, pp. 327–331, 1955.

H. Buckner, A Formula for an Integral Occurring in the Theory of Linear Servomechanisms and Control Systems, *Quart. Appl. Math.*, vol. 10, 1952.

P. Hazenbroek and B. I. Van der Waerden, Theoretical Considerations in the Optimum Adjustment of Regulators, *Trans. Am. Soc. Mech. Engrs.*, vol. 72, 1950.

§6. The results of Hurwitz were obtained in response to the problem of determining when a given polynomial was a stability polynomial; see

A. Hurwitz, Über die Bedingungen unter welchen eine Gleichung nur Wurzeln mit negativen reellen Teilen besitzt, *Math. Ann.*, vol. 46, 1895 (Werke, vol. 2, pp. 533–545).

A large number of different derivations have been given since. A critical review of stability criteria, new results, and an extensive bibliography may be found in

H. Cremer and F. H. Effertz, Über die algebraische Kriterien für die Stabilität von Regelungssystemen, *Math. Annalen,* vol. 137, pp. 328–350, 1959.

A. T. Fuller, Conditions for a Matrix to Have Only Characteristic Roots with Negative Real Parts, *J. Math. Analysis Appl.*, vol. 23, pp. 71–98, 1968.

§7. A great deal of work using this type of argument has been done by R. W. Bass in unpublished papers.

For a discussion of stability problems as they arise in mathematical economics, see

A. C. Enthoven and K. J. Arrow, A Theorem on Expectations and the Stability of Equilibrium, *Econometrica,* vol. 24, pp. 288–293, 1956.

K. J. Arrow and M. Nerlove, A Note on Expectations and Stability, *Econometrica,* vol. 26, pp. 297–305, 1958.

References to earlier work by Metzler and others will be found there.

§9 See the paper

H. R. Schwarz, Ein Verfahren zur Stabilitätsfrage bei Matrizen-Eigenwert-Problemen, *Z. angew. Math. u. Phys.*, vol. 7, pp. 473–500, 1956.

For the problem of the conversion of a given matrix into a Jacobi matrix, see the above paper by Schwarz and

W. Givens, Numerical Computation of the Characteristic Values of a Real Symmetric Matrix, Oak Ridge National Laboratory, *Report* ORNL-1574, 1954.

For closely related results, see

E. Frank, On the Zeros of Polynomials with Complex Coefficients, *Bull. Am. Math. Soc.*, vol. 52, pp. 144–157, 1946.

E. Frank, The Location of the Zeros of Polynomials with Complex Coefficients, *Bull. Am. Math. Soc.*, vol. 52, pp. 890–898, 1946.

Two papers dealing with matrices all of whose characteristic values lie inside the unit circle are

O. Taussky, Analytical Methods in Hypercomplex Systems, *Comp. Math.*, 1937.

O. Taussky and J. Todd, Infinite Powers of Matrices, *J. London Math. Soc.*, 1943.

An interesting and important topic which we have not at all touched upon is that of determining the asymptotic behavior of the solutions of the vector-matrix equation $x(n + 1) = (A + B(n))x(n)$ as $n \to \infty$, where the elements of $B(n)$ tend to zero as $n \to \infty$.

The study of this problem was begun by Poincaré; see

H. Poincaré, Sur les équations linéaires aux différentielles ordinaires et aux différences finies, *Am. J. Math.*, vol. 7, pp. 203–258, 1885.

and continued by Perron,

O. Perron, Über die Poincarésche lineare Differenzgleichung, *J. reine angew. Math.*, vol. 137, pp. 6–64, 1910.

and his pupil Ta Li,

Ta Li, Die Stabilitätsfrage bei Differenzengleichungen, *Acta Math.*, vol. 63, pp. 99–141, 1934.

The most recent result in this field is contained in

G. A. Freiman, On the Theorems of Poincaré and Perron, *Uspekhi Mat. Nauk* (N.S.), vol. 12, no. 3(75), pp. 241–246, 1957. (Russian.)

Other discussion may be found in

R. Bellman, *A Survey of the Theory of the Boundedness, Stability, and Asymptotic Behavior of Solutions of Linear and Non-linear Differ-*

ential and Difference Equations, Office of Naval Research, Department of the Navy, January, 1949.

R. Bellman, *Methods of Nonlinear Analysis*, Academic Press, Inc., New York, 1970.

These questions, as well as the corresponding questions for the differential equation $dx/dt = (A + B(t))x$, belong more to the theory of linear differential and difference equations than to the theory of matrices per se.

For a discussion of the stability of linear differential systems with random coefficients, see

O. Sefl, On Stability of a Randomized Linear System, *Sci. Sinica*, vol. 7, pp. 1027–1034, 1958.

S. Sninivasan and R. Vasodevan, Linear Differential Equations with Random Coefficients, American Elsevier Publishing Company, Inc., New York, 1970 (to appear).

J. C. Samuels, On the Mean Square Stability of Random Linear Systems, *IRE Trans. on Circuit Theory*, May, 1959, Special Supplement (Transactions of the 1959 International Symposium on Circuit and Information Theory), pp. 248–259.

where many additional references to earlier work by Rosenbloom and others will be found.

14

Markoff Matrices and Probability Theory

1. Introduction. In this and the following chapters, which constitute the last third of the book, we wish to study a class of matrices which are generated by some fundamental questions of probability theory and mathematical economics. The techniques we shall employ here are quite different from those utilized in the study of quadratic forms, or in connection with differential equations.

The basic concept is now that of non-negativity, non-negative matrices and non-negative vectors. A matrix $M = (m_{ij})$ is said to be *non-negative* if $m_{ij} \geq 0$ for all i and j. Similarly, a vector x is said to be *non-negative* if all its components are non-negative, $x_i \geq 0$.

The subclass of non-negative matrices for which the stronger condition $m_{ij} > 0$ holds is called *positive.* These matrices possess particularly interesting and elegant properties.

The terms "positive" and "non-negative" are frequently used to describe what we have previously called "positive definite" and "non-negative definite." Since the two types of matrices will not appear together in what follows, we feel that there is no danger of confusion. The adjective "positive" is such a useful and descriptive term that it is understandable that it should be a bit overworked.

We have restricted ourselves in this volume to a discussion of the more elementary properties of non-negative matrices. Detailed discussions, which would require separate volumes, either in connection with probability theory or mathematical economics, will be found in references at the end of the chapter.

This chapter, and the one following, will be devoted to stochastic processes, while the next chapter will cover various aspects of matrix theory and mathematical economics.

2. A Simple Stochastic Process. In order to set the stage for the entrance of Markoff matrices, let us review the concept of a *deterministic* process. Consider a system S which is changing over time in such a way that its state at any instant t can be described in terms of a finite dimensional vector x. Assume further, that the state at any time $s + t$, $s > 0$, can be expressed as a predetermined function of the state at time t, and the elapsed time s, namely,

$$x(s + t) = g(x(t),s) \tag{1}$$

Under reasonable conditions of continuity on the function $g(t)$, we can derive, by expansion of both sides in powers of s, a differential equation for $x(t)$, of the form

$$\frac{dx}{dt} = h(x(t)) \tag{2}$$

Finite-dimensional deterministic systems of the foregoing type are thus equivalent to systems governed by ordinary differential equations. If we introduce functions of more complicated type which depend upon the past history of the process, then more complicated functional equations than (2) result.

The assumption that the present state of the system *completely* determines the future states is a very strong one. It is clear that it must always fail to some degree in any realistic process. If the deviation is slight, we keep the deterministic model because of its conceptual simplicity. It is, however, essential that other types of mathematical models be constructed, since many phenomena cannot, at the present time, be explained in the foregoing terms.

Let us then consider the following *stochastic* process. To simplify the formulation, and to avoid a number of thorny questions which arise otherwise, we shall suppose that we are investigating a physical system S which can exist only in one of a finite number of states, and which can change its state only at discrete points in time.

Let the states be designated by the integers $1, 2, \ldots, N$, and the times by $t = 0, 1, 2, \ldots$. To introduce the random element, we assume that there is a fixed probability that a system in stage j at time t will transform into state i at time $t + 1$. This is, of course, again a very strong assumption of regularity. A "random process" in mathematical parlance is not at all what we ordinarily think of as a "random process" in ordinary verbal terms.

Pursuant to the above, let us then introduce the *transition matrix* $M = (m_{ij})$, where

m_{ij} = the probability that a system in state j at time t will be in state i at time $t + 1$

Observe that we take M to be independent of time. This is the most important and interesting case.

In view of the way in which M has been introduced, it is clear that we wish to impose the following conditions:

$$m_{ij} \geq 0 \tag{3a}$$

$$\sum_{i=1}^{N} m_{ij} = 1 \qquad j = 1, 2, \ldots, N \tag{3b}$$

That m_{ij} is non-negative is an obvious requirement that it be considered a probability. The second condition expresses the fact that a particle in state j at time t must be somewhere, which is to say in one of the allowable states, at time $t + 1$.

3. Markoff Matrices and Probability Vectors. Let us now introduce some notation. A matrix M whose elements satisfy the restrictions of (2.4) will be called a *Markoff matrix*.

A vector x whose components x_i satisfy the conditions

$$x_i \geq 0 \tag{1a}$$

$$\sum_{i=1}^{N} x_i = 1 \tag{1b}$$

will be called a *probability vector*. Generally, a vector all of whose components are positive will be called a *positive* vector.

EXERCISES

1. Prove that $\lambda P + (1 - \lambda)Q$ is a Markoff matrix for $0 \leq \lambda \leq 1$ whenever P and Q are Markoff matrices.

2. Prove that PQ is a Markoff matrix under similar assumptions.

3. Prove that Mx is a probability vector whenever x is a probability vector and M is a Markoff matrix.

4. Prove that Markoff matrices can be characterized in the following way: A matrix M is a Markoff matrix if and only if Mx is a probability vector whenever x is a probability vector.

5. Prove that if M' is the transpose of a Markoff matrix M and $M' = (a_{ij})$, then $\sum_{j=1}^{N} a_{ij} = 1$, for $i = 1, 2, \ldots, N$.

6. Prove that a positive Markoff matrix transforms non-trivial vectors with non-negative components into vectors with positive components.

4. Analytic Formulation of Discrete Markoff Processes. A stochastic process of the type described in Sec. 2 is usually called a *discrete Markoff* process. Let us now see how this process can be described in analytic terms.

Since the state of the system S at any time t is a random variable, assuming any one of the values $i = 1, 2, \ldots, N$, we introduce the N functions of t defined as follows:

$$x_i(t) = \text{the probability that the system is in state } i \text{ at time } t \tag{1}$$

At $t = 0$ we have the relation

$$x_i(0) = \delta_{ik} \tag{2}$$

where k is the initial state of S.

The following relations must then be valid:

$$x_i(t+1) = \sum_{j=1}^{N} m_{ij}x_j(t) \qquad i = 1, 2, \ldots, N \tag{3}$$

One aspect of the problem of predicting regular behavior of S is that of studying the behavior of the solutions of the system of equations in (3), which we may write more compactly in the form

$$x(t+1) = Mx(t) \qquad t = 0, 1, 2, \ldots \tag{4}$$

The full problem is one of great complexity and interest, with analytic, algebraic, topological, and physical overtones. A number of monographs have been written on this topic, and there seems to be an unlimited area for research.

It turns out that particularly simple and elegant results can be obtained in the case in which the elements of M are all positive. Consequently, we shall restrict ourselves principally to a discussion of this case by several methods, and only lightly touch upon the general case.

EXERCISE

1. Prove that $x(t) = M^t x(0)$, $x(t+s) = M^t x(s)$.

5. Asymptotic Behavior. The next few sections will be devoted to two proofs of the following remarkable result:

Theorem 1. *If M is a positive Markoff matrix and if $x(t)$ satisfies* (4) *of Sec.* 4, *then*

$$\lim_{t\to\infty} x(t) = y, \text{ where } y \text{ is a probability vector,} \tag{1a}$$

$$y \text{ is independent of } x(0), \tag{1b}$$

$$y \text{ is a characteristic vector of } M \text{ with associated characteristic root } 1 \tag{1c}$$

That $x(t)$ should settle down to a fixed probability vector y is interesting and perhaps not unexpected in view of the mixing property implied in the assumption $m_{ij} > 0$. That this limit should be *independent* of the initial state is certainly surprising.

6. First Proof. We shall present two proofs in this chapter. A third proof can be derived as a consequence of results obtained for general positive matrices in Chap. 16. The first proof, which we present first because of its simplicity, illustrates a point we have stressed before, namely, the usefulness of considering transformations together with their adjoints.

Since t assumes only the discrete sequence of values 0, 1, 2, . . . , let us replace it by n and speak of $x(n)$ in place of $x(t)$. To establish the fact that $x(n)$ has a limit as $n \to \infty$, we consider the inner product of $x(n)$ with an arbitrary vector b. Since, by virtue of (4.4), $x(n) = M^n x(0)$, we see, upon setting $x(0) = c$, that

$$(x(n),b) = (M^n c,b) = (c,(M^n)'b) = (c,(M')^n b) \tag{1}$$

where M' is the transpose of M.

If we show that $(M')^n b$ converges as $n \to \infty$, for *any* given b, it will follow that $x(n)$ converges as $n \to \infty$, since we can choose for b first the vector all of whose components are zero, except for the first, then the vector all of whose components are zero except for the second, and so on. Let us then introduce the vector

$$z(n) = (M')^n b \tag{2}$$

which satisfies the difference equation

$$z(n+1) = M'z(n) \qquad z(0) = b \tag{3}$$

Let, for each n, $u(n)$ be the component of $z(n)$ of largest value and $v(n)$ the component of $z(n)$ of smallest value. We shall show that the properties of M' imply that $u(n) - v(n) \to 0$ as $n \to \infty$.

Since

$$z_i(n+1) = \sum_{j=1}^{N} m_{ji} z_j(n) \tag{4}$$

and since $\sum_{j=1}^{N} m_{ji} = 1$, $m_{ji} \geq 0$, we see that

$$\begin{aligned} u(n+1) &\leq u(n) \\ v(n+1) &\geq v(n) \end{aligned} \tag{5}$$

Since $\{u(n)\}$ is a monotone decreasing sequence, bounded from below by zero, and $\{v(n)\}$ is a monotone increasing sequence, bounded from above by one, we see that both sequences converge. Let

$$u(n) \to u \qquad v(n) \to v \tag{6}$$

To show that $u = v$, which will yield the desired convergence, we proceed as follows.

Using the component form of (3), we see that

$$\begin{aligned} u(n+1) &\leq (1-d)u(n) + dv(n) \\ v(n+1) &\geq (1-d)v(n) + du(n) \end{aligned} \tag{7}$$

where d is the assumed positive lower bound for m_{ij}, $i, j = 1, 2, \ldots, N$.

From (7) we derive

$$u(n+1) - v(n+1) \leq [(1-d)u(n) + dv(n) - (1-d)v(n) - du(n)]$$
$$\leq (1-2d)(u(n) - v(n)) \quad (8)$$

We see that (8) yields

$$u(n) - v(n) \leq (1-2d)^n(u(0) - v(0)) \quad (9)$$

and hence, since $d \leq \frac{1}{2}$ if $N \geq 2$, that $u(n) - v(n) \to 0$ as $n \to \infty$.

Thus, not only does $z(n)$ converge as $n \to \infty$, but it converges to a vector all of whose components are equal. From the convergence of $z(n)$ we deduce immediately the convergence of $x(n)$, while the equality of the components will yield, as we shall see, the independence of initial value.

Let $\lim_{n\to\infty} z(n) = z$ and $\lim_{n\to\infty} x(n) = y$. As we know, all of the components of z are equal. Let us call each of these components a_1. Then

$$(y,b) = \lim_{n\to\infty} (x(n),b) = (c,z) = a_1[c_1 + c_2 + \cdots + c_n] = a_1 \quad (10)$$

where a_1 is a quantity dependent only upon b. It follows that y is independent of c.

That y is a characteristic vector of M with characteristic root 1 follows easily. We have

$$y = \lim_{n\to\infty} M^{n+1}c = M \lim_{n\to\infty} M^n c = My \quad (11)$$

In other words, y is a "fixed point" of the transformation represented by M.

EXERCISE

1. Use the fact that I is a non-negative Markoff matrix to show that conditions of positivity cannot be completely relaxed.

7. Second Proof of Independence of Initial State. Another method for establishing independence of the initial state is the following. We already know that $\lim_{n\to\infty} M^n c$ exists for any initial probability vector c. Let y and z be two limits corresponding to different initial probability vectors c and d, respectively. Choose the scalar t_1 so that $y - t_1 z$ has at least one zero component, with all other components positive. As we know, $M(y - t_1 z)$ must then be a positive vector if $y - t_1 z$ is nontrivial. However,

$$M(y - t_1 z) = My - t_1 Mz = y - t_1 z \quad (1)$$

a contradiction unless $y - t_1 z = 0$. This means that $t_1 = 1$, since y and z are both probability vectors.

8. Some Properties of Positive Markoff Matrices. Combining the results and techniques of the foregoing sections, we can derive some interesting results concerning positive Markoff matrices.

To begin with, what we have shown by means of the preceding argument is that a positive Markoff matrix cannot possess two linearly independent positive characteristic vectors associated with the characteristic root one.

Furthermore, the same argument shows that a positive Markoff matrix cannot possess any characteristic vector associated with the characteristic root one which is linearly independent of the probability vector y found above. For, let z be such a vector.[1] Then for t_1 sufficiently large, $z + t_1y$ is a positive vector. Dividing each component by the scalar $t_2 = (z + t_1y, c)$, where c is a vector all of whose components are ones, we obtain a probability vector. Hence $(z + t_1y)/t_2 = y$, which means that z and y are actually linearly dependent.

It is easy to see that no characteristic root of M can exceed one in absolute value. For, let x be a characteristic vector of M', λ be a characteristic root, and m be the absolute value of a component of x of greatest magnitude. Then the relation $\lambda x = M'x$ shows that

$$|\lambda|m \leq m \sum_{j=1}^{N} m_{ji} = m \tag{1}$$

whence $|\lambda| \leq 1$.

We can, however, readily show that $\lambda = 1$ is the only characteristic root of absolute value one, provided that M is a positive Markoff matrix.

To do this, let μ be another characteristic root with $|\mu| = 1$, and $w + iz$ an associated characteristic vector with w and z real. Then, choosing c_1 large enough, we can make $w + c_1y$ and $z + c_1y$ both positive vectors.

It follows that

$$M(w + iz + c_1(1 + i)y) = \mu(w + iz) + c_1(1 + i)y \tag{2}$$

Hence, on one hand, we have

$$M^n(w + iz + c_1(1 + i)y) = \mu^n(w + iz) + c_1(1 + i)y \tag{3}$$

On the other hand, as $n \to \infty$, we see that

$$M^n(w + iz + c_1(1 + i)y) = M^n(w + c_1y + i(z + c_1y)) \tag{4}$$

converges, because of the positivity of $w + c_1y$ and $z + c_1y$, to a certain scalar multiple of y.

The vector $\mu^n(w + iz)$, however, converges as $n \to \infty$ only if $\mu = 1$, if μ is constrained to have absolute value one. This completes the proof.

We see then that we have established Theorem 2.

[1] It is readily seen that it suffices to take z real.

Theorem 2. *If M is a positive Markoff matrix, the characteristic root of largest absolute value is 1. Any characteristic vector associated with this characteristic root is a scalar multiple of a probability vector.*

EXERCISE

1. If λ is a characteristic root with an associated characteristic vector which is positive, then $\lambda = 1$.

9. Second Proof of Limiting Behavior. Let us now turn to a second direct proof of the fact that $x(n)$ has a limit as $n \to \infty$. Writing out the equations connecting the components of $x(n)$ and $x(n+1)$, we have

$$x_i(n+1) = \sum_{j=1}^{N} m_{ij}x_j(n) \tag{1}$$

This yields

$$x_i(n+1) - x_i(n) = \sum_{j=1}^{N} m_{ij}[x_j(n) - x_j(n-1)] \tag{2}$$

Since $\sum_{j=1}^{N} x_j(n) = 1$ for all n, it is impossible to have $x_j(n) \geq x_j(n-1)$ for all j, unless we have equality. In this case $x(n) = x(n-1)$, a relation which yields $x(m) = x(n-1)$ for $m \geq n$, and thus the desired convergence.

In any case, for each n let $S(n)$ denote the set of j for which $x_j(n) \geq x_j(n-1)$ and $T(n)$ the set of j for which $x_j(n) < x_j(n-1)$. The comment in the preceding paragraph shows that there is no loss of generality in assuming that $S(n)$ and $T(n)$ possess at least one element for each n.

Referring to (2), we see that

$$\sum_{j \in T(n)} m_{ij}[x_j(n) - x_j(n-1)] \leq x_i(n+1) - x_i(n) \leq \sum_{j \in S(n)} m_{ij}[x_j(n) - x_j(n-1)] \tag{3}$$

Hence

$$\sum_{i \in S(n+1)} [x_i(n+1) - x_i(n)] \leq \sum_{j \in S(n)} \left\{ \sum_{i \in S(n+1)} m_{ij} \right\} [x_j(n) - x_j(n-1)] \tag{4}$$

which yields, since $m_{ij} \geq d > 0$ for all i and j,

$$\sum_{i \in S(n+1)} [x_i(n+1) - x_i(n)] \leq (1-d) \sum_{j \in S(n)} [x_j(n) - x_j(n-1)] \tag{5}$$

Similarly, summing over $i \epsilon T(n+1)$, we have

$$(1-d) \sum_{j \epsilon T(n)} [x_j(n) - x_j(n-1)] \leq \sum_{i \epsilon T(n+1)} [x_i(n+1) - x_i(n)] \quad (6)$$

Combining (5) and (6), we see that

$$\sum_{j=1}^{N} |x_j(n+1) - x_j(n)| \leq (1-d) \sum_{j=1}^{N} |x_j(n) - x_j(n-1)| \quad (7)$$

and hence that the series $\sum_{n=1}^{\infty} \|x(n) - x(n-1)\|$ converges.

This completes the proof of convergence of $x(n)$.

EXERCISES

1. If M is a Markoff matrix and M^2 is a positive Markoff matrix, show that the sequences $\{M^{2n}c\}$ and $\{M^{2n+1}c\}$ converge. Does $\{M^n c\}$ converge?

2. What is the limit of M^n if M is a positive Markoff matrix?

10. General Markoff Matrices. As mentioned above, the study of the general theory of Markoff matrices is quite complex, and best carried on from a background of probability theory.

In order to see how complex roots of absolute value 1 can arise, consider a system with only two states, 1 and 2, in which state 1 is always transformed into state 2 and state 2 into state 1. The corresponding transition matrix is then

$$M = \begin{bmatrix} 0 & 1 \\ 1 & 0 \end{bmatrix} \quad (1)$$

which has the characteristic roots $\lambda_1 = 1$, $\lambda_2 = -1$.

By considering cyclic situations of higher order, we can obtain characteristic roots which are roots of unity of arbitrary order.

Let us, however, discuss briefly an important result which asserts that even in the case where there is no unique limiting behavior, there is an average limiting behavior.

Let us suppose for the moment that M has only simple characteristic roots, so that we may write

$$M = T \begin{bmatrix} \lambda_1 & & & & 0 \\ & \lambda_2 & & & \\ & & \cdot & & \\ & & & \cdot & \\ & & & & \cdot \\ 0 & & & & \lambda_N \end{bmatrix} T^{-1} \quad (2)$$

Then

$$\left(\sum_{k=1}^{n} M^k\right)/n = T \begin{bmatrix} \sum_{k=1}^{n} \lambda_1^k/n & & & & 0 \\ & \sum_{k=1}^{n} \lambda_2^k/n & & & \\ & & \cdot & & \\ & & & \cdot & \\ & & & & \cdot \\ 0 & & & & \sum_{k=1}^{n} \lambda_N^k/n \end{bmatrix} T^{-1} \qquad (3)$$

If $|\lambda_i| < 1$, we have $\sum_{k=1}^{n} \lambda_i^k/n \to 0$ as $n \to \infty$. If $|\lambda_j| = 1$, the limit

$$\lim_{n \to \infty} \sum_{k=1}^{n} \lambda_j^k/n = a_j \qquad (4)$$

exists. If $\lambda_j = e^{i\theta}$, where $\theta/2\pi$ is an integer, then $a_j = 1$; if $\theta/2\pi$ is not an integer, then $a_j = 0$.

In any case, under the assumption of distinct characteristic roots, we can assert the existence of a quasi-limiting state

$$y = \lim_{n \to \infty} \left(\frac{x + Mx + \cdots + M^n x}{n+1}\right) \qquad (5)$$

which is a characteristic vector of M, that is, $My = y$. The limit vector y is a probability vector if x is a probability vector.

The general case can be discussed along similar, but more complicated, lines; see the exercises at the end of the chapter.

EXERCISE

1. If λ is a characteristic root of a Markoff matrix M, with the property that $|\lambda| = 1$, must λ be a root of unity?

11. A Continuous Stochastic Process. Let us now consider the stochastic process described in Sec. 2 under the assumption that the system is observed continuously. We begin with the discrete process in which observations are made at the times $t = 0, \Delta, 2\Delta, \ldots$. Since we want the continuous process to be meaningful, we introduce a continuity hypothesis in the form of a statement that the probability of the system remaining in the same state over any time interval of length Δ is $1 - 0(\Delta)$.

To make this statement precise, we define the following quantities:

$$\begin{aligned} a_{ij}\Delta &= \text{the probability that } S \text{ is in state } i \text{ at } t + \Delta, \text{ given that} \\ &\quad \text{it is in state } j \text{ at time } t, \text{ for } i \neq j \\ 1 - a_{ii}\Delta &= \text{the probability that } S \text{ is in state } i \text{ at } t + \Delta, \text{ given that} \\ &\quad \text{it is in state } i \text{ at time } t \end{aligned} \tag{1}$$

The quantities a_{ij} are assumed to satisfy the following conditions:

$$a_{ij} \geq 0 \tag{2a}$$

$$a_{ii} = \sum_{j \neq i} a_{ji} \tag{2b}$$

Then the equations governing the quantities $x_i(t)$ are

$$x_i(t + \Delta) = (1 - a_{ii}\Delta)x_i(t) + \Delta \sum_{j \neq i} a_{ij}x_j(t) \qquad i = 1, 2, \ldots, N \tag{3}$$

for $t = 0, \Delta, 2\Delta, \ldots$.

Writing, in a purely formal way,

$$x_i(t + \Delta) = x_i(t) + \Delta x_i'(t) + 0(\Delta^2) \tag{4}$$

and letting $\Delta \to 0$ in (2), we obtain the system of differential equations

$$\frac{dx_i}{dt} = -a_{ii}x_i + \sum_{j \neq i} a_{ij}x_j \qquad x_i(0) = c_i \tag{5}$$

$i = 1, 2, \ldots, N$, where c_i, $i = 1, 2, \ldots, N$, are the initial probabilities.

We shall bypass here all the thorny conceptual problems connected with continuous stochastic processes by *defining* our process in terms of this system of differential equations. In order for this to be an operationally useful technique, we must show that the functions generated (5) act like probabilities. This we shall do below.

There are a number of interesting questions arising in this way which we shall not treat here. Some of these are:

1. How does one define a continuous stochastic process directly and derive the differential equations of (5) directly?

2. In what sense can the continuous stochastic process defined by (5) be considered the limit of the discrete stochastic process defined by (3)?

Since these are problems within the sphere of probability theory, we shall content ourselves with mentioning them, and restrain ourselves here to the matrix aspects of (5). As usual, however, we shall not scruple to use these ideas to guide our analysis.

12. Proof of Probabilistic Behavior. Let us now demonstrate that the system of differential equations

$$\frac{dx_i}{dt} = -a_{ii}x_i + \sum_{j \neq i} a_{ij}x_j \qquad x_i(0) = c_i \qquad i = 1, 2, \ldots, N \tag{1}$$

where

$$c_i \geq 0 \qquad \sum_i c_i = 1 \tag{2a}$$

$$a_{ij} \geq 0 \tag{2b}$$

$$\sum_{j \neq i} a_{ji} = a_{ii} \tag{2c}$$

produces a set of functions satisfying the relations

$$x_i(t) \geq 0 \qquad t \geq 0 \tag{3a}$$

$$\sum_{i=1} x_i(t) = 1 \tag{3b}$$

Let us demonstrate (3*b*) first. From (1) we have

$$\frac{d}{dt}\left(\sum_i x_i\right) = \sum_i \left(-a_{ii}x_i + \sum_{j \neq i} a_{ij}x_j\right) = 0 \tag{4}$$

by virtue of (2*c*). Hence, for all t,

$$\sum_i x_i(t) = \sum_i x_i(0) = 1 \tag{5}$$

To show that $x_i \geq 0$, write (1) in the form

$$\frac{d}{dt}(x_i e^{a_{ii}t}) = e^{a_{ii}t} \sum_{j \neq i} a_{ij}x_j$$

Since $x_j \geq 0$ at $t = 0$, this shows that the quantities $x_i e^{a_{ii}t}$ are monotone increasing.

In place of following this procedure, we could use the easily established fact that the solutions of (1) are the limits of the solutions of (11.3) as $\Delta \to 0$. Hence, properties of the solutions of (11.3) valid for all Δ must carry over to the solutions of (1). We recommend that the reader supply a rigorous proof.

13. Generalized Probabilities—Unitary Transformations. Any set of non-negative quantities $\{x_i\}$ satisfying the conditions

$$x_i \geq 0 \tag{1a}$$

$$\sum_{i=1}^{N} x_i = 1 \tag{1b}$$

can be considered to represent a set of probabilities, where x_i is the probability that a system S is in state i.

A set of transformations

$$x_i' = g_i(x_1,x_2, \ldots ,x_N) \qquad i = 1, 2, \ldots , N \tag{2}$$

which preserves the relations in (1) can then be considered to represent the effects of certain physical transformations upon S.

The simplest such transformations are the linear, homogeneous transformations

$$x_i' = \sum_{j=1}^{N} a_{ij}x_j \tag{3}$$

As we know, a necessary and sufficient condition that the relations in (1) be preserved is that $A = (a_{ij})$ be a Markoff matrix.

Let us now consider a quadratic transformation. Let x be a complex vector with components x_i and let

$$x' = Tx \tag{4}$$

where T is a unitary transformation. Then $(x',\overline{x'}) = (x,\bar{x})$.

Hence, if we consider $|x_1|^2, |x_2|^2, \ldots , |x_N|^2$ to represent a set of probabilities, we see that $|x_1'|^2, |x_2'|^2, \ldots , |x_N'|^2$ can also be considered to be a set of probabilities.

We leave it to the reader to establish the relevant limit theorems for the sequence $\{x(n)\}$ defined recurrently by

$$x(n + 1) = Tx(n) \tag{5}$$

where T is unitary and $(x(0),\overline{x(0)}) = 1$

14. Generalized Probabilities—Matrix Transformations. Let us now generalize the concept of probabilities in the following fashion. If $\{X_i\}$ is a finite set of symmetric matrices satisfying the conditions

$$X_i \geq 0 \qquad i = 1, 2, \ldots , N \tag{1a}$$

$$\sum_{i=1}^{N} \operatorname{tr}(X_i) = 1 \tag{1b}$$

we shall call them a set of probabilities. The condition $X_i \geq 0$ here signifies that X_i is non-negative definite.

Consider the transformation

$$Y_i = \sum_{j=1}^{N} A_{ij}'X_jA_{ij} \qquad i = 1, 2, \ldots , N \tag{2}$$

It is easy to establish the following analogue of the result concerning Markoff matrices.

Theorem 3. *A necessary and sufficient condition that the transformation in* (2) *preserve the relations in* (1) *is that*

$$\sum_{i=1}^{N} A'_{ij}A_{ij} = I \qquad j = 1, 2, \ldots, N \tag{3}$$

Although analogues of the limit theorems established for the usual Markoff transformations can be established for the transformations of (2), the results are more difficult to establish, and we shall therefore not discuss these matters any further here.

EXERCISES

1. Let M be a Markoff matrix. Consider a proof of the fact that $\lim_{n\to\infty} \sum_{k=0}^{n} M^k/n$ exists along the following lines. Using the Schur transformation, we can write

$$M = T \begin{bmatrix} e^{i\theta_1} & b_{12} & & & & & & b_{1N} \\ & e^{i\theta_2} & & & & & & \\ & & \ddots & & & & & \\ & & & e^{i\theta_k} & & & & \\ & & & & \lambda_{k+1} & & & \\ & & & & & \ddots & & \\ & 0 & & & & & & \lambda_N \end{bmatrix} T'$$

where the θ_i are real and $|\lambda_i| < 1$, $i = k + 1, \ldots, N$.

It suffices to consider the iterates of the triangular matrix. To do this, consider the system of equations

$$\begin{aligned} x_1(n+1) &= e^{i\theta_1}x_1(n) + b_{12}x_2(n) + \cdots + b_{1N}x_N(n) \\ x_2(n+1) &= \qquad\qquad e^{i\theta_2}x_2(n) + \cdots + b_{2N}x_N(n) \\ &\vdots \\ x_N(n+1) &= \qquad\qquad\qquad\qquad\qquad \lambda_N x_N(n) \end{aligned}$$

$n = 0, 1, 2, \ldots$. Show that $|x_i(n)| \le c_1 r^n$, $0 < r < 1$, $i = k + 1, \ldots, N$. Using the equation $x_k(n+1) = e^{i\theta_k}x_k(n) + y_k(n)$, where $|y_k(n)| \le c_2 r^n$, show that $\lim_{n\to\infty} \sum_{m=0}^{n} x_k(m)/n$ exists. Then continue inductively.

2. Prove the stated result using the Jordan canonical form.

MISCELLANEOUS EXERCISES

1. Consider the matrix

$$A^{(n)} = \begin{bmatrix} -\lambda_0 & \lambda_0 & & & & & \\ \mu_1 & -(\lambda_1+\mu_1) & \lambda_1 & & & & \\ & \mu_2 & -(\lambda_2+\mu_2) & \lambda_2 & & & \\ & & & \cdot & & & \\ & & & & \cdot & & \\ & & & & \mu_{n-1} & -(\lambda_{n-1}+\mu_{n-1}) & \lambda_{n-1} \\ & & & & & \mu_n & -(\lambda_n+\mu_n) \end{bmatrix}$$

where the λ_i and μ_i are positive quantities.

Denote by $\phi_n(\lambda)$ the quantity $|A^{(n)} + \lambda I|$. Show that

$$\phi_n(\lambda) - (\lambda + \lambda_n + \mu_n)\phi_{n-1}(\lambda) + \lambda_{n-1}\mu_n\phi_{n-2}(\lambda) = 0$$

for $n \geq 1$ if $\phi_{-1}(\lambda) = 1$, $\phi_0(\lambda) = \lambda + \lambda_0$ (*W. Ledermann and G. E. Reuter*).

2. Thus, or otherwise, show that the characteristic roots are $A^{(n)}$ are distinct and negative or zero, and separate those of $A^{(n-1)}$ (*W. Ledermann and G. E. Reuter*).

3. In the notation of Exercise 27 of the Miscellaneous Exercises of Chap. 4, is $A^{(n)}$ symmetrizable?

4. By a *permutation matrix* P of order N, we mean a matrix possessing exactly one nonzero element of value one in each row and column. Show that there are $N!$ permutation matrices of order N.

5. Prove that the product of two permutation matrices is again a permutation matrix, and likewise the inverse. What are the possible values for the determinant of a permutation matrix?

6. By a *doubly stochastic matrix* of order N we mean a matrix A whose elements satisfy the following conditions:

(*a*) $a_{ij} \geq 0$

(*b*) $\sum_j a_{ij} = 1 \qquad i = 1, 2, \ldots, N$

(*c*) $\sum_i a_{ij} = 1 \qquad j = 1, 2, \ldots, N$

Is the product of two doubly stochastic matrices doubly stochastic?

7. Prove that any doubly stochastic matrix A may be written in the form

$$A = \sum_r w_r P_r \qquad r = 1, 2, \ldots, N! \qquad w_r \geq 0 \qquad \sum_r w_r = 1$$

where $\{P_r\}$ is the set of permutation matrices of order N. The result is due to Birkhoff.[1]

Another proof, together with a number of applications to scheduling theory, may be found in T. C. Koopmans and M. J. Beckman.[2]

[1] G. Birkhoff, Tres observaciones sobre el algebra lineal, *Rev. univ. nac. Tucumán*, ser. *A*, vol. 5, pp. 147–151, 1946.

[2] T. C. Koopmans and M. J. Beckman, Assignment Problems and the Location of Economic Activities, *Econometrica*, vol. 25, pp. 53–76, 1957.

For some further results concerning doubly stochastic matrices, see S. Schreiber[1] and L. Mirsky.

Mirsky's paper contains a particularly simple proof of the Birkhoff results cited above and of an interesting connection between doubly stochastic matrices and the theory of inequalities.

8. Let A be a 2×2 Markoff matrix. Show that A is a square of a Markoff matrix if and only if $a_{11} \geq a_{12}$.

9. Hence, show that if A is a square of a Markoff matrix, it can be written as the 2^n power of a Markoff matrix for $n = 1, 2, \ldots$.

10. From this, conclude that if A has a square root which is a Markoff matrix, it has a root of every order which is a Markoff matrix.

11. From this, conclude that we can construct a family of Markoff matrices $A(t)$ with the property that $A(s + t) = A(s)A(t)$, for $s, t \geq 0$, $A(0) = I$, and $A(1) = A$.

12. Using this result, or otherwise, show that $A = e^B$, where B has the form

$$B = \begin{bmatrix} -a_1 & a_1 \\ b_1 & -b_1 \end{bmatrix} \qquad a_1, b_1 \geq 0$$

13. If B is a matrix of the type described in Sec. 12, we know that e^B is a Markoff matrix. Let B_1 and B_2 be two matrices of this type and let $A_1 = e^{B_1}$, $A_2 = e^{B_2}$. Is A_1A_2 a matrix of this type?

14. Let A be a Markoff matrix of dimension N with the property that it has a root of any order which is again a Markoff matrix. Let A_n denote a particular 2^n-th root which is a Markoff matrix, and write $A(t) = A_{n_1}A_{n_2} \cdots$, when t has the form $t = 2^{-n_1} + 2^{-n_2} + \cdots$, where the sum is finite. Can we define $A(t)$ for $0 < t < 1$ as a limit of values of $A(t)$ for t in the original set of values? If so, show that $A(s + t) = A(s)A(t)$ for $0 < s, t \leq 1$. Define $A(0)$ to be I. If the foregoing relation holds, does it follow that $A(t)$ is the solution of a differential equation of the form $dX/dt = BX$, $X(0) = I$, where B is a matrix of the type occurring in Sec. 12?

15. Let $p_i(n)$ = the probability of going from state a to state i in n steps, and $q_i(n)$ be similarly defined starting in state b. Consider the 2×2 determinant

$$d_{ij}(n) = \begin{vmatrix} p_i(n) & p_j(n) \\ q_i(n) & q_j(n) \end{vmatrix}$$

Prove that

$$d_{ij}(n) = \sum_{r,s} a_{ir}a_{js}d_{rs}(n - 1)$$

(Cf. the matrices formed in Sec. 14 of Chap. 12.)

16. Let $\phi_{ij}(n)$ = the probability of two particles going from initial distinct states a and b to states i and j at time n without ever being in the same state. Show that

$$\phi_{ij}(n) = \sum_{r,s} a_{ir}a_{js}\phi_{rs}(n - 1)$$

In these two exercises, a_{ij} denotes the probability of going from state j to state i in any particular move.

[1] S. Schreiber, On a Result of S. Sherman Concerning Doubly Stochastic Matrices, *Proc. Am. Math. Soc.*, vol. 9, pp. 350–353, 1953.

L. Mirsky, Proofs of Two Theorems on Doubly-stochastic Matrices, *Proc. Am. Math. Soc.*, vol. 9, pp. 371–374, 1958.

What is the connection between $d_{ij}(n)$ and $\phi_{ij}(n)$? See S. Karlin and J. MacGregor, Coincidence Probabilities, *Stanford University Department of Statistics, Tech. Rept.* 8, 1958, for a detailed discussion.

17. Let M be a positive Markoff matrix and let the unique probability vector x satisfying $Mx = x$ be determined by the relation $x = \lim_{n\to\infty} M^n b$ where b is a probability vector. If we wish to use this result for numerical purposes, should we arrange the calculation so as to obtain M^n first by use of $M_1 = M$, $M_2 = MM_1$, . . . , and then $M_n b$, or should we calculate $M^n b$ by way of $x_1 = b$, $x_2 = Mx_1$, . . . ? Which procedure is more conducive to round-off error?

18. Suppose that $n = 2^k$ and we calculate $M_1 = M$, $M_2 = M_1^2$, $M_3 = M_2^2$, ... Which of the two procedures sketched above is more desirable?

19. What are necessary and sufficient conditions on a complex matrix A that AH be a stability matrix whenever H is positive definite Hermitian? See Carlson[1] and the interesting survey paper by O. Taussky.[2]

Bibliography and Discussion

§1–§12. The results given in this chapter follow the usual lines of the theory of finite Markoff chains, and the proofs are the classical proofs. For detailed discussion and many additional references, see

W. Feller, *An Introduction to Probability Theory and Applications*, John Wiley & Sons, Inc., New York, 1950.

M. S. Bartlett, *An Introduction to Stochastic Processes*, Cambridge University Press, New York, 1955.

N. Arley, *On the Theory of Stochastic Processes and Their Applications to the Theory of Cosmic Radiation*, Copenhagen, 1943.

W. Ledermann and G. E. Reuter, Spectral Theory for the Differential Equations of Simple Birth and Death Processes, *Phil. Trans. Royal Soc. London*, ser. *A*, vol. 246, pp. 321–369, 1953–1954.

W. Ledermann, On the Asymptotic Probability Distribution for Certain Markoff Processes, *Proc. Cambridge Phil. Soc.*, vol. 46, pp. 581–594, 1950.

For an interesting discussion of matrix theory and random walk problems, see

M. Kac, Random Walk and the Theory of Brownian Motion, *Am. Math. Monthly*, vol. 54, pp. 369–391, 1947.

[1] D. Carlson, A New Criterion for H-stability of Complex Matrices, *Lin. Algebra Appl.*, vol. 1, pp. 59–64, 1968.

[2] O. Taussky, Positive Definite Matrices and Their Role in the Study of the Characteristic Roots of General Matrices, *Advances in Mathematics*, vol. 2, pp. 175–186, Academic Press Inc., New York, 1968.

§7. The shuffling of cards is designed to be a Markoff process which makes each new deal independent of the previous hand. In reality, most people shuffle quite carelessly, so that a great deal of information concerning the distribution of honors in the various hands can be deduced by remembering the tricks of the previous deal. Over several hours of play, this percentage advantage can be quite important.

The concept of a Markoff chain is due to Poincaré. See

M. Fréchet, *Traité du calcul des probabilités,* tome 1, fasc. III, 2e livre, Gauthier-Villars & Cie, Paris, 1936.

for a detailed discussion of the application of matrix theory to Markoff chains.

§13. For a discussion of these generalized probabilities and their connection with the Feynman formulation of quantum mechanics, see

E. Montroll, Markoff Chains, Wiener Integrals, and Quantum Theory, *Comm. Pure Appl. Math.*, vol. 5, pp. 415–453, 1952.

§14. This matrix generalization of probabilities was introduced in

R. Bellman, On a Generalization of Classical Probability Theory—I: Markoff Chains, *Proc. Natl. Acad. Sci. U.S.*, vol. 39, pp. 1075–1077, 1953.

See also

R. Bellman, On a Generalization of the Stieltjes Integral to Matrix Functions, *Rend. circ. mat. Palermo,* ser. II, vol. 5, pp. 1–6, 1956.

R. Bellman, On Positive Definite Matrices and Stieltjes Integrals, *Rend. circ. mat. Palermo,* ser. II, vol. 6, pp. 254–258, 1957.

For another generalization of Markoff matrices, see

E. V. Haynsworth, Quasi-stochastic matrices, *Duke Math. J.*, vol. 22, pp. 15–24, 1955.

For some quite interesting results pertaining to Markoff chains and totally positive matrices in the sense of Gantmacher and Krein, see

S. Karlin and J. McGregor, Coincidence Probabilities, *Stanford University Tech. Rept.* 8, 1958.

15

Stochastic Matrices

1. Introduction. In this chapter we wish to discuss in very brief fashion stochastic matrices and a way in which they enter into the study of differential and difference equations. As we shall see, Kronecker products occur in a very natural fashion when we seek to determine the moments of solutions of linear functional equations of the foregoing type with random coefficients.

2. Limiting Behavior of Physical Systems. Consider, as usual, a physical system S, specified at any time t by a vector $x(t)$. In previous chapters, we considered the case in which $x(t + \Delta)$ was determined from a knowledge of $x(t)$ by means of a linear transformation

$$x(t + \Delta) = (I + Z\Delta)x(t) \tag{1}$$

and the limiting form of (1) in which Δ is taken to be an infinitesimal,

$$\frac{dx}{dt} = Zx \tag{2}$$

Suppose that we now assume that Z is a stochastic matrix. By this we mean that the elements of Z are stochastic variables whose distributions depend upon t. Since the concept of a continuous stochastic process, and particularly that of the solution of a stochastic differential equation, is one of great subtlety, with many pitfalls, we shall consider only discrete processes. There is no difficulty, however, in applying the formalism developed here, and the reader versed in these matters is urged to do so.

Returning to (1), let us write $x(k\Delta) = x_k$ to simplify the notation. Then (1) may be written

$$x_{k+1} = (I + Z_k\Delta)x_k \tag{3}$$

Since Z is a stochastic matrix, its value at any time becomes a function of time, and we denote this fact by writing Z_k.

We see then that

$$x_n = \left[\prod_{k=0}^{n-1} (1 + Z_k\Delta) \right] x_0 \tag{4}$$

If Δ is small, we can write this

$$x_n = \left[I + \Delta \prod_{k=0}^{n-1} Z_k + 0(\Delta^2) \right] x_0 \tag{5}$$

Hence, to terms in Δ^2, we can regard the effect of repeated transformations as an additive effect. Additive stochastic processes have been extensively treated in the theory of probability, and we shall in consequence devote no further attention to them here. Instead, we shall examine the problem of determining the limiting behavior of x_n when $Z_k\Delta$ cannot be regarded as a small quantity.

Since this problem is one of great difficulty, we shall consider only the question of the asymptotic behavior of the expected value of the kth powers of the components of x_n as $n \to \infty$ for $k = 1, 2, \ldots$.

3. Expected Values. To illustrate the ideas and techniques, it is sufficient to consider the case where the x_i are two-dimensional matrices and the Z_k are 2×2 matrices. Let us write out the relations of (3) in terms of the components of x_k, which we call u_k and v_k for simplicity of notation.

Then

$$\begin{aligned} u_{k+1} &= z_{11}u_k + z_{12}v_k \\ v_{k+1} &= z_{21}u_k + z_{22}v_k \end{aligned} \tag{1}$$

We shall assume that the Z_i are independent random matrices with a common distribution. More difficult problems involving situations in which the distribution of Z_k depends upon the values assumed by Z_{k-1} will be discussed in a subsequent volume.

The assumption that the Z_k are independent means that we can write

$$\begin{aligned} E(u_{k+1}) &= e_{11}E(u_k) + e_{12}E(v_k) \\ E(v_{k+1}) &= e_{21}E(u_k) + e_{22}E(v_k) \end{aligned} \tag{2}$$

where $E(u_k)$ and $E(v_k)$ represent the expected values of u_k and v_k, respectively, and e_{ij} is the expected value of z_{ij}.

We see then that

$$E(x_{k+1}) = E(Z)E(x_k) = E(Z)^{k+1}x_0 \tag{3}$$

This means that the asymptotic behavior of $E(x_n)$ as $n \to \infty$ depends upon the characteristic roots of $E(Z)$.

EXERCISE

1. Assume that with probability $\frac{1}{2}$ Z is a positive Markoff matrix A, and with probability $\frac{1}{2}$ a positive Markoff matrix B. Prove that $E(x_n) \to y$, a probability vector independent of x_0, assumed also to be a probability vector.

4. Expected Values of Squares. Suppose that we wish to determine the expected values of u_n^2 and v_n^2. From (3.1) we have

$$\begin{aligned} u_{k+1}^2 &= z_{11}^2 u_k^2 + 2z_{11}z_{12}u_k v_k + z_{12}^2 v_k^2 \\ v_{k+1}^2 &= z_{21}^2 u_k^2 + 2z_{21}z_{22}u_k v_k + z_{22}^2 v_k^2 \end{aligned} \tag{1}$$

We see then that in order to determine $E(u_k^2)$ and $E(v_k^2)$ for all k we must determine the values of $E(u_k v_k)$. From (3.1), we have

$$u_{k+1}v_{k+1} = z_{11}z_{21}u_k^2 + (z_{11}z_{22} + z_{12}z_{21})u_k v_k + z_{12}z_{22}v_k^2 \tag{2}$$

Hence

$$\begin{bmatrix} E(u_{k+1}^2) \\ E(u_{k+1}v_{k+1}) \\ E(v_{k+1}^2) \end{bmatrix} = E(Z^{[2]}) \begin{bmatrix} E(u_k^2) \\ E(u_k v_k) \\ E(v_k^2) \end{bmatrix}$$

where $Z^{[2]}$ is the Kronecker square of $Z = (z_{ij})$, as defined in Sec. 8 of Chap. 12.

Observe how stochastic matrices introduce Kronecker products in a very natural fashion.

MISCELLANEOUS EXERCISES

1. Show that $E(Z^{[r]})$ arises from the problem of determining $E(u_k^r)$.

2. Show that the Kronecker square arises in the following fashion. Let Y be a matrix, to be determined, possessing the property that $E(x(n),Yx(n))$ is readily obtained. Determine Y by the condition that

$$E(Z'YZ) = \lambda Y$$

What are the values of λ?

3. If Z has the distribution of Exercise 1 of Sec. 3, do the sequences $\{E(u_n^2)\}$, $\{E(v_n^2)\}$ converge? If so, to what?

Bibliography and Discussion

The problem of studying the behavior of various functions of stochastic matrices, such as the determinants, or the characteristic roots, is one of great interest and importance in the field of statistics and various parts of mathematical physics. Here we wish merely to indicate one aspect of the problem and the natural way in which Kronecker products enter.

For a discussion of the way stochastic matrices enter into statistics, see

S. S. Wilks, *Mathematical Statistics*, Princeton University Press, Princeton, N.J., 1943.

T. W. Anderson, The Asymptotic Distribution of Certain Characteristic Roots and Vectors, *Proc. Second Berkeley Symposium on Mathematical Statistics and Probability*, 1950.

T. W. Anderson, *An Introduction to Multivariate Statistical Analysis*, John Wiley & Sons, Inc., New York, 1958.

and the references in

A. E. Ingham, An Integral Which Occurs in Statistics, *Proc. Cambridge Phil. Soc.*, vol. 29, pp. 271–276, 1933.

See also

H. Nyquist, S. O. Rice, and J. Riordan, The Distribution of Random Determinants, *Quart. Appl. Math.*, vol. 12, pp. 97–104, 1954.

R. Bellman, A Note on the Mean Value of Random Determinants, *Quart. Appl. Math.*, vol. 13, pp. 322–324, 1955.

For some problems involving stochastic matrices arising from mathematical physics, see

E. P. Wigner, Characteristic Vectors of Bordered Matrices with Infinite Dimensions, *Ann. Math.*, vol. 62, pp. 548–564, 1955.

F. J. Dyson, *Phys. Rev.*, (2), vol. 92, pp. 1331–1338, 1953.

R. Bellman, Dynamics of a Disordered Linear Chain, *Phys. Rev.*, (2), vol. 101, p. 19, 1958.

Two books discussing these matters in detail are

U. Grenander, *Probabilities on Algebraic Structures*, John Wiley & Sons, Inc., New York, 1963.

M. L. Mehta, *Random Matrices*, Academic Press Inc., New York, 1967.

For the use of Kronecker products to obtain moments, see

R. Bellman, Limit Theorems for Noncommutative Operations, I, *Duke Math. J.*, vol. 21, pp. 491–500, 1954.

E. H. Kerner, The Band Structure of Mixed Linear Lattices, *Proc. Phys. Soc. London*, vol. 69, pp. 234–244, 1956.

Jun-ichi Hori, On the Vibration of Disordered Linear Lattices, II, *Progr. Theoret. Phys. Kyoto*, vol. 18, pp. 367–374, 1957.

The far more difficult problem of determining asymptotic probability distributions is considered in

H. Furstenberg and H. Kesten, Products of Random Matrices, *Ann. Math. Stat.*, vol. 3, pp. 457–469, 1960.

An expository paper showing the way in which matrices enter into the study of "nearest neighbor" problems is

A. Maradudin and G. H. Weiss, The Disordered Lattice Problem, a Review, *J. Soc. Ind. Appl. Math.*, vol. 6, pp. 302–319, 1958.

See also

D. N. Nanda, Distribution of a Root of a Determinantal Equation, *Ann. Math. Stat.*, vol. 19, pp. 47–57, 1948; *ibid.*, pp. 340–350.

P. L. Hsu, On the Distribution of Roots of Certain Determinantal Equations, *Ann. Eugenics*, vol. 3, pp. 250–258, 1939.

S. N. Roy, The Individual Sampling . . . , *Sankhya*, 1943.

A. M. Mood, On the Distribution of the Characteristic Roots of Normal Second-moment Matrices, *Ann. Math. Stat.*, vol. 22, pp. 266–273, 1951.

M. G. Kendall, *The Advanced Theory of Statistics*, vol. II, London, 1946.

R. Englman, The Eigenvalues of a Randomly Distributed Matrix, *Nuovo cimento*, vol. 10, pp. 615–621, 1958.

Finally, the problem of the iteration of random linear transformations plays an important role in the study of wave propagation through random media, see

R. Bellman and R. Kalaba, Invariant Imbedding, Wave Propagation, and the WKB Approximation, *Proc. Natl. Acad. Sci. U.S.*, vol. 44, pp. 317–319, 1958.

R. Bellman and R. Kalaba, Invariant Imbedding and Wave Propagation in Stochastic Media, *J. Math. and Mech.*, 1959.

16

Positive Matrices, Perron's Theorem, and Mathematical Economics

1. Introduction. In this chapter, we propose to discuss a variety of problems arising in the domain of mathematical economics which center about the themes of linearity and positivity.

To begin our study, however, we shall study some simple branching processes which arise both in the growth of biological entities and in the generation of elementary particles, as in nuclear fission and cosmic-ray cascades.

Oddly enough, the fundamental result concerning positive matrices was established by Perron in connection with his investigation of the multidimensional continued fraction expansions of Jacobi. His result was then considerably extended by Frobenius in a series of papers.

Rather remarkably, a result which arose in number-theoretic research now occupies a central position in mathematical economics, particularly in connection with the "input-output" analysis of Leontieff. The matrices arising in this study were first noted by Minkowski. This result also plays an important role in the theory of branching processes.

What we present here is a very small and specialized part of the modern theory of positive operators, just as the results for symmetric matrices were principally special cases of results valid for Hermitian operators.

Finally, at the end of the chapter, we shall touch quickly upon the basic problem of linear programming and the connection with the theory of games of Borel and von Neumann. In conclusion, we shall mention some Markovian decision processes arising in the theory of dynamic programming.

2. Some Simple Growth Processes. Let us consider the following simple model of the growth of a set of biological objects. Suppose that there are N different types, which we designate by the numbers $1, 2, \ldots, N$, and that at the times $t = 0, 1, 2, \ldots$, each of these different types gives birth to a certain number of each of the other types.

One case of some importance is that where there are only two types, the normal species and the mutant species. Another case of interest is that where we wish to determine the number of females in different age groups. As each year goes by, females of age i go over into females of age $i + 1$, from time to time giving birth to a female of age zero. It is rather essential to attempt to predict the age distribution by means of a mathematical model since the actual data are usually difficult to obtain.

Let us introduce the quantities

$$a_{ij} = \text{the number of type } i \text{ derived at any stage from a single item of type } j,\ i, j = 1, 2, \ldots, N. \tag{1}$$

As usual, we are primarily interested in growth processes whose mechanism does not change over time.

The state of the system at time n is determined by the N quantities

$$x_i(n) = \text{the number of type } i \text{ at time } n \tag{2}$$

$i = 1, 2, \ldots, N$. We then have the relations

$$x_i(n + 1) = \sum_{j=1}^{N} a_{ij}x_j(n) \qquad i = 1, 2, \ldots, N \tag{3}$$

where the relations $x_i(0) = c_i$, $i = 1, 2, \ldots, N$, determine the initial state of the system.

The problem we would like to solve is that of determining the behavior of the components $x_i(n)$ as $n \to \infty$. This depends upon the nature of the characteristic roots of $A = (a_{ij})$, and, in particular, upon those of greatest absolute value.

As we know from our discussion of Markoff matrices, this problem is one of great complexity. We suspect, however, that the problems may be quite simple in the special case where all the a_{ij} are positive, and this is indeed the case.

3. Notation. As in Chap. 14, we shall call a matrix *positive* if all of its elements are positive. If A is positive, we shall write $A > 0$. The notation $A > B$ is equivalent to the statement that $A - B > 0$. Similarly, we may conceive of non-negative matrices, denoted by $A \geq 0$.

This notation definitely conflicts with previous notation used in our discussion of positive definite matrices. At this point, we have the option of proceeding with caution, making sure that we are never discussing both types of matrices at the same time, or of introducing a new notation, such as $A \geq\geq B$, or something equally barbarous. Of the alternatives, we prefer the one that uses the simpler notation, but which requires that the reader keep in mind what is being discussed. This is, after all, the more desirable situation.

A vector x will be called *positive* if all of its components are positive, and *non-negative* if the components are merely non-negative. We shall write $x > 0$, and $x \geq 0$ in the second. The relation $x \geq y$ is equivalent to $x - y \geq 0$.

EXERCISES

1. Show that $x \geq y$, $A \geq 0$, imply that $Ax \geq Ay$.

2. Prove that $Ax \geq 0$ for all $x \geq 0$ implies $A \geq 0$.

4. The Theorem of Perron. Let us now state the fundamental result of Perron.

Theorem 1. *If A is a positive matrix, there is a unique characteristic root of A, $\lambda(A)$, which has greatest absolute value. This root is positive and simple, and its associated characteristic vector may be taken to be positive.*

There are many proofs of this result, of quite diverse origin, structure, and analytical level. The proof we shall give is in some ways the most important since it provides a variational formulation for $\lambda(A)$ which makes many of its properties immediate, and, in addition, can be extended to handle more general operators.

5. Proof of Theorem 1. Our proof of Theorem 1 will be given in the course of demonstrating Theorem 2.

Theorem 2. *Let A be a positive matrix and let $\lambda(A)$ be defined as above. Let $S(\lambda)$ be the set of non-negative λ for which there exist non-negative vectors x such that $Ax \geq \lambda x$. Let $T(\lambda)$ be the set of positive λ for which there exist positive vectors x such that $Ax \leq \lambda x$. Then*

$$\begin{aligned} \lambda(A) &= \max \lambda \qquad \lambda \epsilon S(\lambda) \\ &= \min \lambda \qquad \lambda \epsilon T(\lambda) \end{aligned} \tag{1}$$

Proof of Theorem 2. Let us begin by normalizing all the vectors we shall consider, so that

$$\|x\| = \sum_{i=1}^{N} x_i = 1 \tag{2}$$

This automatically excludes the null vector. Let us once more set $\|A\| = \sum_{i,j=1}^{N} a_{ij}$. If $\lambda x \leq Ax$, we have

$$\lambda \|x\| \leq \|A\| \, \|x\|$$

or

$$0 \leq \lambda \leq \|A\| \tag{3}$$

This shows that $S(\lambda)$ is a bounded set, which is clearly not empty if A is positive.

Let $\lambda_0 = \text{Sup } \lambda$ for $\lambda \epsilon S(\lambda)$; let $\{\lambda_i\}$ be a sequence of λs in $S(\lambda)$ converging to λ_0; and let $\{x^{(i)}\}$ be an associated set of vectors, which is to say

$$\lambda_i x^{(i)} \leq A x^{(i)} \qquad i = 1, 2, \ldots \tag{4}$$

Since $\|x^{(i)}\| = 1$, we may choose a subsequence of the $x^{(i)}$, say $\{x^{(j)}\}$ which converges to $x^{(0)}$, a non-negative, nontrivial vector. Since $\lambda_0 x^{(0)} \leq A x^{(0)}$, it follows that $\lambda_0 \epsilon S(\lambda)$, which means that the supremum is actually a maximum.

Let us now demonstrate that the inequality is actually an equality, that is, $\lambda_0 x^{(0)} = A x^{(0)}$. The proof is by contradiction. Let us suppose, without loss of generality, that

$$\begin{aligned} \sum_{j=1}^{N} a_{1j} x_j - \lambda_0 x_1 &= d_1 > 0 \\ \sum_{j=1}^{N} a_{kj} x_j - \lambda_0 x_k &\geq 0 \qquad k = 2, \ldots, N \end{aligned} \tag{5}$$

where x_i, $i = 1, 2, \ldots, N$ are the components of $x^{(0)}$.

Consider now the vector

$$y_2 x^{(0)} + \begin{bmatrix} d_1/2\lambda_0 \\ 0 \\ \cdot \\ \cdot \\ \cdot \\ 0 \end{bmatrix} \tag{6}$$

It follows from (5) that $Ay > \lambda_0 y$. This, however, contradicts the maximum property of λ_0. Hence $d_1 = 0$, and there must be equality in all the terms in (5).

Consequently λ_0 is a characteristic root of A and $x^{(0)}$ a characteristic vector which is necessarily positive.

Let us now show that $\lambda_0 = \lambda(A)$. Assume that there exists a characteristic root λ of A for which $|\lambda| \geq \lambda_0$, with z an associated characteristic vector. Then from $Az = \lambda z$ we have

$$|\lambda|\, |z| \leq A|z| \tag{7}$$

where $|z|$ denotes the vector whose components are the absolute values of the components of z. It follows from this last inequality and the definition of λ_0, that $|\lambda| \leq \lambda_0$. Since $|\lambda| = \lambda_0$, the argument above shows that the inequality $|\lambda|\, |z| \leq A|z|$ must be an equality. Hence $|Az| = A|z|$; and thus $z = c_1 w$, with $w > 0$ and c_1 a complex number. Consequently, $Az = \lambda z$ is equivalent to $Aw = \lambda w$; whence λ is real and positive and hence equal to λ_0.

To show that w is equivalent to $x^{(0)}$, let us show that apart from scalar multiples there is only one characteristic vector associated with λ_0. Let z be another characteristic vector of A, not necessarily positive, associated with $\lambda(A)$. Then $x^{(0)} + \epsilon z$, for all scalar ϵ, is a characteristic vector of A. Varying ϵ about $\epsilon = 0$, we reach a first value of ϵ for which one or several components of $x^{(0)} + \epsilon z$ are zero, with the remaining components positive, provided that x and z are actually linearly independent. However, the relation

$$A(x^{(0)} + \epsilon z) = \lambda(A)(x^{(0)} + \epsilon z) \tag{8}$$

shows that $x^{(0)} + \epsilon z \geq 0$ implies $x^{(0)} + \epsilon z > 0$. Thus a contradiction.

EXERCISES

1. Show that $A \geq B \geq 0$ implies that $\lambda(A) \geq \lambda(B)$.

2. Show by means of 2×2 matrices that $\lambda(AB) \leq \lambda(A)\lambda(B)$ is not universally valid.

6. First Proof that $\lambda(A)$ Is Simple. Let us now turn to the proof that $\lambda(A)$ is a simple characteristic root. We shall give two demonstrations, one based upon the Jordan canonical form and the other upon more elementary considerations.

Assume that $\lambda(A)$ is not a simple root. The Jordan canonical form shows that there exists a vector y with the property that

$$(A - \lambda(A)I)^k y = 0 \qquad (A - \lambda(A)I)^{k-1} y \neq 0 \tag{1}$$

for some $k \geq 2$.

This means that $(A - \lambda(A)I)^{k-1}y$ is a characteristic vector of A and hence a scalar multiple of $x^{(0)}$. By suitable choice of y we may take this multiple to be one. Hence

$$x^{(0)} = (A - \lambda(A)I)^{k-1}y \tag{2}$$

Now let

$$z = (A - \lambda(A)I)^{k-2}y \tag{3}$$

Then

$$Az = \lambda(A)z + x^0 > \lambda(A)z \tag{4}$$

This yields $A|z| > \lambda(A)|z|$, which contradicts the maximum definition of $\lambda(A)$.

7. Second Proof of the Simplicity of $\lambda(A)$. Let us now present an alternate proof of the simplicity of $\lambda(A)$ based upon the minimum property.

We begin by proving the following lemma.

Lemma. Let A_N be a positive $N \times N$ matrix (a_{ij}), and A_{N-1} any $(N - 1) \times (N - 1)$ matrix obtained by striking out an ith row and jth *column. Then*

$$\lambda(A_N) > \lambda(A_{N-1}) \tag{1}$$

Proof. Let us proceed by contradiction. Write

$$\lambda_N = \lambda(A_N) \qquad \lambda_{N-1} = \lambda(A_{N-1})$$

Assume for the moment that $\lambda_N \leq \lambda_{N-1}$, and take, without loss of generality, $A_{N-1} = (a_{ij})$, $i, j = 1, 2, \ldots, N-1$. We have then the equations

$$\sum_{j=1}^{N-1} a_{ij}y_j = \lambda_{N-1}y_i \qquad i = 1, 2, \ldots, N-1, y_i > 0 \tag{2}$$

and

$$\sum_{j=1}^{N} a_{ij}x_j = \lambda_N x_i \qquad i = 1, 2, \ldots, N, x_i > 0 \tag{3}$$

Using the first $N - 1$ equations in (3), we obtain

$$\begin{aligned}\sum_{j=1}^{N-1} a_{ij}x_j &= \lambda_N x_i - a_{iN}x_N \\ &= \lambda_N(x_i - a_{iN}x_N/\lambda_N) < \lambda_N x_i\end{aligned} \tag{4}$$

which contradicts the minimum property of λ_{N-1}.

Let us now apply this result to show that $\lambda(A)$ is a simple root of $f(\lambda) = |A - \lambda I| = 0$. Using the rule for differentiating a determinant, we obtain readily

$$f'(\lambda) = -|A_1 - \lambda I| - |A_2 - \lambda I| \cdots - |A_N - \lambda I| \tag{5}$$

where by A_k we denote the matrix obtained from A by striking out the kth row and column. Since $\lambda(A) > \lambda(A_k)$ for each k, and each expression $|A_k - \lambda I|$ is a polynomial in λ with the same leading term, each polynomial $|A_k - \lambda I|$ has the same sign at $\lambda = \lambda(A)$. Hence, $f'(\lambda(A)) \neq 0$.

8. Proof of the Minimum Property of $\lambda(A)$. We have given above a proof of the maximum property of $\lambda(A)$, and used, in what has preceded, the minimum property. Let us now present a proof of the minimum property which uses the maximum property rather than just repeating the steps already given. The technique we shall employ, that of using the adjoint operator, is one of the most important and useful in analysis, and one we have already met in the discussion of Markoff matrices.

Let A' be, as usual, the transpose of A. Since the characteristic roots of A and A' are the same, we have $\lambda(A) = \lambda(A')$. As we know, $(Ax,y) = (x,A'y)$. If $Ay \leq \lambda y$ for some $y > 0$, we have for any $z \geq 0$,

$$\lambda(z,y) \geq (z,Ay) = (A'z,y) \tag{1}$$

Let z be a characteristic vector of A' associated with $\lambda(A)$. Then

$$\lambda(z,y) \geq (A'z,y) = \lambda(A)(z,y) \tag{2}$$

Since $(z,y) > 0$, we obtain $\lambda \geq \lambda(A)$. This completes the proof of the minimum property.

9. An Equivalent Definition of $\lambda(A)$. In place of the definition of $\lambda(A)$ given in Theorem 1 above, we may also define $\lambda(A)$ as follows.

Theorem 3

$$\begin{aligned}\lambda(A) &= \max_x \min_i \left\{\sum_{j=1}^{N} a_{ij}x_j/x_i\right\} \\ &= \min_x \max_i \left\{\sum_{j=1}^{N} a_{ij}x_j/x_i\right\}.\end{aligned} \tag{1}$$

Here x varies over all non-negative vectors different from zero.

The proof we leave as an exercise.

10. A Limit Theorem. From Theorem 1, we also derive the following important result.

Theorem 4. *Let c be any non-negative vector. Then*

$$v = \lim_{n\to\infty} A^n c/\lambda(A)^n$$

exists and is a characteristic vector of A associated with $\lambda(A)$, unique up to a scalar multiple determined by the choice of c, but otherwise independent of the initial state c.

This is an extension of the corresponding result proved in Chap. 14 for the special case of Markoff matrices.

EXERCISE

1. Obtain the foregoing result from the special case established for Markoff matrices.

11. Steady-state Growth. Let us now see what the Perron theorem predicts for the asymptotic behavior of the growth process discussed in Sec. 2. We shall suppose that $A = (a_{ij})$ is a positive matrix.

Then, as $n \to \infty$, the asymptotic behavior of the vector $x(n)$, defined by the equation in (3), is given by the relation

$$x(n) \sim \lambda^n \gamma \tag{1}$$

where λ is the Perron root and γ is a characteristic vector associated with λ. We know that γ is positive, and that it is a positive multiple of one particular normalized characteristic vector, say δ. This normalization can be taken to be the condition that the sum of the components totals one. The constant of proportionality connecting γ and δ is then determined by the values of the initial components of c, c_i, the initial population.

What this means is that regardless of the initial population, as long as we have the complete mixing afforded by a positive matrix, asymptotically as time increases we approach a steady-state situation in which the total population grows exponentially, but where the proportions of the various species remain constant.

As we shall see, the same phenomena are predicted by the linear models of economic systems we shall discuss subsequently.

12. Continuous Growth Processes. Returning to the mathematical model formulated in Sec. 2, let us suppose that the time intervals at which the system is observed get closer and closer together. The limiting case will be a continuous growth process.

Let

$$\begin{aligned} a_{ij}\Delta &= \text{the number of type } i \text{ produced by a single item of type } j \text{ in a time interval } \Delta,\ j \neq i \\ 1 + a_{ii}\Delta &= \text{the number of type } i \text{ produced by a single item of type } i \text{ in a time interval } \Delta \\ i &= 1, 2, \ldots, N. \end{aligned} \tag{1}$$

Then, we have, as above, the equations

$$x_i(t + \Delta) = (1 + a_{ii}\Delta)x_i(t) + \Delta \sum_{j \neq i} a_{ij}x_j(t) \qquad i = 1, 2, \ldots, N \tag{2}$$

Observe that the a_{ij} are now rates of production. Proceeding formally to the limit as $\Delta \to 0$, we obtain the differential equations

$$\frac{dx_i}{dt} = a_{ii}x_i + \sum_{j \neq i} a_{ij}x_j \qquad i = 1, 2, \ldots, N \tag{3}$$

We can now define the continuous growth process by means of these equations, with the proviso that this will be meaningful if and only if we can show that this process is in some sense the limit of the discrete process.

The asymptotic behavior of the $x_i(t)$ as $t \to \infty$ is now determined by the characteristic roots of A of largest real part.

13. Analogue of Perron Theorem. Let us now demonstrate Theorem 5.

Theorem 5. *If*

$$a_{ij} > 0 \qquad i \neq j \tag{1}$$

then the root of A with largest real part, which we shall call $\rho(A)$, is real and simple. There is an associated positive characteristic vector which is unique up to a multiplicative constant.

Furthermore,

$$\rho(A) = \max_x \min_i \left\{ \sum_{j=1}^{N} a_{ij}x_j/x_i \right\}$$
$$= \min_x \max_i \left\{ \sum_{j=1}^{N} a_{ij}x_j/x_i \right\} \tag{2}$$

Proof. The easiest way to establish this result is to derive it as a limiting case of Perron's theorem. Let $\rho(A)$ denote a characteristic root of largest real part. It is clear that a root of $e^{\delta A}$ of largest absolute value is given by

$$\lambda(e^{\delta A}) = e^{\delta\rho(A)} \tag{3}$$

for $\delta > 0$. Since $e^{\delta A}$ is a positive matrix under our assumptions, as follows from Sec. 15 of Chap. 10, it follows that $\rho(A)$ is real and positive, and simple.

Using the variational characterization for $\lambda(e^{\delta A})$, we have

$$e^{\delta\rho(A)} = \max_i \min_x \left(1 + \delta \sum_{j=1}^{N} a_{ij}x_j/x_i\right) + 0(\delta^2) \tag{4}$$

Letting $\delta \to 0$, we obtain the desired representation.

EXERCISES

1. If $B \geq 0$, and A is as above, show that $\rho(A + B) \geq \rho(A)$.

2. Derive Theorem 5 directly from Perron's theorem for $sI + A$, where s has been chosen so that $sI + A$ is a positive matrix.

14. Nuclear Fission. The same type of matrix arises in connection with some simple models of nuclear fission. Here the N types of objects may be different types of elementary particles, or the same particle, say a neutron, in N different energy states.

References to extensive work in this field will be found at the end of the chapter.

15. Mathematical Economics. As a simple model of an economic system, let us suppose that we have N different industries, the operation of which is dependent upon the state of the other industries. Assuming for the moment that the state of the system at each of the discrete times $n = 0, 1, \ldots,$ can be specified by a vector $x(n)$, where the ith component, $x_i(n)$, in some fashion describes the state of the ith industry, the statement concerning interrelation of the industries translates into a system of recurrence relations, or difference equations, of the following type:

$$x_i(n + 1) = g_i(x_1(n),x_2(n), \ldots ,x_N(n)) \tag{1}$$

with $x_i(0) = c_i$, $i = 1, 2, \ldots, N$.

This is, however, much too vague a formulation. In order to see how relations of this type actually arise, let us consider a simple model of three interdependent industries, which we shall call for identification purposes the "auto" industry, the "steel" industry, and the "tool" industry.

Using the same type of "lumped-constant" approach that has been so successful in the study of electric circuits, we shall suppose that each of these industries can be specified at any time by its *stockpile* of raw material and by its *capacity* to produce new quantities using the raw materials available.

Let us introduce the following state variables:

$$\begin{aligned} x_1(n) &= \text{number of autos produced up to time } n \\ x_2(n) &= \text{capacity of auto factories at time } n \\ x_3(n) &= \text{stockpile of steel at time } n \\ x_4(n) &= \text{capacity of steel mills at time } n \\ x_5(n) &= \text{stockpile of tools at time } n \\ x_6(n) &= \text{capacity of tool factories at time } n \end{aligned} \tag{2}$$

In order to obtain relations connecting the $x_i(n + 1)$ with the $x_i(n)$, we must make some assumptions concerning the economic interdependence of the three industries:

To increase auto, steel, or tool capacity, we require only steel and tools. (3*a*)
Production of autos requires only auto capacity and steel. (3*b*)
Production of steel requires only steel capacity. (3*c*)
Production of tools requires only tool capacity and steel. (3*d*)

The dynamics of the production process are the following: At the beginning of a time period, n to $n + 1$, quantities of steel and tools, taken from their respective stockpiles, are allocated to the production of additional steel, tools, and autos, and to increasing existing capacities for production.

Let

$$\begin{aligned} z_i(n) &= \text{the amount of steel allocated at time } n \text{ for the purpose of} \\ &\qquad \text{increasing } x_i(n) \qquad i = 1, 2, \ldots, 6 \\ w_i(n) &= \text{the amount of tools allocated at time } n \text{ for the purpose of} \\ &\qquad \text{increasing } x_i(n) \qquad i = 1, 2, \ldots, 6 \end{aligned} \tag{4}$$

Referring to the assumptions in (3), we see that

$$z_3(n) = 0 \tag{5a}$$
$$w_1(n) = w_3(n) = w_5(n) = 0 \tag{5b}$$

In order to obtain relations connecting $x_i(n+1)$ with the $x_j(n)$ and $z_j(n)$ and $w_j(n)$, we must make some further assumptions concerning the relation between input and output. The simplest assumption is that we have a linear production process where the output of a product is directly proportional to input in shortest supply. Thus production is proportional to capacity whenever there is no constraint on raw materials, and proportional to the minimum raw material required whenever there is no limit to capacity.

Processes of this type are called "bottleneck processes."

To obtain the equations describing the process, we use the principle of conservation. The quantity of an item at time $n+1$ is the quantity at time n, less what has been used over $[n, n+1]$, plus what has been produced over $[n, n+1]$.

The constraints on the choice of the z_i and w_i are that we cannot allocate at any stage more than is available in the stockpiles, and further, that there is no point to allocating more raw materials than the productive capacity can accommodate.

The conservation equations are then, taking account of the bottleneck aspects,

$$\begin{aligned}
x_1(n+1) &= x_1(n) + \min\,(\gamma_1 x_2(n), \alpha_1 z_1(n)) \\
x_2(n+1) &= x_2(n) + \min\,(\alpha_2 z_2(n), \beta_2 w_2(n)) \\
x_3(n+1) &= x_3(n) - z_1(n) - z_2(n) - z_4(n) \\
&\qquad - z_5(n) - z_6(n) + \gamma_2 x_4(n) \qquad (6) \\
x_4(n+1) &= x_4(n) + \min\,(\alpha_4 z_4(n), \beta_4 w_4(n)) \\
x_5(n+1) &= x_5(n) - w_2(n) - w_4(n) - w_6(n) + \min\,[\gamma_5 x_6(n), \alpha_5 z_5(n)] \\
x_6(n+1) &= x_6(n) + \min\,(\alpha_6 z_6(n), \beta_6 w_6(n))
\end{aligned}$$

where α_i, β_i, γ_i are constants.

The stockpile constraints on the choice of the z_i and w_i are

$$z_i,\ w_i \geq 0 \qquad (7a)$$
$$z_1 + z_2 + z_4 + z_5 + z_6 \leq x_3 \qquad (7b)$$
$$w_2 + w_4 + w_6 \leq x_5 \qquad (7c)$$

Applying the capacity constraints, we can reduce the equations in (5) to linear form. We see that

$$\alpha_2 z_2 = \beta_2 w_2 \qquad (8a)$$
$$\alpha_4 z_4 = \beta_4 w_4 \qquad (8b)$$
$$\beta_6 z_6 = \beta_6 w_6 \qquad (8c)$$

Using these relations, we can eliminate the w_i from (7) and obtain the linear equations

$$
\begin{aligned}
x_1(n+1) &= x_1(n) + \alpha_1 z_1(n), x_1(0) = c_1 \\
x_2(n+1) &= x_2(n) + \alpha_2 z_2(n), x_2(0) = c_2 \\
x_3(n+1) &= x_3(n) - z_1(n) - z_2(n) - z_4(n) - z_5(n) \\
&\qquad - z_6(n) + \gamma_2 x_4(n), x_3(0) = c_3 \qquad (9) \\
x_4(n+1) &= x_4(n) + \alpha_4 z_4(n), x_4(0) = c_4 \\
x_5(n+1) &= x_5(n) - \epsilon_2 z_2(n) - \epsilon_4 z_4(n) - \epsilon_6 z_6(n) \\
&\qquad + \alpha_5 z_5(n) \qquad \epsilon_i = \alpha_i/\beta_i \qquad x_5(0) = c_5 \\
x_6(n+1) &= x_6(n) + \alpha_6 z_6(n), x_6(0) = c_6
\end{aligned}
$$

The constraints on the choice of the z_i are now

$$
\begin{aligned}
z_i &\geq 0 && (10a) \\
z_1 + z_2 + z_4 + z_5 + z_6 &\leq x_3 && (10b) \\
\gamma_2 z_2 + \gamma_4 z_4 + \gamma_6 z_6 &\leq x_5 && (10c) \\
z_1 &\leq f_2 x_2 && (10d) \\
z_5 &\leq f_6 x_6 && (10e)
\end{aligned}
$$

Suppose that we satisfy these conditions by means of a choice

$$z_2 = e_2 x_5 \qquad z_4 = e_4 x_5 \qquad z_6 = e_6 x_5 \qquad (11)$$

where the scalars e_2, e_4, and e_6 are chosen to satisfy (10c).

$$z_1 = f_2 x_2 \qquad z_5 = f_6 x_6 \qquad (12)$$

and suppose that (10b) is satisfied.

Then the foregoing equations have the form

$$x_i(n+1) = (1 - a_{ii})x_i(n) + \sum_{j \neq i} a_{ij} x_j(n) \qquad i = 1, 2, \ldots, 6 \qquad (13)$$

where

$$a_{ij} \geq 0 \qquad i, j = 1, \ldots, 6 \qquad (14)$$

The continuous version of these equations is

$$\frac{dx_i}{dt} = -a_{ii}x_i + \sum_{j \neq i} a_{ij} x_i \qquad i = 1, 2, \ldots, 6 \qquad (15)$$

What we have wished to emphasize is that a detailed discussion of the economic model shows that a special type of matrix, sometimes called an "input-output" matrix, plays a paramount role. Unfortunately, in the most interesting and important cases we cannot replace (14) by the stronger condition of positivity.

The result is that the study of the asymptotic behavior of the $x_i(n)$ as given by a linear recurrence relation of the form appearing in (13) is quite complex. References to a great deal of research will be found at the end of the chapter.

16. Minkowski-Leontieff Matrices. Let us now discuss a particular class of non-negative matrices specified by the conditions

$$0 \le a_{ij} \tag{1a}$$

$$\sum_{i=1}^{N} a_{ij} \le 1 \tag{1b}$$

They arise in connection with the solution of linear equations of the form

$$x_i = \sum_{j=1}^{N} a_{ij}x_j + y_i \qquad i = 1, 2, \ldots, N \tag{2}$$

which occur in the treatment of models of interindustry production processes quite similar to that formulated above. Let us demonstrate Theorem 6.

Theorem 6. *If* $0 \le a_{ij}$ *and*

$$\sum_{i=1}^{N} a_{ij} < 1 \qquad j = 1, 2, \ldots, N \tag{3}$$

then the equations in (2) *have a unique solution which is positive if the* y_i *are positive.*

If $a_{ij} > 0$, *then* $(I - A)^{-1}$ *is positive.*

Proof. To show that $(I - A)^{-1}$ is a positive matrix under the assumption that $a_{ij} > 0$, it is sufficient to show that its transpose is positive, which is to say that $(I - A')^{-1}$ is positive. Consider then the adjoint system of equations

$$z_i = \sum_{j=1}^{N} a_{ji}z_j + w_i \qquad i = 1, 2, \ldots, N \tag{4}$$

It is easy to see that the solution obtained by direct iteration converges, in view of the condition in (3), and that this solution is positive if $w_i > 0$ for $i = 1, 2, \ldots, N$.

The assumption that $a_{ij} > 0$ shows that z_i is positive whenever w is a nontrivial non-negative vector.

The fact that $(I - A)^{-1}$ exists and is positive establishes the first part of the theorem.

17. Positivity of $|I - A|$. Since the linear system in (16.2) has a unique solution for all y_i under the stated conditions, it follows that $|I - A| \neq 0$. To show that $|I - A| > 0$, let us use the method of continuity as in Sec. 4 of Chap. 4. If $\lambda \ge 0$, the matrix λA satisfies the same conditions as A. Thus $|I - \lambda A|$ is nonzero for $0 \le \lambda \le 1$. Since the determinant is positive at $\lambda = 0$ and continuous in λ, it is positive at $\lambda = 1$.

18. Strengthening of Theorem 6. It is clear that if $\sum_{i=1}^{N} a_{ji} = 1$ for all j, then 1 is a characteristic root of A, whence $|I - A| = 0$. On the other hand, it is reasonable to suppose that the condition $\sum_{i} a_{ij} < 1$ can be relaxed.

It suffices to assume that

$$1 > a_{ij} > 0 \tag{1a}$$

$$\sum_{i} a_{ij} < 1 \text{ for at least one } j \tag{1b}$$

$$\sum_{i} a_{ij} \leq 1 \text{ for all } j \tag{1c}$$

To prove this, we show that A^2 is a matrix for which (1b) holds for all j.

19. Linear Programming. In the first part of this book, we considered the problem of maximizing quadratic forms subject to quadratic constraints, and the problem of maximizing quadratic forms subject to linear constraints. We have, however, carefully avoided any questions involving the maximization of linear forms subject to linear constraints.

A typical problem of this type would require the maximization of

$$L(x) = \sum_{i=1}^{N} c_i x_i \tag{1}$$

subject to constraints of the form

$$\sum_{j=1}^{N} a_{ij} x_j \leq b_i \qquad i = 1, 2, \ldots, M \tag{2a}$$

$$x_i \geq 0 \tag{2b}$$

The reader will speedily find that none of the techniques discussed in the first part of the book, devoted to quadratic forms, are of any avail here. Questions of this type are indeed part of a new domain, the theory of linear inequalities, which plays an important role in many investigations. The theorem of Perron, established above, is closely related, as far as result and method of proof, to various parts of this theory.

In addition to establishing results concerning the existence and nature of solutions of the problem posed above, it is important to develop various algorithms for numerical solution of this problem. This part of the general theory of linear inequalities is called *linear programming*.

Let us now present a very simple example of the way problems of this type arise in mathematical economics. Suppose that we possess quantities $x_1, x_2, \ldots, x_M$ of M different resources, e.g., men, machines, and money, and that we can utilize various amounts of these resources in connection with N different activities, e.g., drilling for oil, building automobiles, and so on.

Let

$$x_{ij} = \text{the quantity of the } i\text{th resource allocated to the } j\text{th activity} \qquad (3)$$

so that

$$\sum_{j=1}^{N} x_{ij} = x_i \qquad i = 1, 2, \ldots, M \qquad (4a)$$

$$x_{ij} \geq 0 \qquad (4b)$$

Suppose, and this is usually a crude approximation, that the utility of allocating x_{ij} is determined by simple proportionality, which is to say that the utility is $a_{ij}x_{ij}$. Making the further assumption that these utilities are additive, we arrive at the problem of maximizing the linear form

$$L(x) = \sum_{i,j} a_{ij}x_{ij} \qquad (5)$$

subject to the constraints of (2).

As another example of the way in which these linear variational problems arise in mathematical economics, consider the model presented in Sec. 15. In place of assigning the quantities z_i and w_i as we did there, we can ask for the values of those quantities which maximize the total output of steel over N stages of the process.

20. The Theory of Games. Suppose that two players, A and B, are engaged in a game of the following simple type. The first player can make any of M different moves and the second player can make any of N different moves. If A makes the ith move and B the jth move, A receives the quantity a_{ij} and B receives $-a_{ij}$.

The matrix

$$A = (a_{ij}) \qquad (1)$$

is called the *payoff* matrix.

We now require that both players make their choices of moves simultaneously, without knowledge of the other's move. An umpire then examines both moves and determines the amounts received by the players according to the rule stated above.

Suppose that this situation is repeated over and over. In general, it will not be advantageous for either player to make the same choice at each play of the game. To guard himself against his opponent taking advantage of any advance information, each player will randomize over choices.

Let

$$x_i = \text{probability that } A \text{ makes the } i\text{th choice}, \ i = 1, 2, \ldots, M$$
$$y_j = \text{probability that } B \text{ makes the } j\text{th choice}, \ j = 1, 2, \ldots, N \quad (2)$$

Then the expected quantity received by A after any particular play is

$$f(x,y) = \sum_{i,j} a_{ij} x_i y_j \quad (3)$$

The problem is to determine how the x_i and y_j should be chosen.

A can reason in the following way: "Suppose that B knew my choice. Then he would choose the y_j so as to minimize $f(x,y)$. Consequently, I will choose the x_i so as to maximize." Proceeding in this fashion, the expected return to A will be

$$v_A = \max_x \min_y f(x,y) \quad (4)$$

where the x and y regions are defined by

$$x_i,\ y_i \geq 0 \quad (5a)$$
$$\sum_i x_i = \sum_j y_j = 1 \quad (5b)$$

Similarly, B can guarantee that he cannot lose more than

$$v_B = \min_y \max_x f(x,y) \quad (6)$$

The fundamental result of the theory of games is that $v_A = v_B$. This is the *min-max theorem* of von Neumann.

What is quite remarkable is that it can be shown that these problems and the problems arising in the theory of linear inequalities, as described in Sec. 19, are mathematically equivalent. This equivalence turns out to be an analytic translation of the duality inherent in N-dimensional Euclidean geometry.

21. A Markovian Decision Process. Let us now discuss some problems that arise in the theory of dynamic programming.

Consider a physical system S which at any of the times $t = 0, 1, \ldots,$ must be in one of the states $S_1, S_2, \ldots, S_N$. Let

$$x_i(n) = \text{the probability that } S \text{ is in state } S_i \text{ at time } n, \ i = 1, 2, \ldots, N \quad (1)$$

For each q, which represents a vector variable, let

$$M(q) = (m_{ij}(q)) \tag{2}$$

represent a Markoff matrix.

Instead of supposing, as in an ordinary Markoff process, that the probability distribution at time n is determined by $M(q)^n x(0)$, for some fixed q, we shall suppose that q can change from stage to stage. Specifically, we shall suppose that this process is being supervised by someone who wants to maximize the probability that the system is in state 1 at any particular time.

In place of the usual equations, we obtain the relations

$$\begin{aligned} x_1(n+1) &= \max_q \sum_{j=1}^{N} m_{ij}(q)x_j(n) \\ x_i(n+1) &= \sum_{j=1}^{N} m_{ij}(q^*)x_j(n) \end{aligned} \tag{3}$$

where q^* is a value of q which maximizes the expression on the first line.

We leave to the reader the quite interesting problem of determining simple conditions under which a "steady-state" or equilibrium solution exists. In Sec. 22, we shall discuss a more general system.

22. An Economic Model. Suppose that we have N different types of resources. Let

$$x_i(n) = \text{the quantity of the } i\text{th resource at time } n \tag{1}$$

Suppose that, as in the foregoing pages, the following linear relations exist:

$$x_i(n+1) = \sum_{j=1}^{N} a_{ij}(q)x_j(n) \tag{2}$$

where q is a vector parameter, as before. If $A(q)$ is a positive matrix, the Perron theorem determines the asymptotic behavior of the system.

Assume, however, as in Sec. 21, that the process is supervised by someone who wishes to maximize each quantity of resource at each time. The new relations are then

$$x_i(n+1) = \max_q \sum_{j=1}^{N} a_{ij}(q)x_j(n) \qquad i = 1, 2, \ldots, N \tag{3}$$

The following generalization of the Perron theorem can then be obtained.

Theorem 7. *If q runs over a set of values $(q_1,q_2, \ldots ,q_M)$ which allow the maximum to be obtained in* (3) (4a)

$0 < m_1 \le a_{ij}(q) \le m_2 < \infty$ (4b)

$\max_q \lambda(A(q))$ *exists and is attained for a* q_i (4c)

then there exists a unique positive λ *such that the homogeneous system*

$$\lambda y_i = \max_q \sum_{j=1}^{N} a_{ij}(q) y_j \tag{5}$$

has a positive solution $y_i > 0$. *This solution is unique up to a multiplicative constant and*

$$\lambda = \max_q \lambda(A(q)) \tag{6}$$

Furthermore, as $n \to \infty$,

$$x_i(n) \sim a_1 y_i \lambda^n \tag{7}$$

where a_1 *depends on the initial values.*

The proof may be found in references cited at the end of the chapter.

MISCELLANEOUS EXERCISES

1. Let $A = (a_{ij})$ be a non-negative matrix. A necessary and sufficient condition that all characteristic roots of A be less than 1 in absolute value is that all the principal minors of $I - A$ be positive (*Metzler*).

2. If A has all negative diagonal elements, and no negative off-diagonal elements, if D is a diagonal matrix, and if the real parts of the characteristic roots of both A and DA are negative, then the diagonal elements of D are positive (*K. D. Arrow and A. C. Enthoven*).

3. Consider a matrix $Z(t) = (z_{ij}(t))$, where the z_{ij} possess the following properties:

(a) $z_{ij} > 0$

(b) $\int_0^\infty z_{ii}\, dt > 1$ for some i

(c) $\int_0^\infty z_{ij} e^{-at}\, dt < \infty$ for some $a > 0$

Then there is a positive vector x and a positive number s_0 for which

$$\left(\int_0^\infty Z e^{-s_0 t}\, dt\right) x = x$$

Furthermore, s_0 is the root of $\left| I - \int_0^\infty Z e^{-st}\, dt \right| = 0$ with greatest real part, and it is a simple root (*Bohnenblust*); see R. Bellman.[1]

4. Let $A = (a_{ij})$, and $s_i = |a_{ii}| - \sum_{j \ne i} |a_{ij}|$. If $s_i > 0$, $1 \le i \le N$, then, as we already know, $A^{-1} = (b_{ij})$ exists, and, furthermore, $|b_{ij}| \le 1/s_j$ (*Ky Fan*).

[1] R. Bellman, *A Survey of the Mathematical Theory of Time-lag, Retarded Control, and Hereditary Processes*, with the assistance of J. M. Danskin, Jr., The RAND Corporation, *Rept.* R-256, March 1, 1954.

5. A real matrix $A = (a_{ij})$ is said to be *maximum increasing* if the relation

$$\max_{1 \leq i \leq N} x_i \leq \max_{1 \leq i \leq N} \sum_{j=1}^{N} a_{ij} x_j$$

holds for any N real numbers x_i. Prove that A is maximum increasing if and only if $A^{-1} = (b_{ij})$ with

(*a*) $b_{ij} \geq 0$

(*b*) $\sum_{j=1}^{N} b_{ij} = 1 \qquad 1 \leq i \leq N$

(*Ky Fan*)

6. Let the x_i be positive, and $f(x_1,x_2, \ldots ,x_N)$, $g(x_1,x_2, \ldots ,x_N)$ denote, respectively, the least and greatest of the $N + 1$ quantities x_1, $x_2 + 1/x_1$, $x_3 + 1/x_2$, . . . , $x_N + 1/x_{N-1}$, $1/x_N$. Prove that

$$\max_{x_i>0} f(x_1,x_2, \ldots ,x_N) = \min_{x_i>0} g(x_1,x_2, \ldots ,x_N) = 2 \cos (\pi/N + 2)$$

(*Ky Fan*)

7. Let $a > 0$. For any arbitrary partition $0 = t_0 < t_1 < \cdots < t_N = 1$ of $[0,1]$ into N subintervals, the approximate Riemann sum $\sum_{i=1}^{N} (t_i - t_{i-1})/(a + t_i)$ to $\int_0^1 dt/(a + t)$ always contains a term $\geq [1 - (a/a + 1)]^{1/N}$ as well as a term less than this quantity (*Ky Fan*).

8. Let A be a matrix of the form $a_{ij} \leq 0, i \neq j, a_{jj} + \sum_{i \neq j} a_{ij} = 0, j = 1, 2, \ldots, N.$ If λ is a characteristic root, then either $\lambda = 0$ or $\text{Re}\,(\lambda) < 0$ (*A. Brauer-O. Taussky*).

9. If $u_i \geq 0$, $a_{ij} \geq 0$, $i \neq j$, $\sum_{i=1}^{N} a_{ij} = a_{jj}$, then

$$D_N(u) = \begin{vmatrix} a_{11} + u_1 & -a_{12} & \cdots & -a_{1N} \\ -a_{21} & a_{22} + u_2 & \cdots & -a_{2N} \\ \vdots & & & \\ -a_{N1} & -a_{N2} & \cdots & a_{NN} + u_N \end{vmatrix} \geq 0$$

(*Minkowski*)

10. If $a_{ij} \geq 0$, $i \neq j$, and $b_j \geq \sum_{i \neq j} a_{ij}$, then all cofactors of

$$D_N = \begin{vmatrix} b_1 & -a_{12} & \cdots & -a_{1N} \\ -a_{21} & b_2 & \cdots & -a_{2N} \\ \vdots & & & \\ -a_{N1} & -a_{N2} & \cdots & b_N \end{vmatrix}$$

are non-negative (*W. Ledermann*).

11. A real matrix X for which $x_{ii} > \sum_{j \neq i} |x_{ij}|$ is called an *Hadamard matrix.* Let A and B both be Hadamard matrices. Then $|A + B| \geq |A| + |B|$ (*E. V. Haynsworth*).[1] See also A. Ostrowski.[2]

Hadamard matrices play an important role in computational analysis.

Matrices of this nature, and of more general type, the *dominant* matrices, enter in the study of n-part networks. See

P. Slepian and L. Weinberg, *Synthesis Applications of Paramount and Dominant Matrices*, Hughes Research Laboratories, 1958.

R. S. Burington, On the Equivalence of Quadrics in m-affine n-space and Its Relation to the Equivalence of $2m$-pole Networks, *Trans. Am. Math. Soc.*, vol. 38, pp. 163–176, 1935.

R. S. Burington, R-matrices and Equivalent Networks, *I'*, *J. Math. and Phys.*, vol. 16, pp. 85–103, 1937.

12. If $|a_{ij}| \leq m|a_{ii}|$, $j < i, i = 1, 2, \ldots, N$, $|a_{ij}| \leq M|a_{ii}|, j > i, i = 1, 2, \ldots, N - 1$, then A is nonsingular if $m/(1 + m)^N < M(1 + M)^N$, and $m < M$. If $m = M$, then $m < 1/(N - 1)$ is sufficient (*A. Ostrowski*).[3]

13. An M matrix is defined to be a real matrix A such that $a_{ij} \leq 0$, $i \neq j$, possessing one of the following three equivalent properties:

(*a*) There exist N positive numbers x_j such that

$$\sum_{j=1}^{N} a_{ij}x_j > 0 \qquad i = 1, 2, \ldots, N$$

(*b*) A is nonsingular and all elements of A^{-1} are non-negative

(*c*) All principal minors of A are positive.

M matrices were first introduced by Ostrowski.[4] See also E. V. Haynsworth, Note on Bounds for Certain Determinants, *Duke Math. J.*, vol. 24, pp. 313–320, 1957, and by the same author, Bounds for Determinants with Dominant Main Diagonal, *Duke Math. J.*, vol. 20, pp. 199–209, 1953, where other references may be found.

14. Let A and B be two positive matrices of order N, with $AB \neq BA$, and c a given positive N-dimensional vector. Consider the vector $x_N = Z_N Z_{N-1} \cdots Z_2 Z_1 c$, where each Z_i is either an A or a B. Suppose that the Z_i are to be chosen to maximize the inner product (x_N,b) where b is a fixed vector. Define the function

$$f_N(c) = \max_{\{Z_i\}} (x_N,b)$$

Then

$$f_1(c) = \max ((Ac,b),(Bc,b))$$
$$f_N(c) = \max (f_{N-1}(Ac),f_{N-1}(Bc)) \qquad N = 2, 3, \ldots$$

[1] E. V. Haynsworth, Bounds for Determinants with Dominant Main Diagonal, *Duke Math. J.*, vol. 20, pp. 199–209, 1953.

[2] A. Ostrowski, Note on Bounds for Some Determinants, *Duke Math. J.*, vol. 22, pp. 95–102, 1955.

A. Ostrowski, Über die Determinaten mit überwiegender Haupt-Diagonale, *Comment. Math. Helveticii*, vol. 10, pp. 69–96, 1937–1938.

[3] A. Ostrowski, On Nearly Triangular Matrices, *J. Research Natl. Bur. Standards*, vol. 52, pp. 319–345, 1954.

[4] A. Ostrowski, *Comment. Math. Helveticii*, vol. 30, pp. 175–210, 1956; *ibid.*, vol. 10, pp. 69–96, 1937.

15. Does there exist a scalar λ such that $f_N(c) \sim \lambda^N g(c)$?

16. Consider the corresponding problem for the case where we wish to maximize the Perron root of $Z_N Z_{N-1} \cdots Z_2 Z_1$.

17. Let A be a positive matrix. There exist two positive diagonal matrices, D_1 and D_2, such that $D_1 A D_2$ is doubly stochastic (*Sinkhorn;* see Menon[1]).

18. Let A be a non-negative matrix. When does there exist a positive diagonal matrix such that DAD^{-1} is positive and symmetric? (The question arose in some work by Pimbley in neutron transport theory. See Parter,[2] Parter and Youngs,[3] and Hearon.[4])

19. Let A be a non-negative matrix, and let $P(A)$ be the finite set of non-negative matrices obtained from A by permuting its elements arbitrarily. Let $B \subset P(A)$.

(*a*) What are the maxima and minima of tr (B^2)?

(*b*) What are the maximum and minimum values of the Perron root of B? (*B. Schwarz, Rearrangements of Square Matrices with Nonnegative Elements, University of Wisconsin, MRC Report* #282, 1961.)

20. Let A be a positive matrix. Prove that the Perron root satisfies the inequalities

$$\max\{\max_i a_{ii}, (N-1)\min_{i \neq j} a_{ij} + \min_i a_{ii}\} \leq \lambda(A) \leq (N-1)\max_{i \neq j} a_{ij} + \max_i a_{ii}$$

(*Szender*)

21. Consider the differential equation $x' = Q(t)x$, $0 < t < t_0$, where $q_{ij}(t) \geq 0$, $i \neq j$. If

$$\int_s^t \psi(t_1)\, dt_1 \to \infty$$

as $s \to 0$, where $\psi(t) = \inf_{i \neq j} [q_{ij}q_{ji}]^{1/2}$, then the equation has a positive solution, unique up to a constant factor. (*G. Birkhoff and L. Kotin, Linear Second-order Differential Equations of Positive Type, J. Analyse Math.*, vol. 18, pp. 43–52, 1967.)

22. Consider the matrix product $C = A \cdot B$ where $c_{ij} = -a_{ij}b_{ij}$, $i \neq j$, $c_{ii} = a_{ii}b_{ii}$. Show that if A and B are M matrices (see Exercise 13), then C is an M matrix. For many further deeper results concerning this matrix product introduced by Ky Fan, see Ky Fan, Inequalities for M-matrices, *Indag. Math.*, vol. 26, pp. 602–610, 1944. M matrices play an important role in the study of iterative methods for solving systems of linear equations, and in the numerical solution of elliptic partial differential equations using difference methods; see Ostrowski[5] and Varga.[6]

The Fan product is analogous to the Schur product ($c_{ij} = a_{ij}b_{ij}$), previously introduced for positive definite matrices. In general, there is an amazing parallel between results for positive definite and M matrices. This phenomenon, noted some time

[1] M. V. Menon, Reduction of a Matrix with Positive Elements to a Doubly Stochastic Matrix, *Proc. Am. Math. Soc.*, vol. 18, pp. 244–247, 1967.

[2] S. V. Parter, On the Eigenvalues and Eigenvectors of a Class of Matrices, *SIAM Journal*, to appear.

[3] S. V. Parter and J. W. T. Youngs, The Symmetrization by Diagonal Matrices, *J. Math. Anal. Appl.*, vol. 4, pp. 102–110, 1962.

[4] J. Z. Hearon, The Kinetics of Linear Systems with Special Reference to Periodic Reactions, *Bull. Math. Biophys.*, vol. 15, pp. 121–141, 1953.

[5] A. M. Ostrowski, Iterative Solution of Linear Systems of Functional Equations, *J. Math. Anal. Appl.*, vol. 2, pp. 351–369, 1961.

[6] R. S. Varga, *Matrix Iterative Analysis*, Prentice-Hall, Inc., Englewood Cliffs, N.J., 1962.

ago by O. Taussky (and also by Gantmacher-Krein, and by D. M. Kotelyanskii, *Math. Rev.*, vol. 12, p. 793), is discussed and analyzed in the following papers:

Ky Fan, An Inequality for Subadditive Functions on a Distributive Lattice, with Applications to Determinantal Inequalities, *Lin. Algebra Appl.*, vol. 1, pp. 33–38, 1966.

Ky Fan, Subadditive Functions on a Distributive Lattice, and an Extension of Szasz's Inequality, *J. Math. Anal. Appl.*, vol. 18, pp. 262–268, 1967.

Ky Fan, Some Matrix Inequalities, *Abh. Math. Sem. Hamburg Univ.*, vol. 29, pp. 185–196, 1966.

23. If A and B are M matrices such that $B - A \geq 0$, then $|A + B|^{1/n} \geq |A|^{1/n} + |B|^{1/n}$, $(tA + (1 - t)B)^{-1} \leq (1 - t)B^{-1}$ for any t in $(0,1)$. Many further results are given in Ky Fan.[1]

24. Call a non-negative matrix A irreducible if $A^N > 0$ for some N. Show that every irreducible A has a positive characteristic root $f(A)$ which is at least as large as the absolute value of any other characteristic root. This characteristic root is simple and the characteristic vector can be taken to be positive (*Frobenius*). See Blackwell[2] for a proof using the min-max theorem of game theory.

25. A matrix A is said to be of *monotone kind* if $Ax \geq 0$ implies that $x \geq 0$. Show that A^{-1} exists if A is of monotone kind, and $A^{-1} \geq 0$. See Collatz,[3] and for an extension to rectangular matrices see Mangasarian.[4]

26. Let $A = (a_{ij})$ be a Markov matrix, and let $A^{-1} = (b_{ij})$. Under what conditions do we have $b_{ii} > 1$, $b_{ij} < 0$? See Y. Uekawa, *P-Matrices and Three Forms of the Stolper-Samuelson Criterion*, Institute of Social and Economic Research, Osaka University, 1968.

27. Let $X \geq X^2 \geq 0$. Then the real characteristic values of X must lie between 1 and $(1 - \sqrt{2})/2$ (*R. E. DeMarr*).

28. Let B be a real square matrix such that $b_{ij} \leq 0$ for $i \neq j$ and the leading minors of B are positive. Then B^{-1} is a non-negative matrix. *Hint:* Use the representation $\int_0^\infty e^{-Bt}\,dt = B^{-1}$. See J. Z. Hearon, The Washout Curve in Tracer Kinetics, *Math. Biosci.*, vol. 3, pp. 31–39, 1968.

Bibliography and Discussion

§1. A very large number of interesting and significant analytic problems arise naturally from the consideration of various mathematical models of economic interaction. For an interesting survey of mathematical questions in this field, see

I. N. Herstein, Some Mathematical Methods and Techniques in Economics, *Quart. Appl. Math.*, vol. 11, pp. 249–261, 1953.

[1] Ky Fan, Inequalities for the Sum of Two M-matrices, *Inequalities*, Academic Press Inc., New York, pp. 105–117, 1967.

[2] D. Blackwell, Minimax and Irreducible Matrices, *J. Math. Anal. Appl.*, vol. 3, pp. 37–39, 1961.

[3] L. Collatz, *Functional Analysis and Numerical Mathematics*, Academic Press., Inc., New York, 1966.

[4] O. L. Mangasarian, Characterization of Real Matrices of Monotone Kind, *SIAM Review*, vol. 10, pp. 439–441, 1968.

Two books that the reader interested in these matters may wish to examine are

T. Koopmans, *Activity Analysis of Production and Allocation*, John Wiley & Sons, Inc., New York, 1951.

O. Morgenstern (ed.), *Economic Activity Analysis*, John Wiley & Sons, Inc., New York, 1954.

For an account of the application of the theory of positive matrices to the study of the dynamic stability of a multiple market system, see

K. J. Arrow and M. Nerlove, A Note on Expectations and Stability, TR No. 41, Department of Economics, Stanford University, 1957.

K. J. Arrow and A. Enthoven, A Theorem on Expectations and the Stability of Expectations, *Econometrica*, vol. 24, pp. 288–293, 1956.

where further references to these questions may be found.

To add to the confusion of terminology, let us note that in addition to the *positive definite* matrices of the first part of the book, and the *positive* matrices of Perron, we also have the important *positive real* matrices mentioned on page 111.

§2. For a discussion of mathematical models of growth processes of this type, see the monograph by Harris,

T. E. Harris, The Mathematical Theory of Branching Processes, *Ergeb. Math.*, 1960,

where many further references will be found.

§3. The concept of a positive operator is a basic one in analysis. For an extensive discussion of the theory and application of these operators, see

M. G. Krein and M. A. Rutman, Lineinye operatory ostavliaiushie invariantnym konus v prostranstve Banakha (Linear Operators Leaving Invariant a Cone in a Banach Space), *Uspekhi Matem. Nauk* (n.s.), vol. 3, no. 1 (23), pp. 3–95, 1948. See *Math. Revs.*, vol. 10, pp. 256–257, 1949.

§4. Perron's result is contained in

O. Perron, Zur Theorie der Matrizen, *Math. Ann.*, vol. 64, pp. 248–263, 1907.

The method of proof is quite different from that given here and extremely interesting in itself. We have emphasized the method in the text since it is one that generalizes with ease to handle infinite dimensional operators. This method was developed by H. Bohnenblust in response to a problem encountered by R. Bellman and T. E. Harris in the study of multidimensional branching processes. See the above cited monograph by T. E. Harris.

Perron's result was considerably extended by Frobenius,

G. Frobenius, Über Matrizen aus nicht-negativen Elementen, *Sitzsbere der kgl. preuss. Akad. Wiss.*, pp. 456–477, 1912,

who studied the much more involved case where only the condition $a_{ij} \geq 0$ is imposed. For many years, the pioneer work of Perron was forgotten and the theorem was wrongly attributed to Frobenius.

For an approach to this theorem by way of the theory of linear differential equations, and some extensions, see

P. Hartman and A. Wintner, Linear Differential Equations and Difference Equations with Monotone Solutions, *Am. J. Math.*, vol. 1, pp. 731–743, 1953.

For a discussion of more recent results, see

G. Birkhoff and R. S. Varga, Reactor Criticality and Nonnegative Matrices, *J. Ind. and Appl. Math.*, vol. 6, pp. 354–377, 1958.

As in the case of Markoff matrices, the problem of studying the behavior of the characteristic roots under merely a non-negativity assumption requires a very careful enumeration of cases. The conditions that there exist a unique characteristic root of largest absolute value are now most readily expressible in probabilistic, economic, or topological terms.

For a thorough discussion of these matters with detailed references to other work, see

Y. K. Wong, Some Mathematical Concepts for Linear Economic Models, *Economic Activity Analysis*, O. Morgenstern (ed.), John Wiley & Sons, Inc., New York, 1954.

Y. K. Wong, An Elementary Treatment of an Input-output System, *Naval Research Logistics Quarterly*, vol. 1, pp. 321–326, 1954.

M. A. Woodbury, Properties of Leontieff-type Input-output Matrices, *Economic Activity Analysis*, O. Morgenstern (ed.), John Wiley & Sons, Inc., New York, 1954.

M. A. Woodbury, Characteristic Roots of Input-output Matrices, *Economic Activity Analysis*, O. Morgenstern (ed.), John Wiley & Sons, Inc., New York, 1954.

S. B. Noble, *Measures of the Structure of Some Static Linear Economic Models*, George Washington University, 1958.

The first proof of the Perron theorem using fixed point theorems is contained in

P. Alexandroff and H. Hopf, *Topologie I*, Berlin, 1935.

For a detailed discussion of this technique, with many extensions and applications, see

Ky Fan, Topological Proofs for Certain Theorems on Matrices with Non-negative Elements, *Monats. Math.*, Bd. 62, pp. 219–237, 1958.

§5. For other proofs and discussion, see

A. Brauer, A New Proof of Theorems of Perron and Frobenius on Non-negative Matrices—I: Positive Matrices, *Duke Math. J.*, vol. 24, pp. 367–378, 1957.

G. Debreu and I. N. Herstein, Non-negative Square Matrices, *Econometrica*, vol. 21, pp. 597–607, 1953.

H. Samelson, On the Perron-Frobenius Theorem, *Michigan Math. J.*, vol. 4, pp. 57–59, 1957.

J. L. Ullman, On a Theorem of Frobenius, *Michigan Math. J.*, vol. 1, pp. 189–193, 1952.

A powerful method for studying problems in this area is the "projective metric" of Birkhoff. See

G. Birkhoff, Extension of Jentzsch's Theorem, *Trans. Am. Math. Soc.*, vol. 85, pp. 219–227, 1957.

R. Bellman and T. A. Brown, Projective Metrics in Dynamic Programming, *Bull. Am. Math. Soc.*, vol. 71, pp. 773–775, 1965.

An extensive generalization of the Perron-Frobenius theory is the theory of monotone processes. See

R. T. Rockafellar, Monotone Processes of Convex and Concave Type, *Mem. Am. Math. Soc.*, no. 77, 1967.

where many further references may be found.

Some papers containing applications of the theory of positive matrices are

S. Kaplan and B. Miloud, *On Eigenvalue and Identification Problems in the Theory of Drug Distribution*, University of Southern California Press, Los Angeles, USCEE-310, 1968.

J. Z. Hearon, Theorems on Linear Systems, *Ann. N.Y. Acad. Sci.*, vol. 108, pp. 36–68, 1963.

P. Mandl and E. Sevieta, *The Theory of Nonnegative Matrices in a Dynamic Programming Problem*, to appear.

P. Novosad, Isoperimetric Eigenvalue Problems in Algebras, *Comm. Pure Appl. Math.*, vol. 21, pp. 401–466, 1968.

Another generalization of the concept of a positive matrix is that of totally positive matrices. See the book

S. Karlin, *Total Positivity*, Stanford University Press, 1968.

for an extensive account and an important bibliography, and

H. S. Price, Monotone and Oscillation Matrices Applied to Finite Difference Approximations, *Math. Comp.*, vol. 22, pp. 489–517, 1968.

See also

P. G. Ciarlet, Some Results in the Theory of Nonnegative Matrices, *Lin. Algebra Appl.*, vol. 1, pp. 139–152, 1968.

R. B. Kellogg, Matrices Similar to a Positive Matrix, *Lin. Algebra Appl.*, to appear.

For the use of graph-theoretic methods, see

B. R. Heap and M. S. Lynn, The Structure of Powers of Nonnegative Matrices, I. The Index of Convergence, *SIAM J. Appl. Math.*, vol. 14, pp. 610–639, 1966; II, *ibid.*, pp. 762–777.

R. B. Marimont, System Connectivity and Matrix Properties, *Bull. Math. Biophysics*, vol. 31, pp. 255–273, 1969.

A. Rescigno and G. Segre, On Some Metric Properties of the Systems of Compartments, *Bull. Math. Biophysics*, vol. 27, pp. 315–323, 1965.

J. Z. Hearon, Theorems on Linear Systems, *Ann. N.Y. Acad. Sci.*, vol. 108, pp. 36–68, 1963.

J. Maybee and J. Quirk, Qualitative Problems in Matrix Theory, *SIAM Review*, vol. 11, pp. 30–51, 1969.

W. R. Spillane and N. Hickerson, Optimal Elimination for Sparse Symmetric Systems as a Graph Problem, *Q. Appl. Math.*, vol. 26, pp. 425–432, 1968.

For further references, see

C. R. Putnam, *Commutators on a Hilbert Space—On Bounded Matrices with Non-negative Elements*, Purdue University, PRF-1421, May, 1958, where it is indicated how Perron's theorem is a consequence of the Vivanti-Pringsheim theorem for power series. This result has been independently discovered by Wintner, Bohnenblust-Karlin, Mullikin-Snow, and Bellman.

Various iterative schemes have been given for obtaining the largest characteristic root. See the above reference to A. Brauer, and also

A. Brauer, *A Method for the Computation for the Greatest Root of a Positive Matrix*, University of North Carolina, December, 1957.

R. Bellman, An Iterative Procedure for Obtaining the Perron Root of a Positive Matrix, *Proc. Am. Math. Soc.*, vol. 6, pp. 719–725, 1955.

§7. This technique is used in

R. Bellman and J. M. Danskin, *A Survey of the Mathematical Theory of Time-lag, Retarded Control, and Hereditary Processes*, The RAND Corporation, *Rept.* R-256, March 1, 1954.

to handle the corresponding question posed in Exercise 3 of the Miscellaneous Exercises.

§9. This variational characterization of $\lambda_N(A)$ appears first to have been implied in

L. Collatz, Einschliessungssatz für die characteristischen Zahlen von Matrizen, *Math. Z.*, vol. 48, pp. 221–226, 1946.

The precise result was first given in

H. Wielandt, Unzerlegbare, nicht negative Matrizen, *Math. Z.*, vol. 52, pp. 642–648, 1950.

It has been rediscovered several times, notably by Bohnenblust as mentioned above, and by von Neumann.

§14. See the above cited monograph by T. E. Harris, and the papers by

R. Bellman, R. E. Kalaba, and G. M. Wing, On the Principle of Invariant Imbedding and One-dimensional Neutron Multiplication, *Proc. Nat. Acad. Sci.*, vol. 43, pp. 517–520, 1957.

R. Bellman, R. E. Kalaba, and G. M. Wing, On the Principle of Invariant Imbedding and Neutron Transport Theory; I, One-dimensional Case, *J. Math. Mech.*, vol. 7, pp. 149–162, 1958.

§15. This model is discussed in detail by the methods of dynamic programming in

R. Bellman, *Dynamic Programming*, Princeton University Press, Princeton, N.J., 1957.

§16. See the papers by Wong and Woodbury referred to above in §4.

§19. See the paper by T. Koopmans referred to above and the various studies in the *Annals of Mathematics* series devoted to linear programming.

A definitive work is the book

G. B. Dantzig, *Linear Programming and Its Extensions,* Princeton University Press, Princeton, N.J., 1963.

An excellent exposition of the theory of linear inequalities is contained in

R. A. Good, Systems of Linear Relations, *SIAM Review,* vol. 1, pp. 1–31, 1959.

where many further references may be found. See also

M. H. Pearl, A Matrix Proof of Farkas' Theorem, *Q. Jour. Math.,* vol. 18, pp. 193–197, 1967.

R. Bellman and K. Fan, On Systems of Linear Inequalities in Hermitian Matrix Variables, *Convexity, Proc. Symp. Pure Math.,* vol. 7, pp. 1–11, 1963.

§20. The classic work in the theory of games is

J. von Neumann and O. Morgenstern, *The Theory of Games and Economic Behavior,* Princeton University Press, Princeton, N.J., 1948.

For a very entertaining introduction to the subject, see

J. D. Williams, *The Compleat Strategyst,* McGraw-Hill Book Company, Inc., New York, 1955.

For a proof of the min-max theorem via an induction over dimension, see

I. Kaplansky, A Contribution to the Von Neumann Theory of Games, *Ann. Math.,* 1945.

The proof that the min-max of $\sum_{i,j} a_{ij}x_iy_j$ over $\sum_i x_i = 1$, $\sum_j y_j = 1$, $x_i,y_j \geq 0$ is equal to the max-min is complicated by the fact that the extrema with respect to x_i and y_j separately are at vertices. One can force the extrema to lie inside the region by adding terms such as $\epsilon \log (1/x_i(1 - x_i)) - \epsilon \log (1/y_j(1 - y_j))$, $\epsilon > 0$.

It is not difficult to establish that min-max = max-min for all $\epsilon > 0$ and to justify that equality holds as $\epsilon \to 0$. This unpublished proof by M. Shiffman (1948), is closely related to the techniques of "regularization" discussed in the books of Lattes-Lions and Lavrentiev referred to in Chap. 19.

§21. See Chap. 11 of the book by R. Bellman referred to in §2, and

R. Bellman, A Markovian Decision Process, *J. Math. Mech.*, vol. 6, pp. 679–684, 1957.

P. Mandl and E. Seneta, The Theory of Non-negative Matrices in a Dynamic Programming Problem, to appear.

§22. For a different motivation, see

R. Bellman, On a Quasi-linear Equation, *Canad. J. Math.*, vol. 8, pp. 198–202, 1956.

An interesting extension of positivity, with surprising ramifications, is that of *positive regions* associated with a given matrix A. We say that R is a positive region associated with A if $x, y \epsilon R$ implies that $(x,Ay) \geq 0$.

See

M. Koecher, Die Geodatischen von Positivitätsbereichen, *Math. Ann.*, vol. 135, pp. 192–202, 1958.

M. Koecher, Positivitätsbereiche im R^n, *Am. J. Math.*, vol. 79, pp. 575–596, 1957.

S. Bochner, Bessel Functions and Modular Relations of Higher Type and Hyperbolic Differential Equations, *Comm. Sem. Math. Univ. Lunde, Suppl.*, pp. 12–20, 1952.

S. Bochner, Gamma Factors in Functional Equations, *Proc. Natl. Acad. Sci. U.S.*, vol. 42, pp. 86–89, 1956.

Another interesting and significant extension of the concept of a positive transformation is that of a *variation-diminishing* transformation, a concept which arises naturally in a surprising number of regions of analysis. For various aspects, see

I. J. Schoenberg, On Smoothing Functions and Their Generating Functions, *Bull. Am. Math. Soc.*, vol. 59, pp. 199–230, 1953.

V. Gantmacher and M. Krein, Sur les matrices complètement non-négatives et oscillatoires, *Comp. math.*, vol. 4, pp. 445–476, 1937.

G. Polya and I. J. Schoenberg, Remarks on de la Vallée Poussin Means and Convex Conformal Maps of the Circle, *Pacific J. Math.*, vol. 8, pp. 201–212, 1958.

S. Karlin, Polya-type Distributions, IV, *Ann. Math. Stat.*, vol. 29, pp. 1–21, 1958 (where references to the previous papers of the series may be found).

S. Karlin and J. L. MacGregor, The Differential Equations of Birth-and-death Processes and the Stieltjes Moment Problem, *Trans. Am. Math. Soc.*, vol. 85, pp. 489–546, 1957.

See also

P. Slepian and L. Weinberg, *Synthesis Applications of Paramount and Dominant Matrices*, Hughes Research Laboratories, Culver City, California.

Finally, let us mention a paper devoted to the study of the domain of the characteristic roots of non-negative matrices.

F. I. Karpelevic, Über die charakterische Wurzeln von Matrizen mit nicht-negativen Elementen, *Isvestja Akad. Nauk., S.S.S.R.*, ser. mat. 15, pp. 361–383, 1951. (Russian.)

17

Control Processes

1. Introduction. One of the major mathematical and scientific developments of the last twenty years has been the creation of a number of sophisticated analytic theories devoted to the feasible operation and control of systems. As might be expected, a number of results concerning matrices play basic roles in these theories. In this chapter we wish to indicate both the origin and nature of some of these auxiliary results and describe some of the methods that are used to obtain them. Our objective is to provide an interface which will stimulate the reader to consult texts and original sources for many further developments in a field of continuing interest and analytic difficulty.

Although we will briefly invade the province of the calculus of variations in connection with the minimization of quadratic functionals, we will not invoke any of the consequences of this venerable theory. The exposition here will be simple and self-contained with the emphasis upon matrix aspects. The theory of dynamic programming will be used in connection with deterministic control processes of both discrete and continuous type and very briefly in a discussion of stochastic control processes of discrete type.

2. Maintenance of Unstable Equilibrium. Consider a system S described by the differential equation

$$\frac{dx}{dt} = Ax \qquad x(0) = 0 \tag{1}$$

where A is not a stability matrix. The solution is clearly $x = 0$ for $t \geq 0$. This solution, however, is unstable as far as disturbances of the initial state are concerned. If y is determined by

$$\frac{dy}{dt} = Ay \qquad y(0) = c \tag{2}$$

and if A possesses some characteristic roots with positive real parts, then unless the vector c is of particular form, y will be unbounded as $t \to \infty$. This can lead to undesirable behavior of the system S, and indeed the linear model in (2) becomes meaningless when $\|y\|$ becomes too large.

One way of avoiding this situation is to redesign the system so as to produce a stability matrix. This approach motivates the great and continuing interest in obtaining usable necessary and sufficient conditions that A be a stability matrix. For many purposes, sufficient conditions are all that are required. This explains the importance of the second method of Lyapunov. Unfortunately, redesign of a system, such as one of biological, engineering, or social type, is not always feasible. Instead, additional influences are exerted on the system to counteract the influences which produce instability. Many interesting and difficult questions of analysis arise in this fashion. This is the principal theme of modern control theory.

One question is the following. Choose a control vector y, a "forcing term," so that the solution of

$$\frac{dx}{dt} = Ax + y \qquad x(0) = c \tag{3}$$

follows a prescribed course over time. For example, we may wish to keep $\|x\|$ small. What rules out a trivial solution, for example $y = -Ax$ (a simple "feedback law"), is the fact that we have not taken into account the cost of control. Control requires effort, time, and resources.

Let us restrict ourselves in an investigation of control processes to a simple case where there are two costs involved, one the cost of deviation of the system from equilibrium and the other the cost involved in a choice of y. Let us measure the first cost by the functional

$$J_1(x) = \int_0^T (x,Bx)\,dt \tag{4}$$

$B > 0$, and the second cost by the functional

$$J_2(y) = \int_0^T (y,y)\,dt \tag{5}$$

We further assume that these costs are commensurable so that we can form a total cost by addition of the individual costs

$$J(x,y) = \int_0^T [(x,Bx) + (y,y)]\,dt \tag{6}$$

The problem we pose ourselves is that of choosing the control function y so as to minimize the functional $J(x,y)$, where x and y are related by (2.3). The variable x is called the *state variable*, y is called the *control variable*, and $J(x,y)$ is called the *criterion function*.

3. Discussion. This problem may be regarded as lying in the domain of the calculus of variations, which suggests a number of immediate questions:

(a) Over what class of functions do we search for a minimum?
(b) Does a minimizing function exist in this class?
(c) If so, is it unique?
(d) Can we develop an analytic technique for the determination of a minimizing function? How many different analytic techniques can we develop? (1)
(e) Can we develop feasible computational techniques which will provide numerical answers to numerical questions? How many such methods can we discover?
(f) Why is there a need for a variety of analytic and computational approaches?

As mentioned previously, we will discuss these questions without the formidable apparatus of the general theory of the calculus of variations. Furthermore, we will refer the reader to other sources for detailed discussions of these questions.

Let us introduce the term "control process" to mean a combination of, first, an equation determining the time history of a system involving state and control variables, and second, a criterion function which evaluates any particular history. In more general situations it is necessary to specify constraints of various types, on the behavior of the system, on the control permissible, on the information available, and so on. Here we will consider only a simple situation where these constraints are absent.

4. The Euler Equation for $A = 0$. Let us consider first the case $A = 0$ where the analysis is particularly simple. Since $x' = y$, the quadratic functional (the criterion function) takes the form

$$J(x) = \int_0^T [(x',x') + (x,Bx)]\, dt \tag{1}$$

As candidates for a minimizing function we take functions x whose derivatives x' are in $L^2(0,T)$—we write $x' \in L^2(0,T)$—and which satisfy the initial condition $x(0) = c$. This is certainly a minimum requirement for an admissible function, and it turns out, fortunately, to be sufficient in the sense that there exists a function in this class yielding the absolute minimum of $J(x)$.

Let x be a function in this class minimizing $J(x)$, assumed to exist for the moment. Let z be a function with the property that $z(0) = 0$, $z' \in L^2(0,T)$, and consider the second admissible function $x + \epsilon z$ where ϵ is a scalar parameter. Then

$$\begin{aligned} J(x + \epsilon z) &= \int_0^T [(x' + \epsilon z',\, x' + \epsilon z') + (x + \epsilon z,\, B(x + \epsilon z))]\, dt \\ &= J(x) + \epsilon^2 J(z) + 2\epsilon \int_0^T [(x',z') + (Bx,z)]\, dt \end{aligned} \tag{2}$$

Since ϵ can assume positive and negative values, for $J(x)$ to be a minimum value we must have the variational condition

$$\int_0^T [(x',z') + (Bx,z)]\, dt = 0 \tag{3}$$

We want to use the fact that this holds for all z of the type specified above to derive a condition on x itself. This condition will be the Euler equation. We permit ourselves the luxury of proceeding formally from this point on since it is the equation which is important for our purposes, not a rigorous derivation.

If we knew that x' possessed a derivative, we would obtain from (3) via integration by parts

$$(x',z)\Big]_0^T + \int_0^T [-(x'',z) + (Bx,z)]\, dt = 0 \tag{4}$$

Since $z(0) = 0$, the integrated term reduces to $(x'(T),z(T))$.

Since (4) holds for all z as described above, we suspect that x must satisfy the differential equation

$$x'' - Bx = 0 \tag{5}$$

subject to the original initial condition $x(0) = c$ and a new boundary condition $x'(T) = 0$.

This, the *Euler equation* for the minimizing function, specifies x as a solution of a two-point boundary-value problem.

EXERCISES

1. What is wrong with the following argument: If $\int_0^T (x'' - Bx, z)\, dt = 0$ for all z such that $z' \in L^2(0,T)$, take $z = x'' - Bx$ and thus conclude that $x'' - Bx = 0$.

2. Integrate by parts in (3) to obtain, in the case where B is constant,

$$\int_0^T \left[(x',z') + \left(B \int_t^T x\, dt_1, z' \right) \right] dt = 0$$

and take $z' = x' + B \int_t^T x\, dt_1$, to conclude that $x' + B \int_t^T x\, dt_1 = 0$ (*Haar*). Why is this procedure rigorous, as opposed to that proposed in the preceding exercise?

3. Carry through the proof for B variable.

5. Discussion. We shall now pursue the following route:

(*a*) Demonstrate that the Euler equation possesses a unique solution. (1)

(*b*) Demonstrate that the function thus obtained yields the absolute minimum of $J(x)$.

If we can carry out this program, we are prepared to overlook the dubious origin of the Euler equation. The procedure is easily carried out as a consequence of the positive character of the "cost function" $J(x)$.

6. Two-point Boundary-value Problem. Let X_1, X_2 be the principal matrix solutions of

$$X'' - BX = 0 \tag{1}$$

which is to say

$$\begin{aligned} X_1(0) &= I \qquad X_1'(0) = 0 \\ X_2(0) &= 0 \qquad X_2'(0) = I \end{aligned} \tag{2}$$

Then every solution of

$$x'' - Bx = 0 \qquad x(0) = c \tag{3}$$

has the representation

$$x = X_1(t)c + X_2(t)d \tag{4}$$

where d is an arbitrary vector. To satisfy the terminal condition $x'(T) = 0$, we set

$$x'(T) = 0 = X_1'(T)c + X_2'(T)d \tag{5}$$

Hence, the vector d is uniquely determined if $X_2'(T)$ is nonsingular.

7. Nonsingularity of $X_2'(T)$. Let us now show that the nonsingularity of $X_2'(T)$ follows readily from the positive definite character of $J(x)$. By this term "positive definite" we mean an obvious extension of the concept for the finite-dimensional case, namely that $J(x) > 0$ for all nontrivial x. This is a consequence of the corresponding property for (x,Bx) under the assumption $B > 0$. As we shall see, we actually need $J(x) > 0$ only for nontrivial x satisfying $x(0) = x'(T) = 0$.

Suppose that $X_2'(T)$ is singular. Then there exists a nontrivial vector b such that

$$X_2'(T)b = 0 \tag{1}$$

Consider the function

$$y = X_2(t)b \tag{2}$$

It is a solution of

$$y'' - By = 0 \tag{3}$$

satisfying the conditions $y(0) = 0$, $y'(T) = 0$. Hence, from (3),

$$\int_0^T (y, y'' - By)\, dt = 0 \tag{4}$$

Integrating by parts, we have

$$\begin{aligned} 0 &= (y,y')\Big]_0^T - \int_0^T [(y',y') + (y,By)]\, dt \\ 0 &= -J(y) \end{aligned} \tag{5}$$

This is a contradiction since y is nontrivial. Hence, $X_2'(T)$ is nonsingular.

EXERCISES

1. Extend the foregoing argument to cover the case where

$$J(x) = \int_0^T [(x',Q(t)x') + (x,B(t)x)]\,dt$$

with $Q(t)$, $B(t) \geq 0$, $0 \leq t \leq T$.

2. Extend the foregoing argument to cover the case where there is a terminal condition $x(T) = d$.

3. Consider the terminal control problem of minimizing

$$J(x,\lambda) = \int_0^T [(x',x') + (x,Bx)]\,dt + \lambda(x(T) - d, x(T) - d)$$

$\lambda \geq 0$, where x is subject solely to $x(0) = c$. Discuss the limiting behavior of $x = x(t,\lambda)$ as $\lambda \to \infty$. (λ is called a *Courant parameter.*)

4. Consider the problem of determining the minimum over c and x of

$$J(x,c) = \int_0^T [(x',x') + (x,Bx)]\,dt + \lambda(c - b, c - b)$$

where $x(0) = c$.

5. Consider the problem of determining the minimum over x and y of

$$J(x,y) = \int_0^T [(y,y) + (x,Bx)]\,dt + \lambda \int_0^T (x' - y, x' - y)\,dt$$

where $\lambda \geq 0$ and $x(0) = c$. What happens as $\lambda \to \infty$?

8. Analytic Form of Solution. Now that we know that (6.5) has a unique solution

$$d = -X_2'(T)^{-1}X_1'(T)c \tag{1}$$

let us examine the analytic structure of the minimizing function and the corresponding minimum value of $J(x)$. We have

$$x = X_1(t)c - X_2(t)X_2'(T)^{-1}X_1'(T)c \tag{2}$$

whence the missing value at $t = 0$ is

$$\begin{aligned} x'(0) &= X_1'(0)c - X_2'(0)X_2'(T)^{-1}X_1'(T)c \\ &= -X_2'(T)^{-1}X_1'(T)c \end{aligned} \tag{3}$$

Let us set

$$R(T) = X_2'(T)^{-1}X_1'(T) \tag{4}$$

for $T > 0$. As we shall see, this matrix plays a fundamental role in control theory. Since X_2' and X_1' commute, we see that R is symmetric.

We can easily calculate the minimum value of $J(x)$ in terms of R. We have, using the Euler equation,

$$\begin{aligned} 0 &= \int_0^T (x, x'' - Bx)\,dt \\ &= (x,x')\Big]_0^T - \int_0^T [(x',x') + (x,Bx)]\,dt \end{aligned} \tag{5}$$

Hence, since $x'(T) = 0$,

$$J(x) = -(x(0),x'(0)) = (c,R(T)c) \tag{6}$$

Thus the minimum value is a positive definite quadratic form in c whose matrix is $R(T)$.

9. Alternate Forms and Asymptotic Behavior. Let $B^{1/2} = D$ denote the positive definite square root of B, where B is assumed to be positive definite. Then we have

$$\begin{aligned} X_1(t) &= (e^{Dt} + e^{-Dt})/2 = \cosh Dt \\ X_2(t) &= D^{-1}(e^{Dt} - e^{Dt})/2 = D^{-1} \sinh Dt \end{aligned} \tag{1}$$

Hence,

$$\begin{aligned} X_2'(T) &= \cosh DT \\ X_1'(T) &= D \sinh DT \\ R(T) &= D \tanh DT \end{aligned} \tag{2}$$

If we had wished, we could have used this specific representation to show that $X_2'(T)$ is nonsingular for $T \geq 0$. A disadvantage, however, of this direct approach is that it fails for variable B, and it cannot be readily used to treat the system mentioned in the exercises following Sec. 18.

As $T \to \infty$, we see from (1) that $\tanh DT$ approaches I, the identity matrix. Hence, we conclude that

$$R(T) \sim D \tag{3}$$

as $T \to \infty$.

Returning to the equation

$$x'' - Bx = 0 \qquad x(0) = c \qquad x'(T) = 0 \tag{4}$$

we see that we can write the solution in the form

$$x = (\cosh DT)^{-1}[\cosh D(t - T)]c \tag{5}$$

10. Representation of the Square Root of B. From the foregoing we obtain an interesting representation of the square root of a positive definite matrix B, namely

$$(c,B^{1/2}c) = \lim_{T\to\infty} \min_{x} \left\{ \int_0^T [(x',x') + (x,Bx)]\, dt \right\} \tag{1}$$

where $x(0) = c$.

EXERCISES

1. Using the foregoing representation, show that $B_1 > B_2 > 0$ implies that $B_1^{1/2} > B_2^{1/2}$.

2. Show that $B^{1/2}$ is a concave function of B, that is,

$$(\lambda B_1 + (1-\lambda)B_2)^{1/2} \geq \lambda B_1^{1/2} + (1-\lambda)B_2^{1/2} \qquad \text{for } B_1, B_2 > 0$$

3. Obtain the asymptotic form of $x(t)$ and $x'(t)$ as $T \to \infty$.
4. Asymptotically as $T \to \infty$, what is the relation between $x'(t)$ and $x(t)$?

11. Riccati Differential Equation for R. Using the expression for $R(T)$ given in (9.2), we see that $R(T)$ satisfies the matrix Riccati equation

$$R'(T) = B - R(T)^2 \qquad R(0) = 0 \tag{1}$$

This means that we can solve the linear two-point boundary-value problem in terms of an initial-value problem. Given a specific value of T, say T_0, we integrate (1) numerically to determine $R(T_0)$. Having obtained $R(T_0)$, we solve the initial-value problem

$$\begin{gathered} x'' - Bx = 0 \\ x(0) = c \qquad x'(0) = -R(T_0)c \end{gathered} \tag{2}$$

again computationally, to determine $x(t)$ for $0 \leq t \leq t_0$.

An advantage of this approach is that we avoid the numerical inversion of the $N \times N$ matrix $X_2'(T_0)$. In return, we must solve a system of N_2 nonlinear differential equations subject to initial conditions.

EXERCISE

1. Why can one anticipate difficulty in the numerical inversion of $X_2'(T_0)$ if T_0 is large or if B possesses characteristic roots differing greatly in magnitude?

12. Instability of Euler Equation. A drawback to the use of (11.2) for computational purposes is the fact that the solution of (11.2) is *unstable* as far as numerical errors incurred in the determination of $R(T_0)c$ and numerical integration are concerned. To see what we mean by this, it is sufficient to consider the scalar version,

$$\begin{gathered} u'' - b^2u = 0 \\ u(0) = c \qquad u'(0) = -b(\tanh bT_0)c \end{gathered} \tag{1}$$

with the solution

$$u = \frac{[\cosh b(t - T_0)]c}{\cosh bT_0} \tag{2}$$

It is easy to see that for fixed t as $T_0 \to \infty$, we have

$$u \sim e^{-bt}c \tag{3}$$

Consider the function

$$v = u + \epsilon e^{bt} \tag{4}$$

satisfying the same differential equation as u, with the initial conditions

$$v(0) = c + \epsilon \qquad v'(0) = u'(0) + b\epsilon \tag{5}$$

If the total time interval is small, the term ϵe^{bt} will cause no trouble. If, however, $bT \gg 1$, routine numerical integration of (11.2) will produce seriously erroneous results. The reason for this is that numerical integration using a digital computer involves the use of an approximate-difference equation, errors in evaluation of functions and round-off errors. Hence, we can consider numerical integration as roughly equivalent to a solution of the differential equation subject to slightly different initial conditions.

There are a number of ways of overcoming or circumventing this instability. One way is to use an entirely different approach to the study of control processes, an approach which extends readily into many areas where the calculus of variations is an inappropriate tool. This we will do as soon as we put the finishing touches on the preceding treatment.

It is interesting to regard numerical solution as a control process in which the aim is to minimize some measure of the final error. A number of complex problems arise in this fashion.

13. Proof that $J(x)$ Attains an Absolute Minimum. Having established the fact that the Euler equation possesses a unique solution x, it is a simple matter to show that this function furnishes the absolute minimum of the quadratic functional J. Let y be another admissible function. Then

$$\begin{aligned} J(y) = J(x + (y - x)) = J(x) + J(y - x) \\ + 2 \int_0^T [(x', y' - x') + (Bx, y - x)]\, dt \end{aligned} \tag{1}$$

as in Sec. 4, where now $\epsilon = 1$, $z = y - x$. Integrating by parts,

$$\begin{aligned} \int_0^T (x', y' - x')\, dt &= (x', y - x)\Big]_0^T - \int_0^T (x'', y - x)\, dt \\ &= -\int_0^T (x'', y - x)\, dt \end{aligned} \tag{2}$$

since $x'(T) = 0$, $x(0) = y(0) = c$. Hence, the third term on the right-hand side of (1) has the form

$$2 \int_0^T [-(x'', y - x) + (Bx, y - x)]\, dt = 0 \tag{3}$$

since x satisfies the Euler equation. Thus,

$$J(y) = J(x) + J(y - x) > J(x) \tag{4}$$

unless $y \equiv x$.

14. Dynamic Programming. In Chap. 9 we pointed out some uses of the theory of dynamic programming in connection with some finite-dimensional minimization and maximization problems. The basic idea was that these could be interpreted as multistage decision processes and thus be treated by means of functional equation techniques. Let us now show that we can equally regard the variational problem posed above as a multistage decision process, but now one of continuous type.

To do this we abstract and extend the engineering idea of "feedback control." We consider the control of a system as the task of determining what to do at any time in terms of the current state of the system. Thus, returning to the problem of minimizing

$$J(x) = \int_0^T [(x',x') + (x,Bx)]\,dt \tag{1}$$

we ask ourselves how to choose x' in terms of the current state x and the time remaining. What will guide us in this choice is the necessity of balancing the effects of an immediate decision against the effects of the continuation of the process.

15. Dynamic Programming Formalism. Let us parallel the procedure we followed in using the calculus of variations. First we shall derive some important results using an intuitive formal procedure, then we shall indicate how to obtain a valid proof of these results. See Fig. 1.

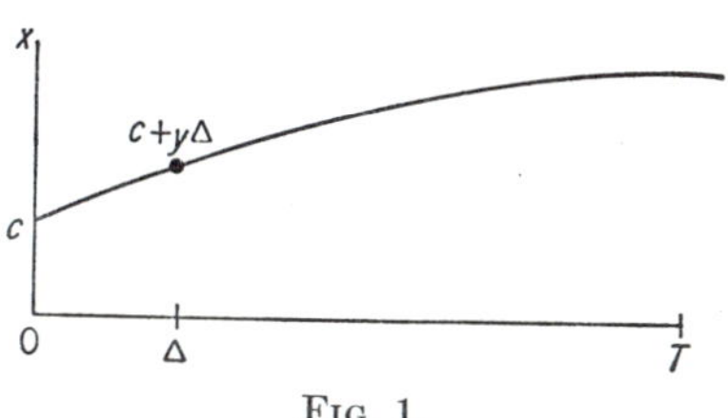

Fig. 1.

To begin with, we imbed the original variational process within a family of processes of similar nature. This is the essential idea of the "feedback control" concept, since we must develop a control policy applicable to any admissible state. Write

$$f(c,T) = \min_x J(x) \tag{1}$$

where x is subject to the constraint $x(0) = c$. This function is defined for $T \geq 0$ and all c. The essential point here is that c and T are the fundamental variables rather than auxiliary parameters.

Referring to Fig. 1, we wish to determine $f(c,T)$ by using the fact that the control process can be considered to consist of a decision over

$[0,\Delta]$ together with a control process of similar type over the remaining interval $[\Delta,T]$. This is a consequence of the additivity of the integral

$$\int_0^T = \int_0^\Delta + \int_\Delta^T \tag{2}$$

If Δ is small, and if we assume sufficient smoothness in the minimizing x (which we already know it possesses from what has preceded), we can regard a choice of $x(t)$ over $[0,\Delta]$ as equivalent to a choice of $x'(0) = y$, an initial slope. Similarly, we may write

$$\int_0^\Delta [(x',x') + (x,Bx)]\,dt = [(y,y) + (c,Bc)]\Delta + O(\Delta^2) \tag{3}$$

Now let us consider the effect of a choice of y over $[0,\Delta]$. The initial state c is changed into $c + y\Delta$ (to terms in $O(\Delta^2)$), and the time remaining in the process is diminished to $T - \Delta$. Hence, for a minimizing choice of x over $[\Delta,T]$, we must have

$$\int_\Delta^T = f(c + y\Delta,\ T - \Delta) + O(\Delta^2) \tag{4}$$

Combining these results, we see that we have

$$f(c,T) = [(y,y) + (c,Bc)]\Delta + f(c + y\Delta,\ T - \Delta) + O(\Delta^2) \tag{5}$$

It remains to choose y. If $f(c,T)$ is to be the minimum value of $J(x)$, y must be chosen to minimize the right side of (5). Thus, we obtain the relation

$$f(c,T) = \min_y\, [[(y,y) + (c,Bc)]\Delta + f(c + y\Delta,\ T - \Delta)] + O(\Delta^2) \tag{6}$$

keeping in mind that we are unreservedly using our formal license.

Finally, we write

$$f(c + y\Delta,\ T - \Delta) = f(c,T) + ((y,\operatorname{grad} f) - f_T)\Delta + O(\Delta^2) \tag{7}$$

where

$$\operatorname{grad} f = \begin{bmatrix} \dfrac{\partial f}{\partial y_1} \\ \dfrac{\partial f}{\partial y_2} \\ \cdot \\ \cdot \\ \cdot \\ \dfrac{\partial f}{\partial y_N} \end{bmatrix} \tag{8}$$

Letting $\Delta - 0$, we are left with the nonlinear partial differential equation

$$f_T = \min_y [(y,y) + (c,Bc) + (y,\text{grad } f)] \tag{9}$$

This is the dynamic programming analogue of the Euler equation as far as the determination of the minimizing function is concerned.

16. Riccati Differential Equation. The minimum with respect to y in (15.9) is readily determined,

$$y = -(\text{grad } f)/2 \tag{1}$$

Substituting this into (15.9), we obtain a nonlinear partial differential equation of conventional form,

$$f_T = (c,Bc) - (\text{grad } f, \text{grad } f)/4 \tag{2}$$

Referring to the definition of $f(c,T)$, we see that the appropriate initial condition is

$$f(c,0) = 0 \tag{3}$$

We can readily dispose of the partial differential equation by observing that thc quadratic nature of the functional (or equivalently the linear nature of the Euler equation) implies that

$$f(c,T) = (c,R(T)c) \tag{4}$$

for some matrix R. Using (4), we have

$$\text{grad } f = 2R(T)c \tag{5}$$

Hence, (2) leads to the relation

$$(c,R'(T)c) = (c,Bc) - (R(T)c,R(T)c) \tag{6}$$

Since this holds for all c, we have

$$R'(T) = B - R^2(T) \qquad R(0) = 0 \tag{7}$$

precisely the result previously obtained in Sec. 11.

The "control law" is given by (1), namely

$$y - -R(T)c \tag{8}$$

a linear relation. We have deliberately used the same notation as in (8.4), since, as (8.6) shows, we are talking about the same matrix.

EXERCISE

1. Show that R^{-1} also satisfies a Riccati equation, and generally that $(CR + D)^{-1}(AR + B)$ satisfies a Riccati equation if R does.

17. Discussion. The path we followed was parallel to that pursued in our treatment involving the calculus of variations. We first obtained a fundamental relation using a plausible intuitive procedure. Once the result was obtained, we provided a valid basis employing entirely different ideas. In this case, we can fall back upon the calculus of variations where the desired results were easily derived. This procedure is typical of mathematical analysis related to underlying physical processes. Scientific insight and experience contribute greatly to mathematical intuition.

It is also not difficult to follow a direct path based upon functional analysis.

18. More General Control Processes. Let us proceed to the consideration of a slightly more general control process where we desire to minimize

$$J(x,y) = \int_0^T [(x,Bx) + (y,y)]\, dt \tag{1}$$

subject to x and y related by the differential equation

$$x' = Ax + y \qquad x(0) = c \tag{2}$$

Once again we suppose that $B > 0$; A, however, is merely restricted to being real.

Writing

$$f(c,T) = \min_y J(x,y) \tag{3}$$

the formal procedure of Sec. 15 leads to the relation

$$f_T = \min_z [(c,Bc) + (z,z) + (Ac + z, \operatorname{grad} f)] \tag{4}$$

where we have set $z = y(0)$. The minimum is attained at

$$z = -(\operatorname{grad} f)/2 \tag{5}$$

and using this value (4) becomes

$$f_T = (c,Bc) + (Ac,\operatorname{grad} f) - (\operatorname{grad} f,\operatorname{grad} f)/4 \tag{6}$$

Setting once again

$$f(c,T) = (c,R(T)c) \tag{7}$$

we obtain the equation

$$(c,R''(T)c) = (c,Bc) + (Ac,2R(T)c) - (R(T)c,R(T)c) \tag{8}$$

Symmetrizing, we have

$$(Ac,2R(T)c) = (\{AR(T) + R(T)A^*\}c,c) \tag{9}$$

where A^* denotes the transpose of A. Hence, (8) yields

$$R' = B + AR + RA^* - R^2 \qquad R(0) = 0 \tag{10}$$

The optimal policy is once again given by

$$z = -R(T)c \tag{11}$$

We use the notation A^* here in place of our usual A' in order not to confuse with the derivative notation.

EXERCISES

1. Obtain the corresponding relations for the control process defined by

$$J(x,y) = \int_0^T [(x,Bx) + (y,Cy)]\,dt \qquad x' = Ax + Dy \qquad x(0) = c$$

2. Consider the control process

$$J(x,y) = \int_0^T [(x,x) + (y,y)]\,dt \qquad x' = Ax + y \qquad x(0) = c$$

Show that the Euler equation satisfied by a desired minimizing y is

$$y' = -A^*y + x \qquad y(T) = 0$$

3. Following the procedures given in Secs. 6–8, show that the system of differential equations possesses a unique solution.

4. Show that this solution provides the absolute minimum of $J(x,y)$.

5. Determine the form of $R(T)$ in the analytic representation of

$$\min J(x,y) = (c,R(T)c)$$

and show that it satisfies the Riccati equation corresponding to (10).

6. Carry through the corresponding analysis for the case where

$$J(x,y) = \int_0^T [(x,Bx) + (y,y)]\,dt \qquad B > 0$$

7. Consider the case where

$$J(x,y) = \int_0^T [(x,Bx) + (y,Cy)]\,dt \qquad x' = Ax + Dy$$

first for the case where D is nonsingular and then where it is singular, under the assumption that $B > 0$ and a suitable assumption concerning C.

8. Show how the nonlinear differential equation in (10) may be solved in terms of the solutions of linear differential equations.

19. Discrete Deterministic Control Processes. A corresponding theory exists for discrete control processes. Let

$$\begin{aligned} x_{n+1} &= Ax_n + y_n \qquad x_0 = c \\ J(\{x_n,y_n\}) &= \sum_{n=0}^{N} [(x_n,x_n) + (y_n,y_n)] \end{aligned} \tag{1}$$

and suppose that we wish to minimize J with respect to the y_n. Write

$$f_N(c) = \min_{\{y_n\}} J(\{x_n,y_n\}) \tag{2}$$

Then, arguing as before, we have

$$f_N(c) = \min_y [(c,c) + (y,y) + f_{N-1}(Ac + y)] \tag{3}$$

$N \geq 1$, with $f_0(c) = (c,c)$.

It is clear once again, or readily demonstrated inductively, that

$$f_N(c) = (c,R_Nc) \tag{4}$$

where R_N is independent of c. Using the representation of (4) in (3), we have

$$(c,R_Nc) = \min_y [(c,c) + (y,y) + (Ac + y, R_{N-1}(Ac + y))] \tag{5}$$

The minimum with respect to y can be obtained readily. The variational equation is

$$y + R_{N-1}(Ac + y) = 0 \tag{6}$$

whence

$$y = -(I + R_{N-1})^{-1}R_{N-1}Ac \tag{7}$$

Substituting this result in (3), we obtain, after some simplification, the recurrence relation

$$R_N = (I + A^*A) - A^*(I + R_{N-1})^{-1}A \tag{8}$$

$N \geq 1$, with $R_0 = I$.

EXERCISES

1. Show that $I \leq R_{N-1} < R_N \leq (I + A^*A)$ and thus that as $N \to \infty$, R_N converges to a matrix R_∞ satisfying

$$R_\infty = (I + A^*A) - A^*(I + R_\infty)^{-1}A$$

2. Obtain the analogue of the Euler equation for the foregoing finite-dimensional variational problem. Prove that the two-point boundary problem for these linear difference equations possesses a unique solution and that this solution yields the absolute minimum of J.

3. Carry out the same program for the case where

$$J(\{x_n,y_n\}) = \sum_{n=0}^{N} [(x_n,Bx_n) + (y_n,y_n)]$$

4. Consider the discrete process $x_{n+1} = (I + A\Delta)x_n + y_n\Delta$, $x_0 = c$,

$$J(\{x_n,y_n\}) = \sum_{n=0}^{N} [(x_n,x_n) + (y_n,y_n)]\Delta$$

Consider the limiting behavior of min J and the minimizing vectors as $\Delta \to 0$ with $N\Delta = T$.

20. Discrete Stochastic Control Processes. Let us now consider the case where we have a stochastic control process of discrete type,

$$x_{n+1} = Ax_n + y_n + r_n \qquad x_0 = c$$
$$J(\{x_n,y_n\}) = \sum_{n=0}^{N} [(x_n,x_n) + (y_n,y_n)] \tag{1}$$

where the r_n are random variables. The objective is to choose the y_n so as to minimize the expected value of $J(\{x_n,y_n\})$ subject to the following provisions:

(a) The r_n are independent random variables with specified probability distribution $dG(r)$.
(b) The vector y_n is chosen at each stage after observation of the state x_n and before a determination of r_n. (2)

Writing

$$f_N(c) = \min_{\{y_n\}} \exp_{\{r_n\}} J(\{x_n,y_n\}) \tag{3}$$

and arguing as before, we obtain the functional equation

$$f_N(c) = \min_y \left[(c,c) + (y,y) + \int_{-\infty}^{\infty} f_{N-1}(Ac + y + r)\, dG(r) \right] \tag{4}$$

$N \geq 1$, with $f_0(c) = (c,c)$. We can then proceed in the same fashion as above to use the quadratic nature of $f_N(c)$ to obtain more specific results.

EXERCISES

1. Show that $f_N(c) = (c,R_Nc) + (b_N,c) + a_N$ and determine recurrence relations for the R_N, b_N, and a_N.

2. In control processes of this particular nature, does it make any difference whether we actually observe the system at each stage or use the expected state without observation?

3. How would one handle the case where the r_i are dependent in the sense that $dG(r_n) = dG(r_n,r_{n-1})$, that is, the probability distribution for r_n depends upon the value of r_{n-1}?

21. Potential Equation. The potential equation

$$u_{xx} + u_{yy} = 0 \qquad (x,y) \in R \tag{1}$$

with u specified on the boundary of a region R, leads to many interesting investigations involving matrices when we attempt to obtain numerical results using a discretized version of (1). Let x and y run through some discrete grid and replace (1) by

$$\frac{u(x+\Delta, y) + u(x-\Delta, y) - 2u(x,y)}{\Delta^2} + \frac{u(x, y+\Delta) + u(x, y-\Delta) - 2u(x,y)}{\Delta^2} = 0 \quad (2)$$

This may be written

$$u(x,y) = \frac{u(x+\Delta, y) + u(x-\Delta, y) + u(x, y+\Delta) + u(x, y-\Delta)}{4} \quad (3)$$

expressing the fact that the value at (x,y) is the mean of the values at the four "nearest neighbors" of (x,y).

Use of (3) inside the region together with the specified values of $u(x,y)$ on the boundary leads to a system of linear algebraic equations for the quantities $u(x,y)$. If an accurate determination is desired, which means small Δ, the system becomes one of large dimension. A number of ingenious methods have been developed to solve systems of this nature, taking advantage of the special structure. We wish to sketch a different approach, using dynamic programming and taking advantage of the fact that (1) is the Euler equation of

$$J(u) = \int_R (u_x^2 + u_y^2)\,dx\,dy \quad (4)$$

22. Discretized Criterion Function. As before, we replace derivatives by differences,

$$u_x \cong \frac{u(x+\Delta, y) - u(x,y)}{\Delta}$$

$$u_y \cong \frac{u(x, y+\Delta) - u(x,y)}{\Delta}$$

For simplicity, suppose that R is rectangular, with $a = N\Delta$, $b = M\Delta$. See Fig. 2. Let $c_1, c_2, \ldots, c_{M-1}$ be the assigned values along $x = 0$

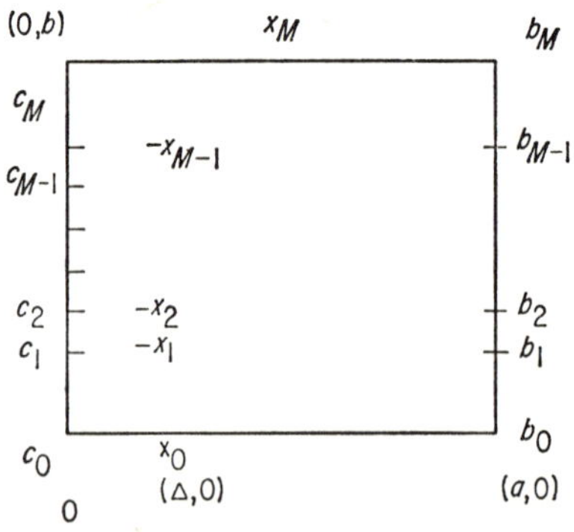

Fig. 2.

and let $x_1, x_2, \ldots, x_{M-1}$ be the values of u at the values (Δ,Δ), $(\Delta,2\Delta), \ldots, (\Delta,(M-1)\Delta)$.

In place of writing a single variational equation for the unknown values at the interior grid points, as in (21.3), we think of this as a multistage decision process in which we must determine first the $M-1$ values along $x = \Delta$, then the $(M-1)$ values along $x = 2\Delta$, and so on. Write

$$f_M(c) = \min \Sigma\Sigma[\{u(x+\Delta, y) - u(x,y)\}^2 + \{u(x, y+\Delta) - u(x,y)\}^2] \quad (2)$$

where c is the vector whose components are the c_i,

$$f_1(c) = (c_0 - b_0)^2 + (c_1 - b_1)^2 + \cdots + (c_M - b_M)^2 + (c_1 - c_0)^2 + (c_2 - c_1)^2 + \cdots + (c_M - c_{M-1})^2 \quad (3)$$

Then

$$f_M(c) = \min_x [(c_0 - x_0)^2 + (c_1 - x_1)^2 + \cdots + (c_M - x_M)^2 + (c_0 - c_1)^2 + (c_1 - c_2)^2 + \cdots + (c_{M-1} - c_M)^2 + f_{M-1}(x)] \quad (4)$$

EXERCISES

1. Use the quadratic nature of $f_M(c)$ to obtain appropriate recurrence relations.

2. What happens if the rectangle is replaced by a region of irregular shape? Consider, for example, a triangular wedge.

3. How would one go about obtaining a solution for a region having the shape in Fig. 3? See Angel.[1]

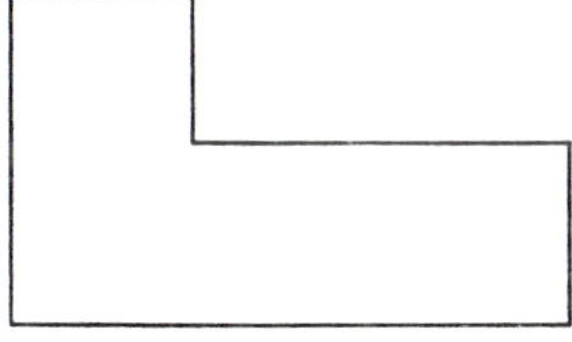

Fig. 3.

[1] E. S. Angel, *Dynamic Programming and Partial Differential Equations*, doctoral dissertation, Electrical Engineering, University of Southern California, Los Angeles, 1968.

E. S. Angel, Dynamic Programming and Linear Partial Differential Equations, *J. Math. Anal. Appl.*, vol. 23, pp. 639ff, 1968.

E. S. Angel, Discrete Invariant Imbedding and Elliptic Boundary-value Problems over Irregular Regions, *J. Math. Anal. Appl.*, vol. 23, pp. 471ff, 1968.

E. S. Angel, A Building Block Technique for Elliptic Boundary-value Problems over Irregular Regions, *J. Math. Anal. Appl.*, to appear.

E. S. Angel, Noniterative Solutions of Nonlinear Elliptic Equations, *Comm. Assoc. Comput. Machinery*, to appear.

E. S. Angel, Invariant Imbedding and Three-dimensional Potential Problems, *J. Comp. Phys.*, to appear.

MISCELLANEOUS EXERCISES

1. Let A be a positive definite matrix and write $A^{1/2} = D + B$, where D is a diagonal matrix whose elements are the square roots of the diagonal elements of A. Consider the recurrence relation $DB_k + B_kD = A - D^2 - B_{k-1}^2$, $B_0 = 0$. Does B_k converge as $k \to \infty$? (*P. Pulag, An Iterative Method for the Determination of the Square Root of a Positive Definite Matrix, Z. angew. Math. Mech., vol.* 46, *pp.* 151*ff*, 1966.)

2. Let $A, B > 0$. Then

$$\left(\frac{A + B}{2}\right) \leq \left(\frac{A^2 + B^2}{2}\right)^{1/2} \leq \cdots \leq \left(\frac{A^{2N} + B^{2N}}{2}\right)^{1/2N}$$

Hence the limit

$$M(A,B) = \lim_{N \to \infty} \left(\frac{A^{2N} + B^{2N}}{2}\right)^{1/2N}$$

exists as $N \to \infty$.

3. Does $\lim_{N \to \infty} \left(\frac{A^N + B^N}{2}\right)^{1/N}$ exist if $A, B > 0$?

4. Every real skew-symmetric matrix is a real, normal square root of a nonpositive definite real symmetric matrix whose nonzero characteristic values have even multiplicities. (*R. F. Rinehart, Skew Matrices as Square Roots, Am. Math. Monthly, vol.* 67, *pp.* 157–161, 1960.)

5. Consider the solution of $R' = A - R^2$, $R(0) = 0$, and the approximation scheme $R'_{n+1} = A - R_n(\infty)R_{n+1}$, $R_{n+1}(0) = 0$, with $R_0(\infty) = B$, $B^2 < A$. Does $R_{n+1}(\infty)$ converge to $A^{1/2}$ with suitable choice of B?

Bibliography and Discussion

§1. For discussions of the modern theory of control processes, see

R. Bellman, *Introduction to the Mathematical Theory of Control Processes,* Academic Press Inc., New York, 1967.

R. Bellman, *Adaptive Control Processes: A Guided Tour,* Princeton University Press, Princeton, New Jersey, 1961.

R. Bellman and R. Kalaba, *Dynamic Programming and Modern Control Theory,* Academic Press Inc., New York, 1965.

M. R. Hestenes, *Calculus of Variations and Optimal Control Theory,* John Wiley & Sons, Inc., New York, 1966.

L. S. Pontryagin, V. G. Boltyanskii, R. V. Gamkrelidze, and E. F. Mishchenko, *The Mathematical Theory of Optimal Processes,* Interscience Publishers, New York, 1962.

§2. For discussions of stability theory and further references, see

R. Bellman, *Stability Theory of Differential Equations*, Dover Publications, New York, 1969.

R. Bellman, *Introduction to Methods of Nonlinear Analysis*, Academic Press, Inc., New York, 1970.

§§3 to 9. We follow the route outlined in the first book cited above. See also

R. Bellman, On Analogues of Poincaré-Lyapunov Theory for Multipoint Boundary-value Problems, I, *J. Math. Anal. Appl.*, vol. 14, pp. 522–526, 1966.

R. Bellman, On Analogues of Poincaré-Lyapunov Theory for Multipoint Boundary-value Problems: Correction, *J. Math. Anal. Appl.*, to appear.

§10. See

R. Bellman, Some Inequalities for the Square Root of a Positive Definite Matrix, *Lin. Algebra Appl.*, vol. 1, pp. 321–324, 1968.

For an interesting way in which $A^{1/2}$ enters, see

E. P. Wigner and M. Y. Yanase, Information Content of Distributions, *Proc. Nat. Acad. Sci. U.S.*, vol. 49, pp. 910–918, 1963.

§12. The equation for $R(T)$, however (11.1), is stable. See

R. Bellman, Functional Equations in the Theory of Dynamic Programming, XIII: Stability Considerations, *J. Math. Anal. Appl.*, vol. 12, pp. 537–450, 1965.

As G. Dahlquist has pointed out, it is essential to distinguish between inherent instability and instability associated with a specific method. Thus, for example, the equation $x'' - Ax = 0$, $x(0) = c$, $x'(T) = 0$ is *stable* as far as small changes in c are concerned, but the solution calculated according to some particular technique may be very sensitive to computational error.

§14. In addition to the first three books cited above in Sec. 1, see the references at the end of Chap. 9.

§15. Geometrically, we can think of dynamic programming as dual to the calculus of variations. A curve is visualized in this theory as an envelope of tangents rather than a locus of points as it is in the calculus of variations. In the derivation of (15.5) we are using the *principle of optimality;* see any of the three books cited above.

§16. For detailed discussions of the ways in which matrix Riccati equations enter into the solution of two-point boundary-value problems, see

W. T. Reid, Solution of a Riccati Matrix Differential Equation as Functions of Initial Values, *J. Math. Mech.*, vol. 8, pp. 221–230, 1959.

W. T. Reid, Properties of Solutions of a Riccati Matrix Differential Equation, *J. Math. Mech.*, vol. 9, pp. 749–770, 1960.

W. T. Reid, Oscillation Criteria for Self-adjoint Differential Systems, *Trans. Am. Math. Soc.*, vol. 101, pp. 91–106, 1961.

W. T. Reid, Principal Solutions of Nonoscillatory Linear Differential Systems, *J. Math. Anal. Appl.*, vol. 9, pp. 397–423, 1964.

W. T. Reid, Riccati Matrix Differential Equations and Nonoscillation Criteria for Associated Linear Differential Systems, *Pacific J. Math.*, vol. 13, pp. 665–685, 1963.

W. T. Reid, A Class of Two-point Boundary Problems, *Illinois J. Math.*, vol. 2, pp. 434–453, 1958.

W. T. Reid, Principal Solutions of Nonoscillatory Self-adjoint Linear Differential Systems, *Pacific J. Math.*, vol. 8, pp. 147–169, 1958.

W. T. Reid, A Class of Monotone Riccati Matrix Differential Operators, *Duke Math. J.*, vol. 32, pp. 689–696, 1965.

See also

R. Bellman, Upper and Lower Bounds for the Solution of the Matrix Riccati Equation, *J. Math. Anal. Appl.*, vol. 17, pp. 373–379, 1967.

M. Aoki, Note on Aggregation and Bounds for the Solution of the Matrix Riccati Equation, *J. Math. Anal. Appl.*, vol. 21, pp. 377–383, 1968.

The first appearance of the Riccati equation in control theory and dynamic programming is in

R. Bellman, On a Class of Variational Problems, *Q. Appl. Math.*, vol. 14, pp. 353–359, 1957.

§§19 to **20.** See the first book cited above. For a detailed discussion of stochastic control processes of discrete type and dynamic programming, see

S. Dreyfus, *Dynamic Programming and the Calculus of Variations,* Academic Press Inc., New York, 1965.

§21. See, for example,

R. S. Varga, *Matrix Iterative Analysis,* Prentice-Hall, Inc., Englewood Cliffs, New Jersey, 1962.

C. G. Broyden and F. Ford, An Algorithm for the Solution of Certain Kinds of Linear Equations, *Numer. Math.,* vol. 8, pp. 307–323, 1966.

18

Invariant Imbedding

1. Introduction. In the previous chapter we considered some two-point boundary-value problems for linear differential systems arising in the theory of control processes. Further, we indicated how the theory of dynamic programming could be used to obtain alternate analytic formulation in terms of a matrix Riccati differential equation. In this chapter, we wish to employ the theory of invariant imbedding—an outgrowth of dynamic programming—to obtain corresponding results for general two-point boundary-value problems of the type

$$\begin{aligned} -x' &= Ax + Dy \qquad & x(a) &= c \\ y' &= Bx + Cy \qquad & y(0) &= d \end{aligned} \tag{1}$$

Why we employ this particular form will become clear as we proceed. We suppose that the matrices A, B, C, D depend on the independent variable t, $0 \le t \le a$. As is to be expected, more can be said when these matrices are constant.

There is little difficulty in establishing existence and uniqueness of the solution of (1) when a is a small quantity. Furthermore, the question of the determination of a_c (the critical value of a), for which existence and uniqueness cease to hold, can be reduced to a Sturm-Liouville problem.

A basic challenge, however, is that of delineating significant classes of equations of this nature for which a unique solution exists for all $a > 0$. We shall approach this question by associating the foregoing system of equations with an idealized physical process, in this case a transport process.

Simple physical principles will then lead us to important classes of matrices $[A,B,C,D]$. Furthermore, the same principles will furnish the clue for a powerful analytic approach.

2. Terminal Condition as a Function of a. Let us examine the system in (1.1) with a view toward determining the missing initial condition $y(a)$. Let (X_1,Y_1) and (X_2,Y_2) be the principal solutions of

$$\begin{aligned} -X' &= AX + DY \\ Y' &= BX + CY \end{aligned} \tag{1}$$

that is to say

$$\begin{aligned} X_1(0) &= I \qquad & Y_1(0) &= 0 \\ X_2(0) &= 0 \qquad & Y_2(0) &= I \end{aligned} \tag{2}$$

The solution of (1.1) may then be written

$$\begin{aligned} x &= X_1 b + X_2 d \\ y &= Y_1 b + Y_2 d \end{aligned} \tag{3}$$

where the vector b is to be determined by the relation

$$c = X_1(a)b + X_2(a)d \tag{4}$$

Since $X_1(0) = I$, it is clear that $X_1(a)$ is nonsingular for a small. We shall operate under this assumption for the moment.

Since the equation in (1.1) is linear, it is permissible to consider the cases $c = 0$ and $d = 0$ separately. Let us then take $d = 0$. The case where $d \neq 0$, $c = 0$ can be treated by replacing t by $a - t$. From (4), we have

$$b = X_1(a)^{-1}c \tag{5}$$

whence we have the representations

$$\begin{aligned} x &= X_1 X_1(a)^{-1}c \\ y &= Y_1 X_1(a)^{-1}c \end{aligned} \tag{6}$$

The missing value of y at $t = a$ is then

$$y(a) = Y_1(a)X_1(a)^{-1}c \tag{7}$$

and at $t = 0$ we have

$$x(0) = X_1(a)^{-1}c \tag{8}$$

Let us call

$$R(a) = Y_1(a)X_1(a)^{-1} \tag{9}$$

the *reflection matrix* and

$$T(a) = X_1(a)^{-1} \tag{10}$$

the *transmission matrix*. We are anticipating the identification of (1.1) with certain equations describing a transport process.

Consider $R(a)$ as a function of a. We have

$$\begin{aligned} R'(a) &= Y_1'(a)X_1(a)^{-1} - Y_1(a)X_1(a)^{-1}X_1'(a)X_1(a)^{-1} \\ &= [B(a)X_1(a) + C(a)Y_1(a)]X_1(a)^{-1} \\ &\quad + Y_1(a)X_1(a)^{-1}[A(a)X_1(a) + D(a)Y_1(a)]X_1(a)^{-1} \\ &= B(a) + C(a)R(a) + R(a)A(a) + R(a)D(a)R(a) \end{aligned} \tag{11}$$

The initial condition is $R(0) = 0$.

Similarly, we have

$$\begin{aligned} T'(a) &= X_1(a)^{-1}[A(a)X_1(a) + D(a)Y_1(a)]X_1(a)^{-1} \\ &= T(a)A(a) + T(a)D(a)R(a) \end{aligned} \tag{12}$$

with $T(0) = I$.

3. Discussion. We observe that we obtain the same type of Riccati differential equation in this general case as in the special case where the system of equations is derived from a control process. Once again an initial-value problem for a nonlinear system of ordinary differential equations replaces a linear system subject to two-point conditions leading to the solution of a linear system of algebraic equations. Sometimes this is desirable, sometimes not. What we want is the flexibility to use whatever approach is most expedient at the moment.

4. Linear Transport Process. Let us consider an idealized transport process where N different types of particles move in either direction along a line of finite length a. They interact with the medium of which the line is composed in ways which will be specified below, but not with each other. The effect of this interaction is to change a particle from one type to another, traveling in the same or opposite direction. We call this a "pure scattering" process.

Let us consider the passage of a particle of type or state i through the interval $[t, t + \Delta]$ where $\Delta > 0$ is a small quantity and i may assume any of the values $i = 1, 2, \ldots, N$; see Fig. 1 and note the reversal of the usual direction. We have done this to preserve consistency of notation with a number of references.

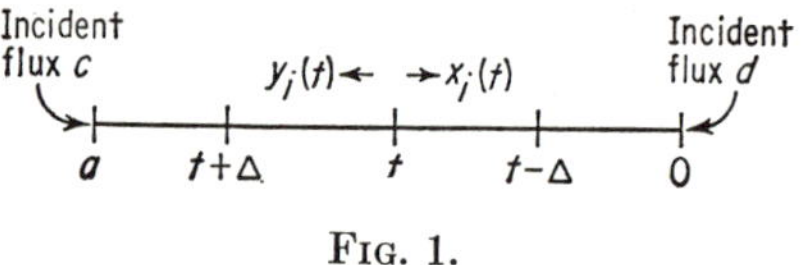

FIG. 1.

Let

$a_{ij}(t)\Delta + o(\Delta)$ = the probability that a particle in state j will be transformed into a particle in state i traveling in the same direction, $j \neq i$, upon traversing the interval $[t + \Delta, t]$ going to the right

$1 - a_{ii}(t)\Delta + o(\Delta)$ = the probability that a particle in state i will remain in state i in the same direction while traversing the interval $[t + \Delta, t]$ going to the right (1)

$b_{ij}(t)\Delta + o(\Delta)$ = the probability that a particle going to the right in state j will be transformed into a particle in state i traveling in the opposite direction upon traversing the interval $[t + \Delta, t]$

Similarly, we introduce the functions $c_{ij}(t)$ and $d_{ij}(t)$, associated with forward scattering and back scattering for a particle going to the left through $[t, t + \Delta]$. We suppose that all of these functions are non-negative for $t \geq 0$, as they should be, of course, to be associated with probabilities. The most important case is that where they are piecewise continuous. Suppose that there are streams of particles of all N types incident at both ends of the line, as indicated in Fig. 1, of constant intensities per unit time. We call this stream of particles a "flux."

5. Classical Imbedding. Let us now introduce the steady-state functions

$$\begin{aligned} x_i(t) &= \text{the expected intensity of flux of particles of type } i \text{ to the} \\ &\quad \text{right at the point } t,\ 0 \leq t \leq a \\ y_i(t) &= \text{the expected intensity of flux of particles of type } i \text{ to the left} \\ &\quad \text{at the point } t,\ 0 \leq t \leq a,\ i = 1, 2, \ldots, N \end{aligned} \tag{1}$$

See Fig. 1. We shall henceforth omit the term "expected" in what follows and proceed formally to obtain some equations for x_i and y_i. There is no need to concern ourselves with rigorous details here, since we are using the stochastic process solely as a guide to our intuition in ascertaining some of the properties of the solution of (1.1). Those interested in a rigorous derivation of the following equations, starting from the continuous stochastic process, may refer to sources cited at the end of the chapter.

An input-output analysis (which is to say, a local application of conservation relations) of the intervals $[t - \Delta, t]$ and $[t, t + \Delta]$ yields the following relations, valid to $o(\Delta)$:

$$\begin{aligned} x_i(t) &= x_i(t + \Delta)(1 - a_{ii}\Delta) + \sum_{j \neq i} a_{ij}\Delta x_j(t + \Delta) + \sum_{j=1}^{N} d_{ij}\,\Delta y_j(t) \\ y_i(t) &= y_i(t - \Delta)(1 - c_{ii}\Delta) + \sum_{j \neq i} c_{ij}\Delta y_j(t - \Delta) + \sum_{j=1}^{N} b_{ij}\,\Delta x_j(t) \end{aligned} \tag{2}$$

Passing to the limit as $\Delta \to 0$, we obtain the differential equations

$$\begin{aligned} -x_i'(t) &= -a_{ii}x_i(t) + \sum_{j \neq i} a_{ij}x_j(t) + \sum_{j=1}^{N} d_{ij}y_j(t) \\ y_i'(t) &= -c_{ii}y_i(t) + \sum_{j \neq i} c_{ij}y_j(t) + \sum_{j=1}^{N} b_{ij}x_j(t) \end{aligned} \tag{3}$$

$i = 1, 2, \ldots, N$. Referring to Fig. 1, we see that the boundary conditions are

$$\begin{aligned} x_i(a) &= c_i \\ y_i(0) &= d_i \qquad i = 1, 2, \ldots, N \end{aligned} \tag{4}$$

We introduce the matrices

$$B = (b_{ij})$$

$$A = \begin{bmatrix} -a_{11} & a_{12} & \cdots & +a_{1N} \\ a_{21} & -a_{22} & \cdots & a_{2N} \\ \cdot & & & \\ \cdot & & & \\ \cdot & & & \\ a_{N1} & a_{N2} & \cdots & -a_{NN} \end{bmatrix}$$

$$D = (d_{ij})$$

$$C = \begin{bmatrix} -c_{11} & c_{12} & \cdots & c_{1N} \\ c_{21} & -c_{22} & \cdots & c_{2N} \\ \cdot & & & \\ \cdot & & & \\ \cdot & & & \\ c_{N1} & c_{N2} & \cdots & -c_{NN} \end{bmatrix} \tag{5}$$

B and D are non-negative matrices, while A and C possess quite analogous properties despite the negative main diagonals; see Chap. 16. The condition of pure scattering imposes certain conditions on the column sums of $A + B$ and $C + D$ which we will use below.

We observe then the special structure of the matrices A, B, C, D required for (1.1) to be associated with a linear transport process. Introducing the vectors x and y, with components x_i and y_i respectively, we obtain a linear vector-matrix system subject to a two-point boundary condition having the form given in (1.1).

If the effects of interaction are solely to change type and direction, we say that the process is one of *pure scattering* as noted above; if particles disappear as a result of interactions, we say *absorption* occurs. We will consider the pure scattering case first. We will not discuss the *fission* case, corresponding to a production of two or more particles as a result of an interaction. Here questions of critical length arise, as well as a number of novel problems.

6. Invariant Imbedding. Let us now present a different analytic formulation of the pure scattering process which will permit us to demonstrate the existence of the reflection and transmission matrices for all $a \geq 0$. This represents a different imbedding of the original process within a family of processes. The preceding was a classical imbedding. To begin with, we introduce the function

$r_{ij}(a)$ = the intensity of flux in state i from a rod of length a emergent at a due to an incident flux of unit intensity in state j incident at a (1)

This function, the intensity of reflected flux, is defined, hopefully, for $a \geq 0$, $i, j = 1, 2, \ldots, N$. It remains to be demonstrated that this is meaningful. See Fig. 2.

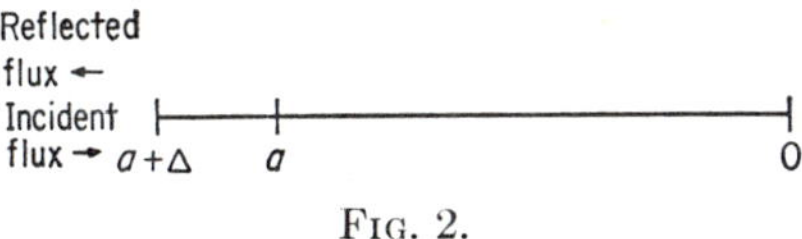

FIG. 2.

To obtain equations for the $r_{ij}(a)$ as functions of a, we keep account of all the possible ways in which interactions in $[a + \Delta, a]$ can contribute to a reflected flux in state i. At first the bookkeeping seems tedious. Subsequently, we shall show that with the aid of matrix notation the multidimensional case can be treated in a simple direct fashion when the physical meaning of the matrices is appreciated. This is typical of mathematical physics when suitable notation is employed. In more complex situations, infinite-dimensional operators replace the finite-dimensional matrices.

Starting with an incident flux in state j, we can obtain a reflected flux in state i in the following fashions:

(a) Interaction in $[a + \Delta, a]$ and immediate reflection resulting in a change of direction.
(b) Interaction in $[a + \Delta, a]$ together with reflection in state i from $[a,0]$.
(c) Passage through $[a + \Delta, a]$ without interaction, reflection from $[a,0]$ in state i, and passage through $[a, a + \Delta]$ without interaction. (2)
(d) Passage through $[a + \Delta, a]$ without interaction, reflection from $[a,0]$ in state $j = i$, and interaction in $[a, a + \Delta]$ resulting in emergence in state i.
(e) Passage through $[a + \Delta, a]$ without interaction, reflection from $[a,0]$ in state k, interaction in $[a, a + \Delta]$ resulting in a state l toward $[a,0]$, and then reflection in state i from $[a,0]$.

All further interactions produce terms of order Δ^2 and can therefore be neglected at this time, since ultimately we let Δ tend to zero. Note that we write both $[a + \Delta, a]$ and $[a, a + \Delta]$ to indicate the direction of the particle.

Taking account of the possibilities enumerated in (2), we may write

$$r_{ij}(a + \Delta) = b_{ij}\Delta + \sum_{k \neq i} a_{kj} r_{ik}(a)\Delta + (1 - a_{jj}\Delta)\Big[r_{ij}(a)(1 - c_{ii}\Delta) + \sum_{k \neq i} r_{kj}(a) c_{ik}\Delta + \sum_{k,l} r_{il}(a) d_{lk} r_{kj}(a)\Delta \Big] \quad (3)$$

to term in $o(\Delta)$. Passing to the limit as $\Delta - 0$, we obtain the nonlinear differential equation

$$r_{ij}'(a) = b_{ij} + \sum_{k \neq i} a_{kj} r_{ik}(a) - (a_{jj} + c_{ii}) r_{ij}(a) + \sum_{k \neq i} r_{kj}(a) c_{ik} + \sum_{k,l} r_{il}(a) d_{kl} r_{kj}(a) \quad (4)$$

for $i, j = 1, 2, \ldots, N$, with the initial condition

$$r_{ij}(0) = 0 \quad (5)$$

This last equation states that the reflection from a rod of zero length is zero.

We are not concerned with the rigorous derivation of (4) in the foregoing fashion at the present time. We do, however, want the difference approximation for an important purpose below; see Sec. 9. A rigorous derivation of (4) from the linear equations has already been given in Sec. 2.

Writing

$$R(a) = (r_{ij}(a)) \quad (6)$$

and letting A, B, C, D have the previous significances, we see that (4) takes the compact form

$$R' = B + RA + CR + RDR \qquad R(0) = 0 \quad (7)$$

This is a matrix differential equation of Riccati type. Similar equations arise in control theory, as we noted in Sec. 3 and in the previous chapter.

7. Interpretation. Once we have obtained the differential equation, we can readily identify each of the terms and then use this identification as a means of writing down the corresponding equation for the transmission matrix. We see that B corresponds to immediate backscattering, that CR corresponds to reflection followed by forward scattering, that RA corresponds to forward scattering followed by reflection, and that RDR corresponds to reflection, backscattering, and then reflection again. The physical interpretation is also useful in obtaining convenient approximations in certain situations.

We expect then the equation for the transmission matrix $T(a)$ to have the form

$$T' = TA + TDR \quad (1)$$

corresponding to forward scattering followed by transmission, plus reflection, backscattering, and then transmission again. The initial condition is $T(0) = I$. We leave this for the reader to verify in various ways.

The importance of the physical approach lies in the fact that it provides an easy way of deriving many important results, and, what is more, of anticipating them.

EXERCISE

1. What differential equations do we get if we use the figure

instead of Fig. 2?

8. Conservation Relation. Since we have assumed a pure scattering process in Sec. 4, we know that no particles are "lost." A particle of type i must transform into a particle of type j, $j = 1, 2, \ldots, N$, for any i. The analytic translations of this observation are the relations

$$M(B + A) = 0 \qquad M(C + D) = 0 \tag{1}$$

where M is the matrix

$$M = \begin{bmatrix} 1 & 1 & \cdots & 1 \\ 0 & 0 & \cdots & 0 \\ \cdot & & & \\ \cdot & & & \\ \cdot & & & \\ 0 & 0 & \cdots & 0 \end{bmatrix} \tag{2}$$

The effect of multiplication by M is to produce column sums.

We expect that the conservation relation

$$M(R + T) = M \tag{3}$$

which states that total input is equal to total output, will hold for $a \geq 0$. Let us begin by showing that it holds for small a. From (6.7) and (7.1), we have

$$\begin{aligned}[M(R + T)]' &= M(B + CR + RA + RDR) + M(TA + TDR) \\ &= MB + MCR + MRA + MTA + M(R + T)DR \\ &= MB + MCR + M(R + T - I)A + MA \\ &\qquad + M(R + T - I)DR + MDR \\ &= M(B + A) + M(C + D)R + M(R + T - I)A \\ &\qquad + M(R + T - I)DR \\ &= M(R + T - I)A + M(R + T - I)DR \end{aligned} \tag{4}$$

Regarding R as a known function, this is a differential equation of the form

$$Z' = ZA + ZDR \qquad Z(0) = 0 \tag{5}$$

for the function $Z = M(R + T - I)$. One solution is clearly $Z = 0$. Uniqueness assures us that it is the only solution within the interval for which $R(a)$ and $T(a)$ exist.

Hence, within the domain of existence of R and T, we have (3). If we can establish what is clearly to be expected, namely that all the elements of R and T are non-negative, (3) will imply that R and T are uniformly bounded for $0 \leq a \leq a_0$, which means that we can continue the solution indefinitely, using the same argument repeatedly.

EXERCISES

1. Show that the relation $M(R + T) = M$ is equivalent to $M(Y_1(a) + I) = MX_1(a)$, provided that $X_1(a)$ is nonsingular.

2. Obtain the foregoing relation in some interval $[0,a_0]$ by showing that

$$MY_1'(a) = MX_1'(a)$$

9. Non-negativity of R and T. Let us now establish the non-negativity of $r_{ij}(a)$ and $t_{ij}(a)$ for $a \geq 0$. We cannot conclude this directly from the differential equation of (6.7) since A and C have negative elements along the main diagonal. Let us present one proof here depending upon the recurrence relation of (6.3). Another, depending upon the properties of A and C as scattering matrices, is sketched in the exercises.

Referring to (6.3), we see that all the terms are non-negative on the right-hand side if Δ is sufficiently small. We can avoid even this restriction by means of the useful device of replacing $(1 - a_{jj}\Delta)$ and $(1 - c_{ii}\Delta)$ by the positive quantities $e^{-a_{jj}\Delta}$ and $e^{-c_{ii}\Delta}$, a substitution valid to terms in $o(\Delta)$. It follows that the sequence $\{r_{ij}(k\Delta)\}$ obtained by use of (6.3) is non-negative, and actually positive if all of the b_{ij} are positive. Exact conditions that guarantee positivity rather than merely non-negativity are not easy to specify analytically, although easy to state in physical terms.

We know from standard theory in differential equations that the functions determined by the difference equation plus linear interpolation converge to the solutions of the differential equation in some interval $[0,a_0]$. Hence, $r_{ij}(a) \geq 0$ in this interval. A similar result holds for the $t_{ij}(a)$. Thus, using the conservation relation (8.3), we can conclude

$$\begin{aligned} &0 \leq r_{ij}(a) \leq 1 \\ &0 \leq t_{ij}(a) \leq 1 \qquad 0 \leq a \leq a_0 \qquad i, j = 1, 2, \ldots, N \end{aligned} \tag{1}$$

It follows that the solution of the differential equation can be continued for another interval $[a_0, a_0 + a_1]$. Using (6.3) starting at $a = a_0$, we establish non-negativity, boundedness, and so on. Thus, the solution of (6.7) can be continued in an interval $[a_0 + a_1, a_0 + 2a_1]$, and so on.

Hence, it exists and is unique for $a \geq 0$. This is a standard argument in the theory of differential equations.

EXERCISES

1. Consider first the case where A and C are constant. Convert the equation $R' = B + CR + RA + RDR$, $R(0) = 0$, into the integral equation

$$R = \int_0^t e^{C(t-t_1)}[B + RDR]e^{A(t-t_1)}\,dt_1$$

2. Use successive approximations, plus the properties of C and A, to deduce that R has non-negative elements in some interval $[0,a_0]$.

3. If A and C are variable, obtain the corresponding integral equation

$$R = \int_0^t X(t)X(t_1)^{-1}[B + RDR]Y(t)Y(t_1)^{-1}\,dt_1$$

where X, Y are respectively solutions of $X' = CX$, $Y' = YA$, $X(0) = Y(0) = I$.

4. Show that $X(t)X(t_1)^{-1}$, $Y(t)Y(t_1)^{-1}$ both have non-negative elements for $0 \leq t_1 \leq t$, and proceed as above.

5. Show that the elements of $R(a)$ are monotone-increasing in a if A, B, C, D are constant.

6. Is this necessarily true if A, B, C, D are functions of a?

7. Show that $R(\infty) = \lim_{a\to\infty} R(a)$ exists and is determined by the quadratic equation $B + CR(\infty) + R(\infty)A + R(\infty)DR(\infty) = 0$.

8. Show that $R(a)$ has the form $R(\infty) + R_1e^{\lambda_1 a} + \cdots$ as $a \to \infty$ where $\text{Re}\,(\lambda_1) < 0$ in the case where A, B, C, D are constant. How does one determine λ_1?

9. Show that the monotonicity of the elements of $R(a)$ implies that $\lambda_1 < 0$.

10. How would one use nonlinear interpolation techniques to calculate $R(\infty)$ from values of $R(a)$ for small a? See R. Bellman and R. Kalaba, A Note on Nonlinear Summability Techniques in Invariant Imbedding, *J. Math. Anal. Appl.*, vol. 6, pp. 465–472, 1963.

10. Discussion. Our objective is the solution of (1.1) in the interval $[0,a]$. Pursuing this objective in a direct fashion we encountered the problem of demonstrating that $X_1(a)$ was nonsingular for $a \geq 0$. To establish this result, we identified $X_1(a)^{-1}$ with the transmission matrix of a linear transport process and thus with the aid of a conservation relation we were able to deduce that $X_1(a)^{-1}$ existed for $a \geq 0$.

This means that the internal fluxes $x(t)$ and $y(t)$ are uniquely determined for $0 \leq t \leq a$, for all $a > 0$.

A basic idea is that of identifying the original equations with a certain physical process and then using a *different* analytic formulation of this process to derive certain important results.

11. Internal Fluxes. Let us now consider a representation for the internal cases in the case where the medium is *homogeneous*, that is, A, B, C, D are constant, and the scattering is *isotropic*, that is,

$$A = C \qquad B = D \tag{1}$$

Consider Fig. 3. Concentrating on the interval $[a,t]$, we regard c and $y(t)$ as the incident fluxes. Hence, we have

$$x(t) = T(a - t)c + R(a - t)y(t) \tag{2}$$

Similarly, considering the interval $[t,0]$, we have

$$y(t) = R(t)x(t) + T(t)d \tag{3}$$

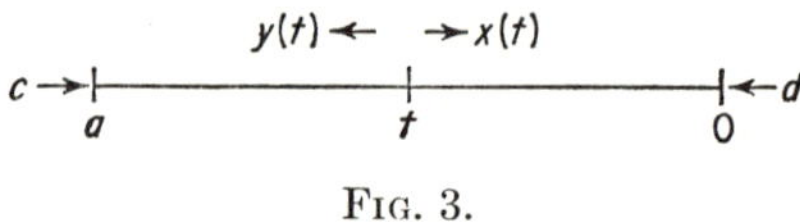

Fig. 3.

Combining (2) and (3), we obtain

$$x(t) = [I - R(a - t)R(t)]^{-1}[T(a - t)c + R(a - t)T(t)d] \tag{4}$$

provided that $[I - R(a - t)R(t)]$ is nonsingular for $0 \leq t \leq a$. This is certainly true for small a. We shall leave as exercises the proof of the result for general a and the demonstration of the non-negativity of x and y for non-negative c and d.

EXERCISES

1. Show that $R(a)$ has largest characteristic root (the Perron root) less than one. *Hint:* Use the conservation relation.

2. Hence, show that $I - R(a/2)^2$ has an inverse and that this inverse is non-negative.

3. Hence, show that $x(a/2)$, $y(a/2)$ are non-negative.

4. Repeating the same type of argument for $[a/2,0]$ and $[a,a/2]$, show that $x(t)$ and $y(t)$ are non-negative for $a/4$, $3a/4$. Proceeding in this fashion and invoking continuity, show that $x(t)$, $y(t)$ are non-negative for $0 \leq t \leq a$.

5. Establish the relations in (2) and (3) purely analytically.

6. From the fact that

$$[I - R(a - t)R(t)]x = T(a - t)c + R(a - t)T(t)d = p$$

possesses *one* solution x for every c and d, conclude that $I - R(a - t)R(t)$ possesses an inverse, and thus that the largest characteristic root of $R(a - t)R(t)$ cannot be one.

7. To show that this largest characteristic root must be less than one in absolute value, iterate the relation

$$\begin{aligned} x &= p + R(a - t)R(t)x \\ &= p + R(a - t)R(t)p + \cdots + [R(a - t)R(t)]^{n-1}p + R(a - t)R(t)^n x \end{aligned}$$

and use the fact that p, x, $R(a - t)R(t)$ are all non-negative.

8. Extend the foregoing considerations to the homogeneous, anisotropic case.

9. Extend the foregoing considerations to the inhomogeneous case.

12. Absorption Processes. Consider the more general case where particles can disappear without producing other particles as a result of an interaction. Call this phenomenon *absorption.* Introduce the quantities $i = 1, 2, \ldots, N$,

$$\begin{aligned} e_{ii}(t)\Delta &= \text{the probability that a particle of type } i \text{ is absorbed in traversing the interval } [t+\Delta, t] \text{ traveling in a right-hand direction} \\ f_{ii}(t)\Delta &= \text{the probability that a particle of type } i \text{ is absorbed traversing the interval } [t, t+\Delta] \text{ traveling in a left-hand direction} \end{aligned} \tag{1}$$

We suppose that $e_{ii}, f_{ii} \geq 0$ and that

$$\begin{aligned} M(A + B + E) &= M \\ M(C + D + F) &= M \end{aligned} \tag{2}$$

for $t \geq 0$. This means that a particle is either forward-scattered, back-scattered, or absorbed.

Introduce the functions

$$l_{ij}(a) = \text{the probability that a particle of type } j \text{ incident upon } [a,0] \text{ from the left results in the absorption of a particle of type } i \text{ in } [a,0],\ i, j = 1, 2, \ldots, N \tag{3}$$

Setting $L(a) = (l_{ij}(a))$, it is easy to see from the type of argument given in Sec. 6 that the matrix L satisfies the differential equation

$$L' = E + LA + FR + LDR \qquad L(0) = 0 \tag{4}$$

where E and F are the diagonal matrices composed of the e_{ii} and f_{ii} respectively. We call L a *dissipation* matrix.

We expect to have the conservation relation

$$M(R + T + L) = M \tag{5}$$

for $a \geq 0$. To demonstrate this we proceed as before. We have, suitably grouping terms,

$$\begin{aligned} [M(R + T + L)]' &= M(B + E) + M(R + T + L)A \\ &\quad + M(C + F)R + M(R + T + L)DR \\ &= M(B + E) + MA + M(C + F)R + MDR \\ &\quad + M(R + T + L - I)A \\ &\quad + M(R + T + L - I)DR \\ &= M(A + B + E) + M(C + D + F)R \\ &\quad + M(R + T + L - I)A \\ &\quad + M(R + T + L - I)DR \\ &= M(R + T + L - I)(A + DR) \end{aligned} \tag{6}$$

This equation clearly has the solution $M(R + T + L - I) = 0$.

As before, we have R and T non-negative and the same is readily demonstrated for L. From here on, the results and methods are as before, and we leave the details as exercises.

MISCELLANEOUS EXERCISES

1. Consider the three-dimensional linear system $x' = Ax$ for which x is specified by the conditions $x_1(0) = c_1$, $(x(t_1),b^{(1)}) = c_2$, $(x(a),b^{(2)}) = c_3$, where x_1 is the first component of x, $b^{(1)}$, $b^{(2)}$ are given vectors and $0 < t_1 < a$. Obtain relations for the missing components of $x(0)$, $x_1(0) = f_1(a,c_2,c_3)$, $x_2(0) = f_2(a,c_2,c_3)$.

2. Obtain corresponding results in the N-dimensional case.

3. Write

$$\begin{pmatrix} f_1(a,c_2,c_3) \\ f_2(a,c_2,c_3) \end{pmatrix} = F(a) \begin{pmatrix} c_2 \\ c_3 \end{pmatrix}$$

and obtain a differential equation satisfied by $F(a)$, See R. Bellman Invariant Imbedding and Multipoint Boundary-value Problems, *J. Math. Anal. Appl.*, vol. 24, 1968.

4. Consider the case where there is only one state and scattering produces a change of direction. Let $r(a)$ denote the intensity of reflected flux. Show by following successive reflections and transmissions that

$$\begin{aligned} r(a+b) &= r(a) + t(a)r(b)t(a) + t(a)r(b)r(a)r(b)t(a) + \cdots \\ &= \frac{r(a) + r(b)}{1 - r(a)r(b)} \end{aligned}$$

5. Obtain the corresponding relation for the transmission function.

6. Obtain the analogous relations in the multitype case where $R(a)$ and $T(a)$ are matrices.

7. Do these functional equations determine $R(a)$ and $T(a)$? See the discussion of Polya's functional equation in Sec. 16 of Chap. 10.

Bibliography and Discussion

§1. For a discussion of the background of invariant imbedding, see

R. Bellman, *Some Vistas of Modern Mathematics,* University of Kentucky Press, Lexington, Kentucky, 1968.

A survey of applications of invariant imbedding to mathematical physics is given in

R. Bellman, R. Kalaba, and G. M. Wing, Invariant Imbedding and Mathematical Physics, I: Particle Processes, *J. Math. Phys.*, vol. 1, pp. 280–308, 1960.

P. B. Bailey and G. M. Wing, Some Recent Developments in Invariant Imbedding with Applications, *J. Math. Phys.*, vol. 6, pp. 453–462, 1965.

See also

R. Bellman, R. Kalaba, and M. Prestrud, *Invariant Imbedding and Radiative Transfer in Slabs of Finite Thickness,* American Elsevier Publishing Company, Inc., New York, 1963.

R. Bellman, H. Kagiwada, R. Kalaba, and M. Prestrud, *Invariant Imbedding and Time-dependent Processes,* American Elsevier Publishing Company, Inc., New York, 1964.

§4. See the foregoing references and

G. M. Wing, *An Introduction to Transport Theory,* John Wiley & Sons, New York, 1962.

R. Bellman and G. Birkhoff (eds.), *Transport Theory,* Proc. Symp. Appl. Math., vol. XIX, Amer. Math. Soc., Providence, R.I., 1968.

R. Preisendorfer, *Radiative Transfer on Discrete Spaces,* Pergamon Press, New York, 1965.

§8. See

R. Bellman, Scattering Processes and Invariant Imbedding, *J. Math. Anal. Appl.*, vol. 23, pp. 254–268, 1968.

R. Bellman, K. L. Cooke, R. Kalaba, and G. M. Wing, Existence and Uniqueness Theorems in Invariant Imbedding, I: Conservation Principles, *J. Math. Anal. Appl.*, vol. 10, pp. 234–244, 1965.

§12. See the references of Sec. 8. The classical theory of semi-groups, discussed in

E. Hille, *Functional Analysis and Semi-groups,* Amer. Math. Soc. Colloq. Publ., vol. XXXI, 1948.

may be thought of as a study of e^{At} where A is an operator, i.e., the study of linear functional equations of the form $x' = Ax + y$, $x(0) = c$. Dynamic programming introduces the study of nonlinear functional equations of the form

$$x' = \max_q [A(q)x + y(q)] \qquad x(0) = c$$

in a natural fashion; see

R. Bellman, *Adaptive Control Processes: A Guided Tour,* Princeton University Press, Princeton, New Jersey, 1961.

Invariant imbedding may be considered as a theory of semi-groups in space and structure, and, particularly, of nonlinear semi-groups of a type different from those just described. See

R. Bellman and T. Brown, A Note on Invariant Imbedding and Generalized Semi-groups, *J. Math. Anal. Appl.*, vol. 9, pp. 394–396, 1964.

A number of interesting matrix results connected with circuits arise from similar applications of invariant imbedding. See, for example,

R. M. Redheffer, On Solutions of Riccati's Equation as Functions of Initial Values, *J. Rat. Mech. Anal.*, vol. 5, pp. 835–848, 1960.

R. M. Redheffer, Novel Uses of Functional Equations, *J. Rat. Mech. Anal.*, vol. 3, pp. 271–279, 1954.

A. Ping and I. Wang, Time-dependent Transport Process, *J. Franklin Inst.*, vol. 287, pp. 409–422, 1969.

For an application of invariant imbedding to wave propagation, see

R. Bellman and R. Kalaba, Functional Equations, Wave Propagation, and Invariant Imbedding, *J. Math. Mech.*, vol. 8, pp. 683–703, 1959.

For discussions of two-point boundary-value problems of unconventional type using functional analysis, see

F. R. Krall, Differential Operators and Their Adjoints Under Integral and Multiple Point Boundary Conditions, *J. Diff. Eqs.*, vol. 4, pp. 327–336, 1968.

W. R. Jones, Differential Systems with Integral Boundary Conditions, *J. Diff. Eqs.*, vol. 3, pp. 191–202, 1967.

R. Bellman, On Analogues of Poincaré-Lyapunov Theory for Multipoint Boundary-value Problems, *J. Math. Anal. Appl.*, vol. 14, pp. 522–526, 1966.

For an extensive application of invariant imbedding to Fredholm integral equation theory, with particular reference to the Wiener-Hopf equation, see

A. McNabb and A. Schumitzky, *Factorization of Operators—I: Algebraic Theory and Examples*, University of Southern California, USCEE-373, July, 1969.

A. McNabb and A. Schumitzky, *Factorization of Integral Operators—II: A Nonlinear Volterra Method for Numerical Solution of Linear Fredholm Equations*, University of Southern California, USCEE-330, March, 1969.

A. McNabb and A. Schumitzky, *Factorization of Operators—III: Initial Value Methods for Linear Two-point Boundary Value Problems*, University of Southern California, USCEE-371, July, 1969.

19

Numerical Inversion of the Laplace Transform and Tychonov Regularization

1. Introduction. In Chap. 11 we briefly indicated some of the advantages of using the Laplace transform to treat linear ordinary differential equations of the form

$$x'' = Ax \qquad x(0) = c \tag{1}$$

where A is a constant matrix. If we set

$$y(s) = L(x) = \int_0^\infty e^{-st}x(t)\,dt \tag{2}$$

we obtain for Re(s) sufficiently large the simple explicit result

$$y = (A - sI)^{-1}c \tag{3}$$

The problem of obtaining x given y poses no analytical difficulty in this case since all that is required is the elementary result

$$\int_0^\infty e^{-st}e^{at}\,dt = \frac{1}{(s - a)} \tag{4}$$

together with limiting forms.

Comparing (1) and (3), we see that one of the properties of the Laplace transform is that of reduction of level of transcendence. Whereas (1) is a linear differential equation for x, (3) is a linear algebraic equation for y. As soon as we think in these terms, we are stimulated to see if the Laplace transform plays a similar role as far as other important types of functional equations are concerned. As we shall readily show, it turns out that it does. This explains its fundamental role in analysis.

This, however, is merely the start of an extensive investigation centering about the use of a readily derived Laplace transform of a function to deduce properties of the function. Many interesting questions of matrix theory arise in the course of this investigation which force us to consider ill-conditioned systems of linear algebraic equations. In this fashion we make contact with some important ideas of Tychonov which lie at the center of a great deal of current work in the analytic and computational study of partial differential equations and equations of more complex type.

EXERCISE

1. Establish the Borel identity

$$L\left(\int_0^t u(t - t_1)v(t_1)\, dt_1\right) = L(u)L(v)$$

under appropriate conditions on u and v by considering integration over the shaded region in Fig. 1 and interchanging the orders of integration.

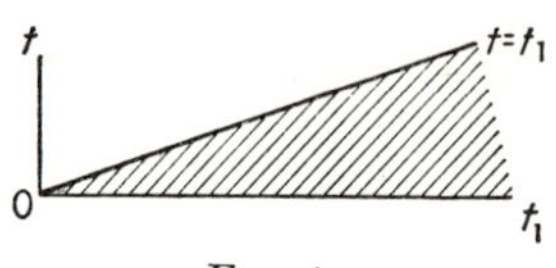

Fig. 1.

2. The Heat Equation. Consider the partial differential equation

$$\begin{aligned} k(x)u_t &= u_{xx} \\ u(0,t) &= u(1,t) = 0 \qquad t > 0 \\ u(x,0) &= g(x) \qquad 0 < x < 1 \end{aligned} \tag{1}$$

Let us proceed formally to obtain an ordinary differential equation for $L(u)$ assuming that all of the operations we perform are justifiable. As mentioned above, our aim is solely to point out some of the uses of the Laplace transform. We have

$$\begin{aligned} L(k(x)u_t) &= L(u_{xx}) \\ k(x)L(u_t) &= L(u)_{xx} \\ -sk(x)g(x) + sk(x)L(u) &= L(u)_{xx} \end{aligned} \tag{2}$$

Hence, setting

$$v = L(u) \equiv v(x,s) \tag{3}$$

we obtain for each fixed s an *ordinary* differential equation for v,

$$\begin{aligned} v_{xx} - sk(x)v &= -k(x)g(x) \\ v(0) &= v(1) = 0 \end{aligned} \tag{4}$$

For any value of s it is an easy matter to solve (4) numerically.

EXERCISES

1. What modifications are necessary if the boundary conditions are $u(0,t) = f_1(t)$, $u(1,t) = f_2(t)$?

2. Obtain an analytic expression for $L(u)$ in the case where $k(x) \equiv k$, a positive constant.

3. The Renewal Equation. Consider the renewal equation

$$u(t) = f(t) + \int_0^t k(t - t_1)u(t_1)\,dt_1 \tag{1}$$

which arises in many parts of pure and applied mathematics. Using the Borel result (Exercise 1 at the end of Sec. 1), we have

$$L(u) = L(f) + L(k)L(u) \tag{2}$$

whence

$$L(u) = \frac{L(f)}{1 - L(k)} \tag{3}$$

an important explicit representation for $L(u)$.

4. Differential-difference Equations. As a third example of the use of the Laplace transform in simplifying equations, consider the linear differential-difference equation

$$\begin{aligned} u'(t) &= au(t) + bu(t-1) \qquad & t \geq 0 \\ u(t) &= g(t) & -1 \leq t \leq 0 \end{aligned} \tag{1}$$

We have

$$L(u') = aL(u) + bL(u(t-1)) \tag{2}$$

Using the relations

$$\begin{aligned} L(u') &= -g(0) + sL(u) \\ \int_0^\infty u(t-1)e^{-st}\,dt &= \int_0^1 + \int_1^\infty \\ &= \int_{-1}^0 g(t_1)e^{-s(1-t_1)}\,dt_1 + e^{-s}\int_0^\infty e^{-st}u(t)\,dt \\ &= \int_{-1}^0 g(t_1)e^{-s(1-t_1)}\,dt_1 + e^{-s}L(u) \end{aligned} \tag{3}$$

we readily obtain the expression

$$L(u) = \frac{\int_{-1}^0 g(t_1)e^{-s(1-t_1)}\,dt_1 + g(0)}{(s - a - be^{-s})} \tag{4}$$

5. Discussion. We have illustrated the point that in a number of important cases it is easy to obtain an explicit representation for $L(u)$. The basic question remains: What is the value of this as far as ascertaining the behavior of u is concerned?

There are two powerful techniques which we can employ. The first is the use of the complex inversion formula

$$u = \frac{1}{2\pi i}\int_C L(u)e^{st}\,ds \tag{1}$$

where C is a suitable contour. Once (1) has been established, we have at our command the powerful resources of the theory of functions of complex variables.

The second general technique involves the use of Tauberian theory, a theory which enables us to relate the behavior of $u(t)$ for large t to the behavior of $L(u)$ at $s = 0$. Both of these methods are discussed in references cited at the end of the chapter.

Here we wish to discuss a numerical inversion technique which yields values of $u(t)$ for $t \geq 0$ in terms of values of $L(u)$ for $s \geq 0$. This numerical work, however, will often be guided by preliminary and partial results obtained using the two methods just mentioned. That there are serious difficulties in this approach to the numerical inversion of the Laplace transform is due essentially to the fact that in its general form the task is impossible.

6. Instability. By this last comment concerning impossibility, we mean that L^{-1} is an unstable operator, which is to say, arbitrarily small changes in $y = L(x)$ can produce arbitrarily large changes in the function x.

Consider as a simple example of this phenomenon the familiar integral

$$\int_0^\infty e^{-st} \sin at \, dt = \frac{a}{(s^2 + a^2)} \qquad a > 0 \tag{1}$$

For real s, we have

$$\frac{a}{(s^2 + a^2)} \leq \frac{1}{a} \tag{2}$$

Hence, by taking a arbitrarily large, we can make $L(\sin at)$ arbitrarily small. Nonetheless, the function $\sin at$ oscillates between values of ± 1 for $t \geq 0$.

Consequently, the question of obtaining the numerical values of u of the linear integral equation

$$\int_0^\infty e^{-st} u(t) \, dt = v(s) \tag{3}$$

given numerical values of $v(s)$, makes sense only if we add to a set of numerical values of $v(s)$ some information concerning the structure of $u(t)$. This is the path we shall pursue.

7. Quadrature. The first step is a simple one, that of approximating to an integral by a sum. There are many ways we can proceed. Thinking of an integral as representing the area under a curve (see Fig. 2), we can write

$$\int_0^1 u(t) \, dt \cong u(0)\Delta + u(\Delta)\Delta + \cdots + u((N-1)\Delta)\Delta \tag{1}$$

where $N\Delta = 1$. More accurate results can be obtained by using a trapezoidal approximation to the area $\int_{k\Delta}^{(k+1)\Delta} u(t) \, dt$, and so forth.

As usual in computational analysis, we are torn between two criteria, one in terms of accuracy, the other in the coordinates of time. A method designed to yield a high degree of accuracy can require a correspondingly large amount of computation; a procedure requiring a relatively small number of calculations can produce undesirably large errors. Let us consider, to illustrate this Scylla and Charybdis of numerical analysis, a situation encountered, say, in connection with the heat equation discussed in Sec. 2, where the evolution of each functional value $u(k\Delta)$ consumes an appreciable time. The price of desired accuracy using the quadrature relation in (1), namely choosing sufficiently small Δ, may thus be too high.

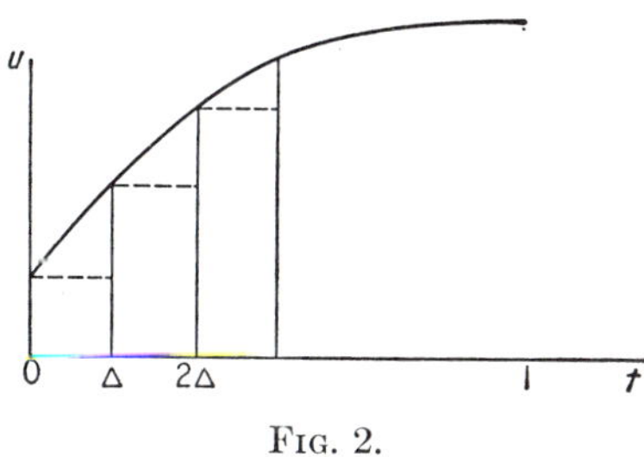

Fig. 2.

Let us then provide ourselves with a bit more flexibility by calculating values of $u(t)$ at irregularly spaced points $\{t_i\}$, called *quadrature points*, and suitably weighting the results. We write

$$\int_0^1 u(t)\,dt \cong \sum_{i=1}^{N} w_i u(t_i) \tag{2}$$

where the weights w_i are called the *Christoffel numbers*.

There are clearly many ways of choosing the t_i and w_i, each convenient according to some preassigned criterion of convenience. We shall follow a procedure due to Gauss.

8. Gaussian Quadrature. Let us show that the w_i and t_i are uniquely determined by the condition that (7.2) is exact for polynomials of degree less than or equal to $2N - 1$. To demonstrate this, let us recall some facts concerning the Legendre polynomials, $\{P_n(t)\}$.

These polynomials constitute an orthogonal set over $[-1,1]$,

$$\int_{-1}^{1} P_m P_n\,dt = 0 \qquad m \neq n \tag{1}$$

which is equivalent to the statement that

$$\int_{-1}^{1} t^k P_n\,dt = 0 \qquad k = 0, 1, 2, \ldots, n-1 \tag{2}$$

Furthermore, they are a complete set over $L^2(-1,1)$ in the sense that any function $u(t)$ belonging to $L^2(-1,1)$ can be arbitrarily closely approximated by a linear combination $\sum_{k=0}^{N} a_k P_k(t)$ in the L^2 norm for N sufficiently large. If we take $u(t)$ to be continuous, we can replace the L^2 norm by the maximum norm $\|u\| = \max_{-1 \le t \le 1} |u|$.

We begin our derivation of the Gaussian quadrature formula by considering the shifted Legendre polynomial

$$P_n^*(t) = P_n(1 - 2t) \qquad n = 0, 1, 2, \ldots \tag{3}$$

These polynomials constitute a complete orthogonal set over [0,1] and are determined by the analogue of (2),

$$\int_0^1 t^k P_n^*(t)\, dt = 0 \qquad k = 0, 1, 2, \ldots, n - 1 \tag{4}$$

up to a normalizing factor which is chosen sometimes one way and sometimes another depending upon our convenience.

Using the functions $t^k P_N^*$, $k = 0, 1, \ldots, N - 1$ as "test functions," we see that the required equality in (7.2) yields the equations

$$0 = \int_0^1 t^k P_N^*(t)\, dt = \sum_{i=1}^{N} w_i t_i^k P_N^*(t_i) \tag{5}$$

$k = 0, 1, \ldots, N - 1$. Regard this as a system of N simultaneous linear algebraic equations for the N quantities $w_i P_N^*(t_i)$, $i = 1, 2, \ldots, N$. The matrix of coefficients is

$$V(t_1, t_2, \ldots, t_N) = (t_i^k) \tag{6}$$

the Vandermonde matrix, which, as we know, is nonsingular if the t_i are distinct. Naturally, we assume that they are. Hence, (5) implies that

$$w_i P_N^*(t_i) = 0 \qquad i = 1, 2, \ldots, N \tag{7}$$

Since the w_i are assumed nonzero, we see that the quadrature points are the N zeros of $P_N^*(t)$, which we know lie in [0,1].

EXERCISES

1. Show that

$$w_i = \int_0^1 \frac{P_N^*(t)\, dt}{(t - t_i) P_N^{*\prime}(t_i)} \qquad i = 1, 2, \ldots, N$$

How does one show that the w_i are positive?

2. With the foregoing determination of t_i and w_i, show that

$$\int_0^1 u(t)\,dt = \sum_{i=1}^{N} w_i u(t_i)$$

for any polynomial of degree $2N - 1$ or less.

3. Determine the possible sets of quadrature points and weights if we assume that (7.2) holds for all polynomials of degree less than or equal to M which are zero at $t = 0$. What relation holds between M and N if we wish the w_i and t_i to be uniquely determined?

4. Obtain similar results for the case where conditions hold at both $t = 0$ and $t = 1$ on $u(t)$ and its derivatives.

9. Approximating System of Linear Equations. With this as background, let us return to the numerical solution of

$$\int_0^\infty u(t)e^{-st}\,dt = v(s) \tag{1}$$

To employ the preceding results, we first make a change of variable

$$e^{-t} = r \tag{2}$$

obtaining

$$\int_0^1 u\left(\log \frac{1}{r}\right) r^{s-1}\,dr = v(s) \tag{3}$$

Write

$$g(r) = u\left(\log \frac{1}{r}\right) \tag{4}$$

and employ the quadrature approximation of Sec. 8. The result is the relation

$$\sum_{i=1}^{N} w_i g(r_i) r_i^{s-1} \cong v(s) \tag{5}$$

We now set $s = 1, 2, \ldots, N$, and consider the approximate system of linear algebraic equations

$$\sum_{i=1}^{N} w_i r_i^k g(r_i) = v(k+1) \qquad k = 0, 1, \ldots, N-1 \tag{6}$$

These N equations uniquely determine the N quantities $w_i g(r_i)$ since the determinant of the coefficients is the Vandermonde determinant which is nonzero.

10. A Device of Jacobi. There is strong temptation at this point to relegate the further steps of the numerical solution to the sturdy keepers of algorithms for solving linear systems of algebraic equations. There are important reasons for resisting this apparently easy way out. In the first place, there is nothing routine about the procedure; in the second place, some important analytic concepts are involved.

To begin with, let us show that we can obtain an explicit representation for the inverse of the Vandermonde matrix, which means that we can obtain an explicit solution of (9.6) for arbitrary v.

Introduce the new variables $y_i = w_i g(x_i)$, $i = 1, 2, \ldots, N$, and set $a_k = v(k+1)$. The system in (9.6) takes the simpler form

$$\sum_{i=1}^{N} r_i^k y_i = a_k \qquad k = 0, 1, \ldots, N-1 \tag{1}$$

Multiply the kth equation by the parameter q_k, to be determined in a moment, and add. The result is

$$\sum_{i=1}^{N} y_i \left(\sum_{k=0}^{N-1} q_k r_i^k \right) = \sum_{k=0}^{N-1} a_k q_k \tag{2}$$

Let the polynomial $f(r)$ be defined by

$$f(r) = \sum_{k=0}^{N-1} q_k r^k \tag{3}$$

Then (2) may be written

$$\sum_{i=1}^{N} y_i f(r_i) = \sum_{k=0}^{N-1} a_k q_k \tag{4}$$

It remains to choose f in an expeditious fashion. Suppose that f is chosen to satisfy the conditions

$$\begin{aligned} f(r_j) &= 1 \\ f(r_i) &= 0 \qquad i \neq j \end{aligned} \tag{5}$$

Call this polynomial obtained in this fashion f_j and write

$$f_j(r) = \sum_{k=0}^{N-1} q_{kj} r^k \tag{6}$$

Then (4) reduces to

$$y_j = \sum_{k=0}^{N-1} a_k q_{kj} \tag{7}$$

The polynomial satisfying (5) is determined in principle by the Lagrange interpolation formula. In this case, it is readily obtained, namely

$$f_j(r) = \frac{P_N^*(r)}{(r - r_j) P_N^{*\prime}(r_j)} \qquad j = 1, 2, \ldots, N \tag{8}$$

With the aid of the recurrence relations for the Legendre polynomials, the q_{kj} can be calculated to any desired degree of accuracy.

EXERCISE

1. Why does it make any difference computationally that we can use (8) rather than the general formula for the Lagrange interpolation polynomial?

11. Discussion. With the aid of analytic and numerical expressions for the explicit inverse of $V(r_1,r_2, \ldots ,r_N)$ we have apparently completed our self-imposed task, that of obtaining an approximate inversion formula. Nonetheless, the results of Sec. 6, which tell us that L^{-1} is an unstable operator warn us that there must be a catch somewhere. There is!

The unboundedness of L^{-1} manifests itself in the fact that $V(r_1,r_2, \ldots ,r_N)$ is an *ill-conditioned matrix*. This is equivalent to saying that $x = V^{-1}b$, the solution of

$$Vx = b \tag{1}$$

is highly sensitive to small changes in b. To see this, let us note the values of the inverse of $(w_i r_i^{j-1})$ for the $N = 10$. It is sufficient to present the values of the last row of this matrix to illustrate the point we wish to make.

−6.2972078903048367	−1
6.8631241594216109	1
−1.8007322514163832	3
1.9787480981636393	4
−1.1232455134290224	5
3.6273559299309234	5
−6.9145818794948088	5
7.6901847430849219	5
−4.6080521705455609	5
1.1482677561426949	5

The last digits indicate the appropriate power of 10 by which the preceding seventeen-digit value is to be multiplied.

We see from an examination of the foregoing table that small changes in the values of the a_k in (9.7) can result in large changes in the value of the y_j. For the case where the values of the a_k can be generated analytically, the problem is not overwhelming since we can calculate the desired values, at some small effort, to an arbitrary degree of accuracy, single precision, double precision, triple precision, etc. In many cases, however, the values are obtained experimentally with a necessarily limited accuracy. In other cases, V, the Vandermonde matrix, is replaced by another ill-conditioned matrix whose inverse cannot be readily calculated with any desired degree of accuracy; see Sec. 16 for an illustration of this.

The problem of solving unstable equations ("improperly posed" problems in the terminology of Hadamard) is fundamental in modern science, and feasible solution represents a continual challenge to the ingenuity and art of the mathematician. It is clear, from the very nature of the problem, that there can never be a uniform or definitive solution to a question of this nature. Hence, we have the desirable situation of a never-ending supply of almost, but not quite, impossible problems in analysis.

12. Tychonov Regularization. The problem we pose is that of devising computational algorithms for obtaining an acceptable solution of

$$Ax = b \tag{1}$$

where A^{-1} exists, but where A is known to be ill-conditioned.

A first idea is to regard (1) as the result of minimizing the quadratic form

$$Q(x) = (Ax - b, Ax - b) \tag{2}$$

and thus replace (1) by

$$A^T A x = A^T b \tag{3}$$

The matrix of coefficients is now symmetric.

Unfortunately, it turns out that $A^T A$ may be more ill-conditioned than A and thus that (3) may result in even more chaotic results. A slight, but fundamental, modification of the foregoing procedure does, however, lead to constructive results. Let us replace the solution of (1) by the problem of minimizing

$$R(x) = (Ax - b, Ax - b) + \varphi(x) \tag{4}$$

where $\varphi(x)$ is a function carefully chosen to ensure the stability of the equation for the minimizing x. This is called *Tychonov regularization.*

The rationale behind this is the following. We accept the fact that the general problem of solving (1), where A is ill-conditioned and b is imprecisely given, is impossible. We cannot guarantee accuracy. Rather than a single solution, (1) determines an equivalence class, the set of vectors x such that $\|Ax - b\| \leq \epsilon$, where $\|\cdot\cdot\cdot\|$ is some suitable norm. Due to the ill-conditioning of A we possess no way of choosing the vector we want in this equivalence class without some further information concerning x. The function φ represents the use of this additional information.

Our thesis is that in any particular analytic or scientific investigation there is always more data concerning x than that present in (1). The challenge is to make use of it.

Let us give two examples which we will expand upon. Suppose that as a result of preliminary studies we know that $x \cong c$, a given vector.

Then we can choose

$$\varphi(x) = \lambda(x - c, x - c) \tag{5}$$

where $\lambda > 0$, and consider the problem of minimizing

$$R_1(x,\lambda) = (Ax - b, Ax - b) + \lambda(x - c, x - c) \tag{6}$$

Alternatively, we may know that the components of x, the x_i, viewed as functions of the index i, vary "smoothly" with i. By this we mean, for example, that we can use the values of $x_1, x_2, \ldots, x_{i-1}$ to obtain a reasonable estimate for the value of x_i. Alternately, we can say that there is a high degree of correlation among the values of the x_i. This is an intrinsic condition, as opposed to the first example, where we have invoked external data. In this case we may take

$$\varphi(x) = \lambda[(x_2 - x_1)^2 + (x_3 - x_2)^2 + \cdots + (x_N - x_{N-1})^2] \tag{7}$$

$\lambda > 0$, and consider the problem of minimizing

$$R_2(x,\lambda) = (Ax - b, Ax - b) + \lambda[(x_2 - x_1)^2 + \cdots + (x_N - x_{N-1})^2] \tag{8}$$

How to choose λ will be discussed below.

13. The equation $\varphi(x) = \lambda(x - c, x - c)$. Let us consider the minimization of

$$R(x) = (Ax - b, Ax - b) + \lambda(x - c, x - c) \tag{1}$$

where $\lambda > 0$. It is easy to see that the minimum value is given by

$$x = (A^TA + \lambda I)^{-1}(A^Tb + \lambda c) \tag{2}$$

The question arises as to the choice of λ. If λ is "small," we are close to the desired value $A^{-1}b$; if λ is "small," however, $A^TA + \lambda I$ will still be ill-conditioned and we may not be able to calculate x effectively. How then do we choose λ?

In general, there is no satisfactory answer since, as we mentioned above, there is no *uniform* method of circumventing the instability of L^{-1}. We must rely upon trial and error and experience. In some cases we can apply various analytic techniques to circumvent partially the uncertainty concerning a suitable choice of λ. We shall indicate some of these devices in the exercises.

EXERCISES

1. Show that, for small λ, $x(\lambda) = x_0 + \lambda y_1 + \lambda^2 y_2 + \cdots$, where the y_i are independent of λ.

2. Discuss the possibility of using "deferred passage to the limit" to calculate x_0 from $x(\lambda)$ calculated using convenient values of λ, that is, set $x_0 = 2x(\lambda) - x(2\lambda) + O(\lambda^2)$, etc.

3. Consider the method of successive approximations,

$$x_{n+1} = (A^TA + \lambda I)^{-1}(A^Tb + \lambda x_n) \qquad n \geq 0$$

$x_0 = c$. Show that x_n converges to x for any $\lambda > 0$ and any c.

4. What is the rate of convergence? How does this rate of convergence affect a choice of λ? Discuss the possibility of using nonlinear extrapolation techniques.

5. Consider an application of dynamic programming to the minimization of $R(x)$ along the following lines. Write

$$R_1(x) = \lambda_1(x_1 - c_1)^2 + \sum_{i=1}^{N} (a_{i1}x_1 - b_i)^2$$

$$R_M(x) = \left[\lambda(x_1 - c_1)^2 + \lambda(x_2 - c_2)^2 + \cdots + \lambda(x_M - c_M)^2 + \sum_{i=1}^{N}\left(\sum_{j=1}^{M} a_{ij}x_j - b_i\right)^2\right]$$

and let $b = (b_1,b_2, \ldots ,b_N)$. Write

$$f_M(b) = \min_x R_M(x)$$

Show that

$$f_M(b) = \min_{x_M} [\lambda(x_M - c_M)^2 + f_{M-1}(b - x_M a^{(M)})]$$

where $a^{(M)} = (a_{1M},a_{2M}, \ldots ,a_{NM})$.

6. Use the foregoing functional equation and the fact that

$$f_M(b) = (b,Q_M b) + 2(p_M,b) + r_M$$

to obtain recurrence relations for Q_M, p_M, r_M.

7. Is there any particular advantage in numbering the unknowns $x_1, x_2, \ldots, x_N$ in one order rather than another? (What we are hinting at here is that computing may be regarded as an adaptive control process where the objective is to minimize the total error, or to keep it within acceptable bounds.)

14. Obtaining an Initial Approximation. One way to obtain an initial approximation as far as the numerical inversion of the Laplace transform is concerned is to begin with a low-order quadrature approximation. Although the accuracy of the results obtained is low, we encounter no difficulty in calculating the solution of the associated linear equations.

Suppose, for example, we use initially a five-point Gaussian quadrature (see Fig. 3).

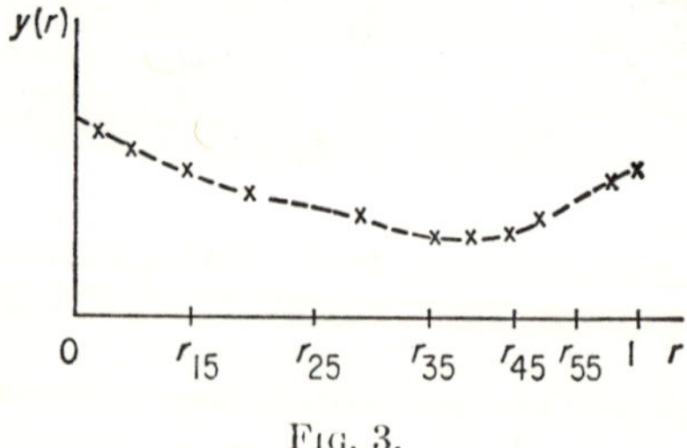

FIG. 3.

Using the Lagrange interpolation formula, we pass a polynomial of degree four through these points, $\{r_{i5},g(r_{i5})\}$, and then use this polynomial to obtain starting values for the values of g at the seven quadrature points for a quadrature formula of degree seven. The seven values constitute our initial vector c in the successive approximation scheme of the exercises of Sec. 13. We can continue in this way, increasing the dimension of the process step-by-step until we obtain sufficient agreement.

One measure of agreement would be the simultaneous smallness of the quantities $(Ax - b, Ax - b)$ and $(x - c, x - c)$.

15. Self-consistent $\varphi(x)$. The numerical inversion of the Laplace transform furnishes us with an example of a case where the x_i vary smoothly with i. Let us then consider the function introduced in (12.7). One interesting approach to the problem of minimizing

$$R(x) = (Ax - b, Ax - b) + \lambda[(x_2 - x_1)^2 + (x_3 - x_2)^2 + \cdots (x_N - x_{N-1})^2] \quad (1)$$

is by way of dynamic programming.

Introduce the sequence of functions

$$f_k(z,c) = \min_x \Big[\sum_{i=1}^{M} \Big(\sum_{j=k}^{N} a_{ij}x_j - z_i \Big)^2 + \lambda[(x_N - x_{N-1})^2 + \cdots (x_k - c)^2] \quad (2)$$

$k = 2, 3, \ldots, N$. Then

$$f_k(z,c) = \min_{x_k} [\lambda(x_k - c)^2 + f_{k+1}(z - a^{(k)}x_k, x_k)] \quad (3)$$

where $a^{(k)} = (a_{1k},a_{2k}, \ldots ,a_{Mk})$.

Setting

$$f_k(z,c) = (z,Q_kz) + 2(z,p_k)c + r_kc^2 \quad (4)$$

we can, as in Chap. 17, use (3) to obtain recurrence relations for Q_k, p_k, and r_k. Once the quadratic $f_2(b,x_1)$ has been determined in this fashion, we minimize over x. We leave the nontrivial details as a set of exercises.

EXERCISE

1. Carry out a similar procedure for the minimization of

$$R(x) = (Ax - b, Ax - b) + \lambda[(x_3 - 2x_2 + x_1)^2 + \cdots + (x_N - 2x_{N-1} + x_{N-2})^2]$$

16. Nonlinear Equations. Suppose that we attempt to apply the same techniques to nonlinear systems. Consider, as a simple case illustrating the overall idea, the equation

$$u' = -u + u^2 \qquad u(0) = c \quad (1)$$

Take $|c| < 1$ which ensures the existence of $u(t)$ for $t \geq 0$. We have

$$sL(u) - c = -L(u) + L(u^2)$$
$$L(u) = \frac{c}{(s+1)} + \frac{L(u^2)}{(s+1)} \tag{2}$$

We can approach the problem of calculating $L(u)$ by means of successive approximations, namely

$$L(u_0) = \frac{c}{(s+1)}$$
$$L(u_{n+1}) = \frac{c}{(s+1)} + \frac{L(u_n{}^2)}{(s+1)} \tag{3}$$

If we apply quadrature techniques, we see that we encounter the same coefficient matrix at each stage which means that we can use the explicit inverse previously obtained. Note that $L(u_n{}^2)$ is evaluated using the values of u_n at the fixed quadrature points. A serious difficulty with the foregoing approach, however, is that the convergence is slow. Since the evaluation of $L(u_n{}^2)$ by means of quadrature can introduce non-negligible errors, it is plausible that repeated use of the foregoing techniques can lead to considerable and unacceptable error.

Let us employ then in place of (3) a quasilinearization technique. Writing

$$u'_{n+1} = -u_{n+1} + 2u_n u_{n+1} - u_n{}^2 \qquad u_{n+1}(0) = c \tag{4}$$

we have

$$L(u'_{n+1}) = -L(u_{n+1}) + 2L(u_n u_{n+1}) - L(u_n{}^2) \tag{5}$$
$$(s+1)L(u_{n+1}) = c + 2L(u_n u_{n+1}) - L(u_n{}^2)$$

The advantage of quasilinearization is that it provides quadratic convergence. Now, however, when we employ the quadrature technique described in the preceding pages, we face a different ill-conditioned matrix at each stage of the calculation. This is due to the presence of the term $L(u_n u_{n+1})$ in (5). It is this which motivates the discussion and utilization of Tychonov regularization.

17. Error Analysis. Let us now critically examine the validity of the foregoing procedures. Essentially what we have been assuming, in treating the integral equation

$$\int_0^1 r^{s-1} g(r)\, dr = v(s) \tag{1}$$

as we did, is that we can approximate to $g(r)$ sufficiently accurately by means of a polynomial of degree N, namely

$$\|g(r) - p(r)\| \leq \epsilon \tag{2}$$

For the moment, let us think in terms of the form

$$\|\cdot\cdot\cdot\| = \max_{0\le r\le 1} |\cdot\cdot\cdot| \tag{3}$$

To see the effect of this approximation, let us write

$$\int_0^1 r^{s-1}p(r)\,dr = v(s) + \int_0^1 r^{s-1}(p(r) - g(r))\,dr \tag{4}$$

Since $s = 1, 2, \ldots, N$, and $p(r)$ is a polynomial of degree N, the term on the left may be written

$$\sum_{i=1}^{N} w_i r_i^{s-1} p(r_i) \tag{5}$$

Then (4) is equivalent to the system of linear equations

$$\sum_{i=1}^{N} w_i r_i^{s-1} p(r_i) = v(s) + \int_0^1 r^{s-1}(p(r) - g(r))\,dr \tag{6}$$

where $s = 1, 2, \ldots, N$. It remains to estimate the discrepancy term. Using (2), we have, crudely,

$$\left|\int_0^1 r^{s-1}(p(r) - g(r))\,dr\right| \le \|g(r) - p(r)\| \int_0^1 r^{s-1}\,dr \le \frac{\epsilon}{s} \tag{7}$$

We face a stability problem in comparing the solution of (6), given the estimate in (7), with the solution of

$$\sum_{i=1}^{N} w_i r_i^{s-1} g_i = v(s) \qquad s = 1, 2, \ldots, N \tag{8}$$

The problem is, as we have indicated, a serious one since $(w_i r_i^j)$ is ill-conditioned. Nevertheless, it is clear that $\max_i |g_i - p(r_i)|$ will be small, provided that ϵ is sufficiently small. The question in practice is an operational one, depending upon the relation between the degree of approximation of g by a polynomial of relatively low degree and the magnitude of the coefficients in V^{-1}. In view of the size of these terms (see Sec. 10), it is surprising that the method works so well in practice.

One explanation may lie in the fact that we can estimate the term $\int_0^1 r^{s-1}(p(r) - g(r))\,dr$ in a far sharper fashion than that given in (7). We have

$$\begin{aligned}\left|\int_0^1 r^{s-1}(p(r) - g(r))\,dr\right| &\le \int_0^1 r^{s-1}|p(r) - g(r)|\,dr \\ &\le \left(\int_0^1 r^{2s-2}\,dr\right)^{1/2}\left(\int_0^1 (p(r) - g(r))^2\,dr\right)^{1/2} \\ &\le \frac{1}{(2s-1)^{1/2}}\left(\int_0^1 (p(r) - g(r))^2\,dr\right)^{1/2}\end{aligned} \tag{9}$$

upon applying the Cauchy-Schwarz inequality. It is well known that for the same degree polynomial we can obtain a considerably better mean-square approximation than that available in the Cebycev norm.

EXERCISE

1. What determines whether we choose $s = 1, 2, \ldots, N$ or $s = 1, 1 + a, 1 + 2a, \ldots, 1 + (N - 1)a$?

MISCELLANEOUS EXERCISES

1. Consider the problem of minimizing the quadratic form

$$Q(x) = \int_0^1 (f(t) - x_0 - x_1 t - \cdots - x_N t^N)^2 \, dt$$

where $f(t)$ is a given function. Carry out some numerical experiments to ascertain whether the matrix of the associated system of linear equations is ill-conditioned or not.

2. Discuss the advantages or disadvantages of replacing $1, t, \ldots, t^N$ by the shifted Legendre polynomials.

3. Show that

$$\begin{vmatrix} 1 & -M & 0 & \cdots & 0 \\ 0 & 1 & -M & \cdots & 0 \\ \cdot & & & & \\ \cdot & & & & \\ \cdot & & & & \\ 0 & \cdots & & 1 & -M \\ -m & 0 & \cdots & 0 & 1 \end{vmatrix} = 1 - mM^{n-1}$$

where n is the order of the determinant.

4. Hence, discuss the stability of solutions of triangular systems subject to small changes in the zero elements below the main diagonal. See A. M. Ostrowski, On Nearly Triangular Matrices, *J. Res. Nat. Bur. Standards*, vol. 52, pp. 319–345, 1954.

5. Consider the equation $x + \lambda Ax = b$ where A is a positive matrix. Show that we can find a parameter k such that $x = b + \mu b_1 + \mu^2 b_2 + \cdots$ converges for all $\mu \geq 0$, where $\mu = \lambda/(\lambda + k)$.

6. Consider the equation $x = y + sAx$ where s is a scalar parameter. Show that

$$x = \frac{y + s(A + k_1 I)y + s^2(A^2 + k_1 A + k_2 I) + \cdots}{1 + k_1 s + k_2 s^2 + \cdots}$$

is a solution for any sequence of values $k_1, k_2, \ldots$. What different sets of criteria are there for choosing various sets of values?

7. Consider the solution of $x = y + Ax$ by means of continued fractions using the fact that $A^n x = A^n y + A^{n+1} x$. See Bellman and Richardson,[1] Brown,[2] and Drobnies.[3]

[1] R. Bellman and J. M. Richardson, A New Formalism in Perturbation Theory using Continued Fractions, *Proc. Nat. Acad. Sci. USA*, vol. 48, pp. 1913–1915, 1962.

[2] T. A. Brown, Analysis of a New Formalism in Perturbation Theory Using Continued Fractions, *Proc. Nat. Acad. Sci. USA*, vol. 50, pp. 598–601, 1963.

[3] S. I. Drobnies, Analysis of a Family of Fraction Expansions in Linear Analysis, *J. Math. Anal. Appl.*, to appear in 1970.

Bibliography and Discussion

§1. For a detailed discussion of the Laplace transform and various inversion techniques, see

G. Doetsch, *Handbuch der Laplace-Transformation* (3 volumes), Basel, 1950–1956.

R. Bellman and K. L. Cooke, *Differential-difference Equations*, Academic Press Inc., New York, 1963.

For some interesting and detailed accounts of the computational solution of $Ax = b$, see

J. Todd, The Problem of Error in Digital Computation, in L. B. Roll (ed.), *Error in Digital Computations*, vol. 1, pp. 3–41, 1965.

G. E. Forsythe, Today's Computational Methods of Linear Algebra, *SIAM Review*, vol. 9, pp. 489–515, 1967.

A. S. Householder, *Theory of Matrices in Numerical Analysis*, Blaisdell Publishing Company, New York, 1964.

A. M. Ostrowski, *Solution of Equations and Systems of Equations*, Academic Press Inc., New York, 1960.

§§2 to **4.** See the books cited above for detailed discussion of these three classes of equations.

§7. For a discussion of quadrature techniques and for further discussion of most of the results of this chapter, see

R. Bellman, R. Kalaba, and J. Lockett, *Numerical Inversion of the Laplace Transform*, American Elsevier Publishing Company, Inc., New York, 1966.

For a different approach, see

L. Weiss and R. N. McDonough, Prony's Method, *Z*-transforms and Padé Approximation, *SIAM Review*, vol. 5, pp. 145–149, 1963.

§10. For extensive numerical applications of the foregoing techniques, see the book cited immediately above. For a careful discussion of the ill-conditioning of Vandermonde matrix and the impossibility of completely circumventing it, see

W. Gautschi, On Inverses of Vandermonde and Confluent Vandermonde Matrices, *Numer. Math.*, vol. 4, pp. 117–123, 1962.

W. Gautschi, On the Conditioning of a Matrix Arising in the Numerical Inversion of the Laplace Transform, *Math. Comp.*, to appear.

§11. See

J. Todd, On Condition Numbers, *Programmation en Mathématiques Numériques,* Colloques Internationaux du Centre National de la Recherche Scientifique, no. 165, pp. 141–159, 1968.

A novel approach to solving $Ax = b$ where A is ill-conditioned is given in

Ju. V. Vorobev, A Random Iteration Process, II, *Z. Vycisl. Mat. i Mat. Fiz.*, vol. 5, pp. 787–795, 1965. (Russian.)
J. N. Franklin, *Well-posed Stochastic Extensions of Ill-posed Linear Problems,* Programming Rept. 135, California Institute of Technology, 1969.

§12. See the books

R. Lattes and J. L. Lions, *Theory and Application of the Method of Quasi-reversibility,* American Elsevier Publishing Company, Inc., New York, 1969.

M. M. Lavrentiev, *Some Improperly Posed Problems of Mathematical Physics,* Springer-Verlag, Berlin, 1967.

for a general discussion of "regularization," one of the important concepts of modern analysis.

See the book cited in Sec. 7 for all that follows.

§13. See

R. S. Lehman, Dynamic Programming and Gaussian Elimination, *J. Math. Anal. Appl.*, vol. 5, pp. 1–16, 1962.

§16. For a detailed discussion of quasilinearization, see

R. Bellman and R. Kalaba, *Quasilinearization and Nonlinear Boundary-value Problems,* American Elsevier Publishing Company, Inc., New York, 1965.

Appendix A

Linear Equations and Rank

1. Introduction. In this appendix, we wish to recall some elementary results concerning determinants and the solution of linear systems of equations of the form

$$\sum_{j=1}^{N} a_{ij}x_j = c_i \qquad i = 1, 2, \ldots, N \tag{1}$$

by their means. The coefficients a_{ij} and c_i are assumed to be complex numbers.

Using these results, we wish to determine when a system of homogeneous linear equations of the form

$$\sum_{j=1}^{N} a_{ij}x_j = 0 \qquad i = 1, 2, \ldots, M \tag{2}$$

possesses a solution distinct from $x_1 = x_2 = \cdots = x_N = 0$, a solution we shall call the *trivial* solution.

For our purposes in this volume, the most important of these linear systems is that where $M = N$. However, we shall treat this case in the course of studying the general problem.

Finally, we shall introduce a function of the coefficients in (2) called the *rank* and study some of its elementary properties.

2. Determinants. We shall assume that the reader is acquainted with the elementary facts about determinants and their relation to the solution of linear systems of equations. The two following results will be required.

Lemma 1. *A determinant with two rows or two columns equal has the value zero.*

This result is a consequence of Lemma 2.

Lemma 2. *The determinant*

$$|a_{ij}| = \begin{vmatrix} a_{11} & a_{12} & \cdots & a_{1N} \\ a_{21} & a_{22} & \cdots & a_{2N} \\ \cdot & & & \\ \cdot & & & \\ \cdot & & & \\ a_{N1} & a_{N2} & \cdots & a_{NN} \end{vmatrix} \tag{1}$$

may be expanded in the elements of the first row,

$$|a_{ij}| = a_{11}A_{11} + a_{12}A_{12} + \cdots + a_{1N}A_{1N} \tag{2}$$

where A_{ij} is $(-1)^{j+i}$ times the determinant of order $N - 1$ formed by striking out the ith row and jth column of $|a_{ij}|$.

The quantity A_{ij} is called the *cofactor* of a_{ij}.

3. A Property of Cofactors. Combining Lemmas 1 and 2, we see that the cofactors A_{ij}, $j = 1, 2, \ldots, N$, satisfy the following systems of linear equations

$$\sum_{j=1}^{N} a_{1j}A_{ij} = 0 \qquad i = 2, \ldots, N \tag{1}$$

since this expresses the fact that the determinant with first and ith rows equal is zero.

4. Cramer's Rule. If $|a_{ij}| \neq 0$, we may express the solution of (1.1) in the form

$$x_1 = \frac{\begin{vmatrix} c_1 & a_{12} & \cdots & a_{1N} \\ c_2 & a_{22} & \cdots & a_{2N} \\ \cdot & \cdot & & \\ \cdot & \cdot & & \\ \cdot & \cdot & & \\ c_N & a_{2N} & \cdots & a_{NN} \end{vmatrix}}{|a_{ij}|} \tag{1}$$

with x_i given by a similar expression with the ith column of $|a_{ij}|$ replaced by the column of coefficients $c_1, c_2, \ldots, c_N$.

5. Homogeneous Systems. Let us now use the foregoing facts to establish the following result.

Theorem 1. *The system of linear equations*

$$\sum_{j=1}^{N} a_{ij}x_j = 0 \qquad i = 1, 2, \ldots, M \tag{1}$$

possesses a nontrivial solution if $N > M$.

If $N \le M$, a necessary and sufficient condition that there be a nontrivial solution is that every $N \times N$ determinant $|a_{ij}|$ formed from N rows be zero.

Furthermore, we can find a nontrivial solution in which each of the x_i are polynomials in the a_{ij}.

The proof will be by way of a long, but elementary, induction over M.

Proof. Let us begin with the case $M = 1$ of the first statement where the statement is clearly true. Next, consider the first statement for $M = 2$. If $N = 2$, we have the equations

$$\begin{aligned} a_{11}x_1 + a_{12}x_2 &= 0 \\ a_{21}x_1 + a_{22}x_2 &= 0 \end{aligned} \tag{2}$$

which, if $|a_{ij}| = 0$, have the solutions

$$x_1 = a_{22} \qquad x_2 = -a_{21} \tag{3}$$

If this is a trivial solution, that is, $x_1 = x_2 = 0$, let us use the cofactors of the coefficients in the second equation,

$$x_1 = a_{11} \qquad x_2 = -a_{12} \tag{4}$$

If both of these solutions are trivial, then all the coefficients a_{ij} are zero, which means that any set of values x_1, x_2 constitute a solution.

If $N > 2$, we have a set of equations

$$\begin{aligned} a_{11}x_1 + a_{12}x_2 + \cdots + a_{1N}x_N &= 0 \\ a_{21}x_1 + a_{22}x_2 + \cdots + a_{2N}x_N &= 0 \end{aligned} \tag{5}$$

To obtain nontrivial solutions, we use an obvious approach. Let us solve for x_1 and x_2 in terms of the remaining x_i. This is easily done if the determinant

$$\begin{vmatrix} a_{11} & a_{12} \\ a_{21} & a_{22} \end{vmatrix} \tag{6}$$

is nonzero. However, this may not be the case. Let us then see if we can find two variables, x_i and x_j, whose determinant is nonzero. If one such 2×2 determinant,

$$\begin{vmatrix} a_{1i} & a_{1j} \\ a_{2i} & a_{2j} \end{vmatrix} \tag{7}$$

exists, we can carry out the program of solving for x_i and x_j in terms of the remaining variables.

Suppose, however, that all such 2×2 matrices are zero. We assert then that one of the equations in (5) is redundant, in the sense that any set of x_i-values satisfying one equation necessarily satisfies the other. Specifically, we wish to show that there exist two parameters λ_1 and λ_2, one at least of which is nonzero, such that

$$\lambda_1(a_{11}x_1 + a_{12}x_2 + \cdots + a_{1N}x_N) \\ + \lambda_2(a_{21}x_1 + a_{22}x_2 + \cdots + a_{2N}x_N) = 0 \quad (8)$$

identically in the x_i. If $\lambda_1 \neq 0$, it is clear that it is sufficient to satisfy the second equation in (5) in order to satisfy the first. We have thus reduced the problem of solving two equations in N unknowns to that of one equation in N unknowns.

The equation in (8) is equivalent to the statement that the following equations hold:

$$\begin{aligned} \lambda_1 a_{11} + \lambda_2 a_{21} &= 0 \\ \lambda_1 a_{12} + \lambda_2 a_{22} &= 0 \\ &\vdots \\ \lambda_1 a_{1N} + \lambda_2 a_{2N} &= 0 \end{aligned} \quad (9)$$

Consider the first two equations:

$$\begin{aligned} \lambda_1 a_{11} + \lambda_2 a_{21} &= 0 \\ \lambda_1 a_{12} + \lambda_2 a_{22} &= 0 \end{aligned} \quad (10)$$

These possess a nontrivial solution since we have assumed that we are in the case where all 2×2 determinants are equal to zero. Without loss of generality, assume that a_{11} or a_{12} is different from zero, since the stated result is trivially true for both coefficients of x_1 equal to zero; take $a_{11} \neq 0$. Then

$$\lambda_1 = a_{21} \qquad \lambda_2 = -a_{11} \quad (11)$$

is a nontrivial solution.

This solution, however, must satisfy the remaining equations in (9) since our assumption has been that all 2×2 determinants of the type appearing in (7) are zero.

The interesting fact to observe is that the proof of the first statement, involving $N \times 2$ systems, has been made to depend upon the second statement involving $2 \times N$ systems. We have gone into great detail in this simple case since the pattern of argumentation is precisely the same in the general $N \times M$ case.

Let us establish the general result by means of an induction on M. If we have the M equations

$$\sum_{j=1}^{N} a_{ij}x_j = 0 \qquad i = 1, 2, \ldots, M \quad (12)$$

with $N > M$, we attempt to find nontrivial solutions by solving for M of the x_i in terms of the remaining $N - M$ variables. We can accomplish this if one of $M \times M$ determinants formed from M different columns in the matrix (a_{ij}) is nonzero.

Suppose, however, that *all* such $M \times M$ determinants are zero. Then we assert that we can find quantities $y_1, y_2, \ldots, y_M$, not all equal to zero, such that

$$y_1 \left(\sum_{j=1}^{N} a_{1j}x_j \right) + y_2 \left(\sum_{j=1}^{N} a_{2j}x_j \right) + \cdots + y_M \left(\sum_{j=1}^{N} a_{Mj}x_j \right) = 0 \qquad (13)$$

identically in the x_i. Suppose, for the moment, that we have established this, and let, without loss of generality, $y_1 \neq 0$. Then, whenever the equations

$$\sum_{j=1}^{N} a_{ij}x_j = 0 \qquad i = 2, 3, \ldots, M \qquad (14)$$

are satisfied, the first equation $\sum_{j=1}^{N} a_{1j}x_j = 0$ will also be satisfied. We see then that we have reduced the problem for M equations to the problem for $M - 1$ equations.

It remains, therefore, to establish (13), which is equivalent to the equations

$$\begin{aligned} y_1a_{11} + y_2a_{21} + \cdots + y_Ma_{M1} &= 0 \\ y_1a_{12} + y_2a_{22} + \cdots + y_Ma_{M2} &= 0 \\ &\vdots \\ y_1a_{1N} + y_2a_{2N} + \cdots + y_Ma_{MN} &= 0 \end{aligned} \qquad (15)$$

This is the type of system of linear equations appearing in the second statement of Theorem 1. Again, we use an obvious technique to find nontrivial solutions.

Consider the first M equations. Choose y_1 to be the cofactor of a_{11} in the determinantal expansion, y_2 to be the cofactor of a_{21}, and so on. These values must satisfy all of the other equations, since our assumption has been that all $M \times M$ determinants are zero. Hence, if some cofactor A_{i1} is nonzero, we have our desired nontrivial solution.

If all the A_{i1} are zero, we consider another set of M equations and proceed similarly. This procedure can fail only if all $(M - 1) \times (M - 1)$ determinants formed by striking out one column and $N - M + 1$ rows of (a_{ji}) are zero.

In that case, we assert that we can set $y_M = 0$ and still obtain a nontrivial set of solutions for $y_1, y_2, \ldots, y_{M-1}$. For, the remaining system is now precisely of the type described by the second statement of Theorem 1 with M replaced by $M - 1$. Consequently, in view of our inductive hypothesis, there does exist a nontrivial solution of these equations. This completes the proof.

6. Rank. Examining the preceding proof, we see that an important role is played by the nonzero determinants of highest order, say r, formed by striking out $N - r$ columns and $M - r$ rows from the $N \times M$ matrix $A = (a_{ij})$.

This quantity r is called the rank of A.

As far as square matrices are concerned, the most important result is Theorem 2.

Theorem 2. *If T is nonsingular, the rank of TA* is the same as the rank of A.*

Proof. If A has rank r, we can express $N - r$ columns as linear combinations of r columns, which are linearly independent. Let these r columns be denoted by $a^1, a^2, \ldots, a^r$, and without loss of generality, let A be written in the form

$$A = (a^1a^2 \cdots a^r(c_{11}a^1 + c_{12}a^2 + \cdots + c_{1r}a^r) \cdots (c_{N-1,1}a^1 + \cdots + c_{N-r,r}a^r)) \qquad (1)$$

If the rows of T are $t^1, t^2, \ldots, t^N$, we have

$$B = TA = \begin{bmatrix} (t^1,a^1)(t^1,a^2) \cdots (t^1,a^r)[c_{11}(t^1,a^1) + c_{12}(t^1,a^2) + \cdots + c_{1r}(t^1,a^r)] \cdots \\ (t^2,a^1)(t^2,a^2) \cdots (t^2,a^r)[c_{11}(t^2,a^1) + c_{12}(t^2,a^2) + \cdots + c_{1r}(t^2,a^r)] \cdots \\ \vdots \end{bmatrix} \qquad (2)$$

Hence the rank of B is at most r. But, since we can write $A = T^{-1}B$, it follows that if the rank of B were less than r, the rank of A would have to be less than r, a contradiction.

EXERCISES

1. What is the connection between the rank of A and the number of zero characteristic roots?

2. Prove the invariance of rank by means of the differential equation $dx/dt = Ax$.

3. The rank of $A - \lambda_1 I$ is $N - k$ where k is the multiplicity of the characteristic root λ_1.

7. Rank of a Quadratic Form. The rank of a quadratic form (x,Ax) is defined to be the rank of the matrix A. It follows that the rank of (x,Ax) is preserved under a nonsingular transformation $x = Ty$.

8. Law of Inertia (*Jacobi-Sylvester*). If (x,Ax) where A is real is transformed by means of $x = Ty$, T nonsingular and real, into a sum of squares

$$(x,Ax) = \sum_{i=1}^{r} b_i y_i^2 \tag{1}$$

where none of the b_i are zero, it follows that r must be the rank of A.

In addition, we have Theorem 3.

Theorem 3. *If a real quadratic form of rank r, (x,Ax), is transformed by two real nonsingular transformations $x = Ty$, $x = Sz$, into a sum of r squares of the form*

$$\begin{aligned}(x,Ax) &= b_1y_1^2 + b_2y_2^2 + \cdots + b_ry_r^2 \\ &= c_1z_1^2 + c_2z_2^2 + \cdots + c_rz_r^2\end{aligned} \tag{2}$$

then the number of positive b_i is equal to the number of positive c_i.

9. Signature. Let $p(A)$ denote the number of positive coefficients in the foregoing representation and $n(A)$ the number of negative coefficients. The quantity $p(A) - n(A)$ is called the *signature* of A, $s(A)$. It follows from the above considerations that

$$s(A) = s(TAT') \tag{1}$$

if T is nonsingular.

EXERCISES

1. Show that $r(AB) + r(BC) \leq r(B) + r(ABC)$, whenever ABC exists (*G. Frobenius*).

2. Derive the Jacobi-Stieltjes result from this.

MISCELLANEOUS EXERCISES

1. For any $N \times N$ matrix $A = (a_{ij})$ of rank r, we have the relations

$$\sum_{i=1}^{N} \frac{|a_{ii}|}{\sum_{j=1}^{N} |a_{ij}|} \leq r \qquad \sum_{i=1}^{N} \frac{|a_{ii}|^2}{\sum_{j=1}^{N} |a_{ij}|^2} \leq r$$

Whenever $0/0$ occurs, we agree to interpret it as 0. (*Ky Fan-A. J. Hoffman*)

In the paper by Ky Fan and A. J. Hoffman,[1] there is a discussion of the problem of finding lower bounds for the rank of A directly in terms of the elements a_{ij}, and the relation between this problem and that of determining estimates for the location of characteristic values.

2. For complex A, write $A = B + iC$, where $B = (A + \bar{A})/2$, $C = -(A - \bar{A})i/2$. Then if A is Hermitian semidefinite, the rank of B is not less than the rank of C (*Broeder-Smith*).

3. If A is Hermitian semidefinite, the rank of B is not less than the rank of A (*Broeder-Smith*).

4. Let A be a complex matrix, and write $A = S + iQ$, where $S = (A + \bar{A})/2$, $Q = -i(A - \bar{A})/2$. If A is Hermitian positive-indefinite, then $r(S) \geq r(Q)$ and $r(S) \geq r(A)$.

Bibliography and Discussion

An excellent discussion of the many fascinating questions arising in the course of the apparently routine solution of linear systems is given in

G. E. Forsythe, Solving Linear Algebraic Equations Can Be Interesting, *Bull. Am. Math. Soc.*, vol. 59, pp. 299–329, 1953.

See also the interesting article

C. Lanczos, Linear Systems in Self-adjoint Form, *Am. Math. Monthly*, vol. 65, pp. 665–679, 1958.

For an interesting discussion of the connection between rank and a topological invariant, see

D. G. Bourgin, Quadratic Forms, *Bull. Am. Math. Soc.*, vol. 51, pp. 907–908, 1945.

For some further results concerning rank, see

R. Oldenburger, Expansions of Quadratic Forms, *Bull. Am. Math. Soc.*, vol. 49, pp. 136–141, 1943.

For a generalization of linear dependence, see

H. C. Thatcher, Jr., Generalization of Concepts Related to Linear Dependence, *J. Ind. Appl. Math.*, vol. 6, pp. 288–300, 1958.

G. Whaples, A Note on Degree-n Independence, *ibid.*, pp. 300–301.

[1] Ky Fan and A. J. Hoffman, Lower Bounds for the Rank and Location of the Eigenvalues of a Matrix, *Contributions to the Solution of Systems of Linear Equations and the Determination of Eigenvalues, Natl. Bur. Standards Applied Math.*, ser. 39, 1954.

Appendix B

The Quadratic Form of Selberg

As another rather unexpected use of quadratic forms, let us consider the following method of Selberg, which in many parts of number theory replaces the sieve method of Brun.

Suppose that we have a finite sequence of integers $\{a_k\}$, $k = 1, 2, \ldots, N$, and we wish to determine an upper bound for the number of the a_k which is not divisible by any prime $p \leq z$.

Let $\{x_v\}$, $1 \leq v \leq z$, be a sequence of real numbers such that $x_1 = 1$, while the other x_v are arbitrary. Consider the quadratic form

$$Q(x) = \sum_{k=1}^{N} \Big(\sum_{v|a_k} x_v\Big)^2 \tag{1}$$

Whenever a_k is not divisible by a prime $\leq z$, the sum $\sum_{v|a_k} x_v$ yields $x_1 = 1$. Hence, if we denote the number of a_k, $k = 1, 2, \ldots, N$, divisible by any prime $p \leq z$ by $f(N,z)$ we have

$$f(N,z) \leq \sum_{k=1}^{N} \Big(\sum_{v|a_k} x_v\Big)^2 \tag{2}$$

for *all* x_v with $x_1 = 1$. Hence

$$f(N,z) \leq \min_{x_v} \sum_{k=1}^{N} \Big(\sum_{v|a_k} x_v\Big)^2 \tag{3}$$

To evaluate the right-hand side, we write

$$Q(x) = \sum_{v_1,v_2<z} x_{v_1}x_{v_2} \left\{\sum_{\frac{v_1v_2}{K}|a_k} 1\right\} \tag{4}$$

where a_k runs over the sequence $a_1, a_2, \ldots, a_N$. Here K denotes the greatest common divisor of v_1 and v_2.

Now suppose that the sequence $\{a_k\}$ possesses sufficient regularity properties so that there exists a formula of the type

$$\sum_{\substack{\rho | a_k \\ k=1,2,\ldots,N}} 1 = \frac{N}{f(\rho)} + R(\rho) \tag{5}$$

where R_ρ is a remainder term and the "density function" $f(\rho)$ is multiplicative, i.e.,

$$f(\rho_1\rho_2) = f(\rho_1)f(\rho_2) \text{ for } (\rho_1,\rho_2) = 1 \tag{6}$$

Then

$$\sum_{\frac{v_1v_2}{K} | a_k} 1 = \frac{N}{f(v_1v_2/K)} = \frac{f(K)N}{f(v_1)f(v_2)} + R\left(\frac{v_1v_2}{K}\right) \tag{7}$$

Using this formula in (4), the result is

$$f(N,z) \leq N \sum_{v_1,v_2 \leq z} \frac{x_{v_1}x_{v_2}f(K)}{f(v_1)f(v_2)} + \sum_{v_1,v_2 \leq z} x_{v_1}x_{v_2}R\left(\frac{v_1v_2}{K}\right) \tag{8}$$

Consider now the problem of determining the x_v, $2 \leq v \leq z$, for which the quadratic form

$$P(x) = \sum_{v_1,v_2 \leq z} \frac{x_{v_1}x_{v_2}f(K)}{f(v_1)f(v_2)} \tag{9}$$

is a minimum.

To do this, we introduce the function

$$f_1(n) = \sum_{d|n} \mu(d)f(n/d) \tag{10}$$

If n is squarefree, we have

$$f_1(n) = f(n) \prod_{p|n} \left(1 - \frac{1}{f(p)}\right) \tag{11}$$

Using the Möbius inversion formula, (10) yields

$$f(K) = \sum_{n|K} f_1(n) = \sum_{\substack{n|v_1 \\ n|v_2}} f_1(n) \tag{12}$$

Using this expression for $f(K)$ in (9), we see that

$$P(x) = \sum_{n \leq z} f_1(n) \left\{ \sum_{\substack{n|v \\ v \leq z}} \frac{x_v}{f(v)} \right\}^2 \tag{13}$$

Now perform a change of variable. Set

$$y_n = \sum_{\substack{n|v \\ v \leq z}} \frac{x_v}{f(v)} \tag{14}$$

Then, using the Möbius inversion formula once more, we have

$$\frac{x_v}{f(v)} = \sum_{n \leq \frac{z}{v}} \mu(n) y_{nv} \tag{15}$$

The problem is thus that of minimizing

$$P = \sum_{n<z} f_1(n) y_n{}^2 \tag{16}$$

over all $\{y_k\}$ such that

$$\sum_{n<z} \mu(n) y_n = \frac{\lambda_1}{f(1)} = 1 \tag{17}$$

It is easily seen that the minimizing $\{y_k\}$ is given by

$$y_n = \frac{\mu(n)}{f_1(n)} \frac{1}{\displaystyle\sum_{\rho' \leq z} \frac{\mu^2(\rho')}{f_1(\rho')}} \tag{18}$$

and that the minimum value of the form P is

$$\frac{1}{\displaystyle\sum_{n \leq z} \frac{\mu^2(n)}{f_1(n)}} \tag{19}$$

The corresponding λ_n are determined by the relations

$$\begin{aligned} \lambda_n &= \frac{f(n)}{\displaystyle\sum_{\rho \leq z} \frac{\mu^2(\rho)}{f_1(\rho)}} \sum_{\rho \leq z/v} \frac{\mu(\rho)\mu(\rho v)}{f_1(\mu v)} \\ &= \mu(v) \prod_{p|v} \left(1 - \frac{1}{f(p)}\right)^{-1} \cdot \frac{1}{\displaystyle\sum_{\rho \leq z} \frac{\mu^2(\rho)}{f_1(\rho)}} \cdot \sum_{\substack{\rho \leq z/v \\ (\rho, v) = 1}} \frac{\mu^2(\rho)}{f_1(\rho)} \end{aligned} \tag{20}$$

Inserting these values in (8), we obtain the inequality

$$f(N,z) \leq \frac{N}{\sum_{\rho \leq z} \frac{\mu^2(\rho)}{f_1(\rho)}} + \sum_{v_1, v_2 \leq z} |\lambda_{v_1} \lambda_{v_2} R(v_1 v_2/K)| \tag{21}$$

If z is chosen properly, say $N^{1/2-\epsilon}$, the upper bound is nontrivial.

A particularly simple case is that where $a_k \equiv k$, so that $f(\rho) \equiv \rho$. We then obtain a bound for the number of primes less than or equal to N.[1]

Bibliography and Discussion

For a treatment of the Selberg form by means of dynamic programming, see

R. Bellman, Dynamic Programming and the Quadratic Form of Selberg, *J. Math. Anal. Appl.*, vol. 15, pp. 30–32, 1966.

[1] A. Selberg, On an Elementary Method in the Theory of Primes, *Kgl. Norske Videnskab. Selskabs Forh.*, Bd 19, Nr. 18, 1947.

Appendix C

A Method of Hermite

Consider the polynomial $p(x)$ with real coefficients and, for the moment, unequal real roots $r_1, r_2, \ldots, r_N$, and suppose that we wish to determine the number of roots greater than a given real quantity a_1.

Let

$$y_i = x_1 + r_i x_2 + r_i^2 x_3 + \cdots + r_i^{N-1} x_N \tag{1}$$

where the x_i are real quantities and $Q(x)$ denotes the quadratic form

$$Q(x) = \sum_{i=1}^{N} y_i^2/(r_i - a_1) \tag{2}$$

It now follows immediately from the "law of inertia" that if $Q(x)$ is reduced to another sum of squares in any fashion by means of a real transformation, then the number of terms with positive coefficients will be equal to the number of real roots of $p(x) = 0$ greater than a_1.

This does not seem to be a constructive result immediately, since the coefficients in $Q(x)$ depend upon the unknown roots. Observe, however, that $Q(x)$ is a symmetric rational function of the roots and hence can be expressed as a rational function of the known coefficients of $p(x)$.

The more interesting case is that where $p(x)$ has complex roots. This was handled by Hermite in the following fashion. Let r_1 and r_2 be conjugate complex roots and write

$$\begin{aligned} r_1 &= \rho(\cos \alpha + i \sin \alpha) \qquad r_2 = \rho(\cos \alpha - i \sin \alpha) \\ y_1 &= u + iv \qquad y_2 = u + iv \end{aligned} \tag{3}$$

where u_i and v_i are real. Similarly, set

$$\frac{1}{r_1 - a_1} = r(\cos \phi + i \sin \phi) \qquad \frac{1}{r_2 - a_1} = r(\cos \phi - i \sin \phi) \tag{4}$$

It follows that

$$\begin{aligned} \frac{y_1^2}{r_1 - a_1} + \frac{y_2^2}{r_2 - a_1} &= 2r(u \cos \phi/2 - v \sin \phi/2)^2 \\ &\quad - 2r(u \sin \phi/2 + v \cos \phi/2)^2 \end{aligned} \tag{5}$$

From this, Hermite's result may be concluded:

Theorem. *Let the polynomial $p(x)$ have real coefficients and unequal roots. Let*

$$Q(x) = \sum_{i=1}^{N} y_i^2/(r_i - a_1) \tag{6}$$

be reduced to a sum of squares by means of a real nonsingular substitution. Then the number of squares with positive coefficients will be equal to the number of imaginary roots of the equation $p(x) = 0$ plus the number of real roots greater than a_1.

On the other hand, the result in Sec. 3 of Chap. 5 also yields a method for determining the number of roots in given intervals by way of Sturmian sequences. The connection between these two approaches and some results of Sylvester is discussed in detail in the book on the theory of equations by Burnside and Panton to which we have already referred, and whose presentation we have been following here.

An extension of the Hermite method was used by Hurwitz[1] in his classic paper on the determination of necessary and sufficient conditions that a polynomial have all its roots with negative real parts.

Further references and other techniques may be found in Bellman, Glicksberg, and Gross.[2]

What we have wished to indicate in this appendix, and in Appendix B devoted to an account of a technique due to Selberg, is the great utility of quadratic forms as "counting devices." Another application of this method was given by Hermite[3] himself in connection with the reduction of quadratic forms.

[1] A. Hurwitz, Über die Bedingungen unter welchen eine Gleichung nur Wurzeln mit negativen reellen Teilen besitzt, *Math. Ann.*, vol. 46, 1895 (*Werke*, vol. 2, pp. 533–545).

E. J. Routh, *The Advanced Part of a Treatise on the Dynamics of a System of Rigid Bodies*, Macmillan & Co., Ltd., pp. 223–231, London, 1905.

[2] R. Bellman, I. Glicksberg, and O. Gross, *Some Aspects of the Mathematical Theory of Control Processes*, The RAND Corporation, *Rept.* R-313, January 16, 1958.

[3] C. Hermite, *Oeuvres*, pp. 284–287, pp. 397–414; *J. Crelle*, Bd. LII, pp. 39–51, 1856.

Appendix D

Moments and Quadratic Forms

1. Introduction. The level of this appendix will be somewhat higher than its neighbors, since we shall assume that the reader understands what is meant by a Riemann-Stieltjes integral. Furthermore, to complete both of the proofs below, we shall employ the Helly selection theorem.

2. A Device of Stieltjes. Consider the sequence of quantities $\{a_k\}$, $k = 0, 1, \ldots$, defined by the integrals

$$a_k = \int_0^1 t^k \, dg(t) \tag{1}$$

where $g(t)$ is a monotone increasing function over $[0,1]$, with $g(1) = 1$, $g(0) = 0$. These quantities are called the moments of dg.

It was pointed out by Stieltjes that the quadratic forms

$$Q_N(x) = \sum_{k,l=0}^{N} a_{k+l} x_k x_l = \int_0^1 \left(\sum_{k=0}^{N} x_k t^k\right)^2 dg(t) \tag{2}$$

are necessarily non-negative definite. Let us call the matrices

$$A_N = (a_{k+l}) \qquad k, l = 0, 1, \ldots, N \tag{3}$$

Hankel matrices.

The criteria we have developed in Chap. 5, Sec. 2, then yield a number of interesting inequalities such as

$$\begin{vmatrix} \int_0^1 t^k \, dg & \int_0^1 t^{k+l} \, dg \\ \int_0^1 t^{k+l} \, dg & \int_0^1 t^{k+2l} \, dg \end{vmatrix} \geq 0 \tag{4}$$

with strict inequality if g has enough points of increase.

If, in place of the moments in (1), we employ the trigonometric moments

$$c_k = \int_0^1 e^{2\pi i k t} \, dg \qquad k = 0, \pm 1, \pm 2, \ldots \tag{5}$$

then the quadratic form obtained from

$$\int_0^1 \left| \sum_{k=0}^{N} x_k e^{2\pi i k t} \right|^2 dg \tag{6}$$

is called the Toeplitz form, and the corresponding matrix a Toeplitz matrix.

It follows from these considerations that a necessary condition that a sequence $\{c_k\}$ be derived from (5) for some monotone increasing function $g(t)$, normalized by the condition that $g(1) = g(0) = 1$, is that

$$C = (c_{k-l}) \tag{7}$$

be a non-negative definite Hermitian matrix. It is natural to ask whether or not the condition is sufficient.

As we shall see, in one case it is; in the other, it is not.

3. A Technique of E. Fischer. Let us consider an infinite sequence of real quantities $\{a_n\}$ satisfying the condition that for each N, the quadratic form

$$Q_N(x) = \sum_{i,j=0}^{N} a_{i+j} x_i x_j \tag{1}$$

is positive definite.

In this case we wish to establish Theorem 1.

Theorem 1. *If the $2N + 1$ real quantities $a_0, a_1, \ldots, a_{2N}$ are such that $Q_N(x)$ is positive definite, then they can be represented in the form*

$$a_k = b_0 r_0^k + b_1 r_1^k + \cdots + b_N r_N^k \qquad k = 0, 1, \ldots, 2N \tag{2}$$

where $b_i > 0$ and the r_i are real and different.

Proof. Consider the associated quadratic form

$$P_N(x) = \sum_{i,j=0}^{N} a_{1+i+j} x_i x_j \tag{3}$$

Since, by assumption, $Q_N(x)$ is positive definite, we can find a simultaneous representation

$$\begin{aligned} Q_N(x) &= \sum_{j=0}^{N} (m_{j0}x_0 + \cdots + m_{jN}x_N)^2 \\ P_N(x) &= \sum_{j=0}^{N} r_i (m_{j0}x_0 + \cdots + m_{jN}x_N)^2 \end{aligned} \tag{4}$$

where the r_i are real.

The expression for $Q_N(x)$ yields

$$a_{i+(j+1)} = m_{0i}m_{0,j+1} + \cdots + m_{Ni}m_{N,j+1} \qquad i = 0, 1, \ldots, N, \quad j = 0, \ldots, N-1 \tag{5}$$

On the other hand, from the equation for $P_N(x)$ we obtain

$$a_{1+i+j} = r_0 m_{0i} m_{0j} + \cdots + r_N m_{Ni} m_{Nj} \tag{6}$$

Combining (5) and (6), we obtain the system of linear homogeneous equations

$$m_{0i}(m_{0,j+1} - r_0 m_{0j}) + \cdots + m_{Ni}(m_{N,j+1} - r_N m_{N,j}) = 0 \tag{7}$$

for $j = 0, 1, \ldots, N$.

Since $Q_N(x)$ is positive definite, the determinant $|m_{ij}|$ must be nonzero. Hence, (7) yields the relations

$$m_{k,j+1} = r_k m_{k,j} \qquad k = 0, \ldots, N \tag{8}$$

or

$$m_{k,j} = r_k{}^j m_{k,0} \tag{9}$$

It follows that we can write $Q_N(x)$ in the form

$$Q_N(x) = \sum_{j=0}^{N} m_{j0}{}^2(x_0 + r_1 x_1 + \cdots + r_1{}^N x_N)^2 \tag{10}$$

from which the desired representation follows. Since $Q_N(x)$ is positive definite, $r_i \neq r_j$ for $i \neq j$.

4. Representation as Moments. Using Theorem 1, it is an easy step to demonstrate Theorem 2.

Theorem 2. *If a real sequence $\{a_n\}$ satisfies the two conditions*

$$\sum_{i,j=0}^{N} a_{i+j}x_i x_j \text{ positive definite for all } N \tag{1a}$$

$$\sum_{i,j=0}^{N} (a_{i+j} - a_{i+j+1})x_i x_j \text{ positive definite for all } N \tag{1b}$$

it is a moment sequence, i.e., *there exists a bounded monotone increasing function $g(t)$ defined over $[-1,1]$ such that*

$$a_n = \int_{-1}^{1} t^n \, dg(t) \tag{2}$$

Proof. The representation of Theorem 1, combined with condition (1b) above, shows that $|r_i| \leq 1$ for $i = 0, 1, \ldots, N$. Hence, we can find a monotone increasing function $g_N(t)$, bounded by a_0, such that

$$a_n = \int_{-1}^{1} t^n \, dg_M(t) \qquad n = 0, 1, \ldots, N \tag{3}$$

To complete the proof, we invoke the Helly selection theorem which assures us that we can find a monotone increasing function $g(t)$, bounded by a_0, with the property that

$$\lim_{M\to\infty} \int_{-1}^{1} t^n \, dg_M(t) = \int_{-1}^{1} t^n \, dg(t) \tag{4}$$

as M runs through some sequence of the N's.

5. A Result of Herglotz. Using the same techniques, it is not difficult to establish the following result of Herglotz.

Theorem 3. *A necessary and sufficient condition that a sequence of complex numbers* $\{a_n\}$, $n = 0, \pm 1, \ldots$, *with* $a_{-n} = \overline{a_n}$, *have the property that*

$$\sum_{h=1}^{N} \sum_{k=1}^{N} a_{h-k} x_h \bar{x}_k \geq 0 \tag{1}$$

for all complex numbers $\{x_n\}$ *and all* N *is that there exist a real monotone nondecreasing function* $u(y)$, *of bounded variation such that*

$$a_n = \int_0^{2\pi} e^{iny} \, du(y) \tag{2}$$

Fischer's method is contained in his paper.[1]

Bibliography and Discussion

Quadratic forms of the type $\sum\limits_{i,j=0} a_{i-j} x_i x_j$ are called L forms after the connection discovered between these forms and the Laurent expansion of analytic functions and thus the connection with Fourier series, in general. They play an extremely important role in analysis. See, for example, the following papers:

O. Szasz, Über harmonische Funktionen und L-formen, *Math. Z.*, vol. 1, pp. 149–162, 1918.

O. Szasz, Über Potenzreihen und Bilinearformen, *Math. Z.*, vol. 4, pp. 163–176, 1919.

L. Fejer, Über trigonometrische Polynome, *J. Math.*, vol. 146, pp. 53–82, 1915.

[1] E. Fischer, Über die Caratheodorysche Problem, Potenzreihen mit positivem reellen Teil betreffend, *Rend. circ. mat. Palermo*, vol. 32, pp. 240–256, 1911.

I. Schur, Bemerkungen zur Theorie der Beschränkten Bilinearformen mit unendlich vielen Veränderlichen, *J. Math.*, vol. 140, pp. 1–28, 1911.

A. Caratheodory and L. Fejer, Über den Zusammenhang der Extremen . . . , *Rend. circ. mat. Palermo*, vol. 32, pp. 1–22, 1911.

Further references will be found in the monograph,

E. F. Beckenbach and R. Bellman, *Inequalities*, Springer-Verlag, 1961.

and in the extensive bibliography of

M. Kac, L. Murdock, and G. Szego, On the Eigenvalues of Certain Hermitian Forms, *J. Rat. Mech. Analysis*, vol. 2, pp. 767–800, 1953.

Indexes

Name Index

Subject Index